Exploring Culture and Heritage Through Experience Tourism

Yashwant Singh Rawal
Parul Institute of Hotel Management, Parul University, India

Rupa Sinha
Amity Institute of Travel and Tourism, Amity University, India

Sankar Kumar Mukherjee
Amity Institute of Travel and Tourism, Amity University, India

Debasish Batabyal
Amity Institute of Travel and Tourism, Amity University, India

IGI Global
PUBLISHER of TIMELY KNOWLEDGE

A volume in the Advances in Hospitality, Tourism, and the Services Industry (AHTSI) Book Series

Published in the United States of America by
IGI Global
Business Science Reference (an imprint of IGI Global)
701 E. Chocolate Avenue
Hershey PA, USA 17033
Tel: 717-533-8845
Fax: 717-533-8661
E-mail: cust@igi-global.com
Web site: http://www.igi-global.com

Library of Congress Cataloging-in-Publication Data

Names: Rawal, Yashwant Singh, 1983- editor. | Sinha, Rupa, 1991- editor. | Mukherjee, Sankar Kumar, 1976- editor. | Batabyal, Debasish, 1978- editor.
Title: Exploring culture and heritage through experience tourism / edited by Yashwant Singh Rawal, Rupa Sinha, Sankar Kumar Mukherjee, and Debasish Batabyal.
Description: Hershey, PA : Business Science Reference, [2023] | Includes bibliographical references and index. | Summary: "The book covers a wide range of topics. It explores the interplay between culture, heritage, and tourism, examining their dynamic relationship and current trends, such as event tourism, creative tourism, and indigenous tourism. Through case studies and insightful analysis, it provides an in-depth understanding of the positive and negative aspects of cultural heritage tourism, shedding light on the significance of traditional cultural practices and the management of intangible heritage"-- Provided by publisher.
Identifiers: LCCN 2023030276 (print) | LCCN 2023030277 (ebook) | ISBN 9781668499573 (h/c) | ISBN 9781668499580 (s/c) | ISBN 9781668499597 (eISBN)
Subjects: LCSH: Heritage tourism--Case studies. | Culture and tourism--Case studies.
Classification: LCC G156.5.H47 E935 2023 (print) | LCC G156.5.H47 (ebook) | DDC 338.4/791--dc23/eng/20230817
LC record available at https://lccn.loc.gov/2023030276
LC ebook record available at https://lccn.loc.gov/2023030277

This book is published in the IGI Global book series Advances in Hospitality, Tourism, and the Services Industry (AHTSI) (ISSN: 2475-6547; eISSN: 2475-6555)

British Cataloguing in Publication Data
A Cataloguing in Publication record for this book is available from the British Library.

For electronic access to this publication, please contact: eresources@igi-global.com.

Advances in Hospitality, Tourism, and the Services Industry (AHTSI) Book Series

Maximiliano Korstanje
University of Palermo, Argentina

ISSN:2475-6547
EISSN:2475-6555

Mission

Globally, the hospitality, travel, tourism, and services industries generate a significant percentage of revenue and represent a large portion of the business world. Even in tough economic times, these industries thrive as individuals continue to spend on leisure and recreation activities as well as services.

The Advances in Hospitality, Tourism, and the Services Industry (AHTSI) book series offers diverse publications relating to the management, promotion, and profitability of the leisure, recreation, and services industries. Highlighting current research pertaining to various topics within the realm of hospitality, travel, tourism, and services management, the titles found within the AHTSI book series are pertinent to the research and professional needs of managers, business practitioners, researchers, and upper-level students studying in the field.

Coverage

- Service Design
- Health and Wellness Tourism
- Destination Marketing and Management
- International Tourism
- Casino Management
- Sustainable Tourism
- Cruise Marketing and Sales
- Customer Service Issues
- Hotel Management
- Service Management

IGI Global is currently accepting manuscripts for publication within this series. To submit a proposal for a volume in this series, please contact our Acquisition Editors at Acquisitions@igi-global.com or visit: http://www.igi-global.com/publish/.

The Advances in Hospitality, Tourism, and the Services Industry (AHTSI) Book Series (ISSN 2475-6547) is published by IGI Global, 701 E. Chocolate Avenue, Hershey, PA 17033-1240, USA, www.igi-global.com. This series is composed of titles available for purchase individually; each title is edited to be contextually exclusive from any other title within the series. For pricing and ordering information please visit http://www.igi-global.com/book-series/advances-hospitality-tourism-services-industry/121014. Postmaster: Send all address changes to above address.

Titles in this Series

For a list of additional titles in this series, please visit: http://www.igi-global.com/book-series/advances-hospitality-tourism-services-industry/121014

Management and Practices of Pilgrimage Tourism and Hospitality
S.K. Gupta (Centre for Mountain Tourism and Hospitality Management, HNB Garhwal University, India) Lilibeth C. Aragon (College of International Tourism and Hospitality Management, Lyceum of the Philippines University, Philippines) Pankaj Kumar (Department of Tourism and Hospitality Management, Mizoram University, India) Madhurima S. (Department of Environmental Science, Mizoram University, India) and Rajesh Ramasamy (Department of Tourism and Hospitality Management, Mizoram Universit, India)
Business Science Reference • © 2023 • 264pp • H/C (ISBN: 9798369314142) • US $250.00

Critical Social Challenges in the Tourism Industry Labor, Commodification, and Drugs
Barış Çıvak (Anadolu University, Turkey) and Engin Bayraktaroğlu (Anadolu University, Turkey)
Business Science Reference • © 2023 • 298pp • H/C (ISBN: 9781668492550) • US $250.00

Prospects and Challenges of Global Pilgrimage Tourism and Hospitality
S.K. Gupta (Centre for Mountain Tourism and Hospitality Management, HNB Garhwal University, India) Lilibeth C. Aragon (College of International Tourism and Hospitality Management, Lyceum of the Philippines University, Philippines) Pankaj Kumar (Department of Tourism and Hospitality Management, Mizoram University, India) Madhurima S. (Department of Environmental Science, Mizoram University, India) and Rajesh Ramasamy (Department of Tourism and Hospitality Management, Mizoram University, India)
Business Science Reference • © 2023 • 291pp • H/C (ISBN: 9781668448175) • US $250.00

Handbook of Research on Innovation, Differentiation, and New Technologies in Tourism, Hotels, and Food Service
Gonçalo Poeta Fernandes (CITUR, Polytechnic Institute of Guarda, Portugal & CICS, Universidade NOVA de Lisboa, Portugal) and António Silva Melo (CiTUR, Polytechnic Institute of Porto, Portugal & CIDTFF, University of Aveiro, Portugal)
Business Science Reference • © 2023 • 389pp • H/C (ISBN: 9781668469859) • US $295.00

Exploring Niche Tourism Business Models, Marketing, and Consumer Experience
Maria Antónia Rodrigues (CEOS.PP, ISCAP, Polytechnic of Porto, Portugal) and Maria Amélia Carvalho (ISCAP, Polytechnic of Porto, Portugal)
Business Science Reference • © 2023 • 397pp • H/C (ISBN: 9781668472422) • US $250.00

IGI Global
PUBLISHER of TIMELY KNOWLEDGE

701 East Chocolate Avenue, Hershey, PA 17033, USA
Tel: 717-533-8845 x100 • Fax: 717-533-8661
E-Mail: cust@igi-global.com • www.igi-global.com

Table of Contents

Detailed Table of Contents

A historic town centers local economic activities and has a significant concentration of heritage structures. Such a place or historic core generally has a definite identity—a distinct townscape character worthy of preservation. It has been proven through several studies that cultural heritage has value and the ability to be a significant component of community development, provided it is properly maintained. This chapter aims to identify the prime movers of the heritage economy and utilize the potential of heritage towns to retain and enhance the role of heritage in its stakeholders' social and economic lives. Based on an extensive systematic literature review, the role of entrepreneurship in the heritage economy and various parameters that contribute to the economic assessment of heritage towns are explored. The proposed methodological framework contributes to identifying techniques for assessing the economic value of heritage towns for local economic development.

This book chapter explores the use of AR and VR in cultural heritage tourism, enhancing visitor experiences with historical locations and artefacts. AR and VR allow for immersive interactions, enabling visitors to comprehend and appreciate heritage places better. These technologies revive old structures and artefacts, offering personalized experiences and interactive storytelling. However, challenges like cost, hardware limitations, and maintaining authenticity need to be addressed. Future opportunities

Religion plays an important role in tourism that ultimately contributes to the human development. Religious attractions belong to the cultural elements and are an important component of economic and social activities. The basic and essential elements religious tourism areas considered as more than pilgrimages. Pilgrimage has material and spiritual values which voluntarily engages with religious significance. In religious tourism, it is not necessarily involving the tourist's belief in a certain religion. The concept of culture entails the need to treat religious tourism as a manifestation of fundamental motivations. Religion, as a part of culture, can contribute to cultural conservation.

Religious tourism is a rapidly growing global phenomenon that involves the travel of individuals or groups to sacred destinations for spiritual or religious purposes. This research paper aims to explore the complex interplay between harmony and holiness in religious tourism contexts. It explores the various issues and challenges associated with religious tourism. It examines the socio-economic impacts, environmental concerns, cultural clashes, and management challenges that arise from the influx of pilgrims and tourists visiting sacred sites. By analyzing case studies and examining best practices, this paper seeks to propose sustainable strategies to navigate the challenges of religious tourism while preserving the sanctity and cultural integrity of sacred sites. Promoting harmony and holiness in religious tourism requires a collective effort from all stakeholders. Responsible tourism practices that prioritize sustainability, cultural preservation, and interfaith dialogue are essential for ensuring a positive impact on both visitors and local communities.

The study attempts to analyze the temple Corridor Project in detail and understand its impact on cultural tourism. It focuses on the scope and challenges of heritage preservation and tourism infrastructure development, had been conducted through field visits and observations. During the investigation, different sources have been taken into account, including document review, personal interviews with the experts, local community stakeholders, and those who are/were directly or indirectly involved or affected by the SKVTC project. The study proposed unique way of using different but related set of factors of different natures in cultuaral heritage tourism, and hence offers an important empirical contribution to the literature of heritage management.

This research aimed to examine the Travel and Tourism Competitiveness Index (TTCI) as an evaluation tool for heritage sites with limited tourism development. The authors aimed to assess how well the TTCI aligns with stakeholders' perceptions when applied to sites facing conservation challenges. The research revealed that stakeholder perceptions significantly impact heritage planning strategies. The study highlighted the potential of stakeholder engagement in enhancing a destination's competitiveness by addressing visitor satisfaction, overall tourism appeal, and the likelihood of repeat visits. The findings emphasized the importance of a collaborative governance model and suggested potential modifications to the TTCI to better suit underserved heritage sites. The research aimed to transform heritage planning into an engaging and profitable endeavor, emphasizing the need for conservation improvements at sites lacking proper attention and promoting successful heritage planning for other sites by making them more appealing to tourists and visitors.

Tourism, as well as ICH, boosts the country's economy. Humans are the most significant aspect of intangible cultural assets and tourism, because other resources cannot meet goals and many factors influence institution development. The purposes of the current study are twofold, i.e., to explore the role of human resources in the preservation of intangible cultural heritage and to examine the process of maintaining and developing skilled and efficient human resources in the sustenance of ICH. To collect data, the articles from databases were searched using keywords and ultimately sixteen articles were reviewed and analyzed. The study findings suggested that unlike tangible heritage, which can be physically touched, intangible cultural heritage is more difficult to preserve. It is being neglected due to a lack of government initiative as well as other social or financial factors. Due to the host community's lack of skill, as a result of inadequate training, the potential for human resource development is also being overlooked. Training guidelines can be prepared for maintaining ICH.

In this part of the book, the importance of virtual museums and their place in the conservation of cultural heritage and cultural heritage tourism will be emphasized. In this part, first, the concept of museum, the emergence and development of virtual museums, the importance of museums and virtual museums in the conservation and cultural heritage tourism and transfer of cultural heritage to future generations, the importance of virtual museums in terms of education will be conceptually explained. The increasing importance of virtual museums during the Covid-19 pandemic will also be mentioned by giving examples around the world. Within the context of the current study, students attending Aksaray University (Turkey) will be made to visit virtual museums of various museums around Turkey and the world. After the visit, the Virtual Museum Satisfaction Scale will be administered to the students. The data to be collected through the scale will be evaluated by using descriptive analysis and difference analyses applied in relation to different variables.

The rural area of Sriniketan in West Bengal, India is full of cultural embodiments that can not only serve as a base to develop tourism but also generate sustainable livelihoods. However, the Sriniketan region suffers from chronic poverty and its unique culture is getting depleted thanks to the lack of awareness and interest among locals. With the help of the DFID Sustainable Livelihood Framework (DFID-SLF), this study tries to analyse the contributions that culture-based tourism can make towards generating sustainable livelihoods at Sriniketan. A modified SLF has been prepared with an added element of cultural capital as a contribution to the existing livelihood literature and guiding sheet for future practitioners. Based on the primary (in-depth interviews of fifteen households and five key respondents) and secondary data collected, this paper concludes that tourism development in Sriniketan can not only aid its cultural preservation but also generate an alternate source of livelihood and thereby, making both (culture and livelihoods) sustainable

The travel industry globally involves the movement of people from one place to another due to various reasons. Every human travelling has different needs and aspirations that they expect to get fulfilled at the places they travel. The involvement of humans as the key factor of travel acts as the indicators for the most complex and dynamic factors. It is not only the attraction that fulfills the note of satisfaction in the traveler, but every interaction at various places during the travel that creates an experience. Human Resource Management involves people at the centre of its study and helps develop the quality of the

Foreword

This book unlocks the potential of culture and heritage tourism to enrich our lives and connect us with the diverse tapestry of human history. There are nineteen chapters in this collection and all the authors claim responsibility of handling their subjects in a remarkable manner. I'm happy to introduce this collection to the series of books on advances in Tourism Industry, IGI Global, USA. Through the book, readers will have chance to find crucial and different insights for assessment and management of culture and heritage tourism. In this work all the authors have investigated on contemporary trends and issues of culture and heritage tourism management and practices around the world.

This book is composed of nineteen chapters with a wide range of diversified contemporary culture and heritage tourism trends and issues with managerial and strategic implications. The first chapter of this book, aims to deal with the economic value of heritage towns and local economic phenomena. The second chapter is on how AR and VR enhance visitor experiences by providing immersive interactions with historical locations and artifacts. The author in the third chapter investigated the role of ancient temples in promoting tourism and preserving cultural heritage in Odisha. While the forth chapter has dealt with the challenges and best practices in balancing historical preservation with current expectations. The fifth chapter seeks to understand why tourists are drawn to historical, cultural, and heritage-related sites. By addressing this gap in research, the authors offer insights into the factors influencing cultural heritage tourism in Oman. In the sixth chapter the authors studied the attitudes of communities towards the development of cultural and heritage tourism and discusses the emergence of cultural and heritage tourism, challenges, and future trends. The seventh chapter discusses how religion, as a part of culture, contributes to cultural conservation. By examining the connection between religious tourism and cultural preservation, this chapter sheds light on the complex dynamics of cultural heritage in religious contexts. Eighth chapter examines the socio-economic impacts, environmental concerns, cultural clashes, and management challenges that arise from the influx of pilgrims and tourists visiting sacred sites. The chapter proposes sustainable strategies to navigate these challenges while preserving the sanctity and cultural integrity of sacred sites. Responsible tourism practices are emphasized for a positive impact on visitors and local communities. Chapter nine offers a unique perspective on using a combination of factors to improve cultural heritage tourism and provides important insights into heritage management.

Chapter ten addresses the issue of animal cruelty and sacrifice in religious and mythological tourism in South Asia. They reviewed the literature surrounding the ritualistic killing and cruelty of animals in religious tourism and highlight region-specific cruelty-conservation paradoxes. The chapter also discusses responsible and sustainable orientations in animal utilization in tourism. It concludes with recommendations for policy issues and business scopes in this context.

The next chapter eleventh focuses on community participation and the government's role in the development of religious tourism. The study assesses the potential of religious tourism in the area and evaluates the level of community participation in tourism development. Twelfth chapter explore the hedonic experiences of tourists at a UNESCO heritage site in India the study provides valuable insights into improving tourist experiences and connecting with heritage monuments

The chapter thirteen discusses the impact of cultural tourism on community development in Oman's Dhofar governorate. It highlights the role of cultural heritage tourism, especially its intangible component, in the region's economic growth. It offers recommendations for maximizing the benefits of cultural heritage tourism while addressing challenges. The next fourteen explores the role of folk music in creating memorable tourism experiences in this chapter. Fifteen chapter examine the Travel and Tourism Competitiveness Index (TTCI) as an evaluation tool for heritage sites in this chapter. They emphasize the impact of stakeholder perceptions on heritage planning strategies and competitiveness. Sixteen chapter focuses on the critical role of human resources in the preservation and sustenance of Intangible Cultural Heritage (ICH) within the context of tourism. Chapter seventeen emphasizes the importance of virtual museums in conserving cultural heritage in this chapter. The author discusses the concept of virtual museums and their significance in education and heritage conservation, especially during the Covid-19 pandemic. Chapter eighteen focus on the potential of culture-based tourism to generate sustainable livelihoods and preserve culture in Sriniketan, West Bengal, India. Using the DFID Sustainable Livelihood Framework, the chapter assesses the contributions of cultural tourism to local livelihoods and cultural preservation

The last chapter explores the significance of human resource management in enhancing the tourist experience at cultural destinations. The chapter emphasizes the role of human interaction in creating a positive experience for tourists.

Therefore, I congratulate the editor and the contributors who had put in a lot of effort and time in bringing out this book to address many vital concerns for a better understanding of the issues surrounding our economy.

Bivraj Bhushan Parida
Mahatma Gandhi University, India & University of Burdwan, India

Preface

Welcome to *Exploring Culture and Heritage Through Experience Tourism*. In this preface, we want to emphasize the crucial role that research plays in addressing the challenges and complexities outlined in this comprehensive reference book. Our aim is to shed light on how research can be a powerful tool for understanding, mitigating, and ultimately solving the pressing issues within culture and heritage tourism.

The world of culture and heritage tourism is vast and intricate, with numerous facets that demand our attention. As the editors of this volume, we recognize that the issues surrounding cultural and heritage tourism are not only evolving but also interwoven with various socio-economic and environmental factors. This is where research becomes paramount.

1. Identification and Recognition: The first goal of this book is to identify and recognize new areas of culture and heritage tourism. Research allows us to explore the uncharted territories of this field, discover hidden gems, and shed light on aspects that have been overlooked. By conducting thorough research, we can uncover the richness of cultural experiences waiting to be shared with travelers.
2. Understanding and Conceptualization: To truly understand culture and heritage tourism, we must delve deep into its concepts, scope, prospects, and challenges. Research provides the foundation for this understanding. It enables us to gather data, analyze trends, and draw meaningful conclusions. Through research, we can conceptualize innovative ideas and strategies that will shape the future of cultural and heritage tourism.
3. Inter-linkages: Culture and heritage tourism are not isolated domains; they have profound effects on various sectors and the overall economy. Research allows us to map out these inter-linkages. By studying the data and conducting rigorous analyses, we can gain insights into how culture and heritage tourism impact local communities, economies, and the global landscape. This knowledge is essential for informed decision-making and policy development.
4. Community and Society: Cultural and heritage tourism cannot thrive without the active participation of communities and cooperation at a transnational level. Research is the tool that helps us understand the intricate dynamics between cultural tourism, community development, social economy, and international collaboration. Through research, we can identify best practices, challenges, and opportunities for fostering sustainable relationships.
5. Opportunities and Integration: Identifying opportunities for cultural and heritage tourism to integrate with other diverse areas, such as events, resort development, and community orientation, requires thorough research. Research enables us to assess market trends, consumer preferences, and emerging technologies that can be harnessed to create unforgettable cultural experiences.

6. Solving Complex Issues: The challenges outlined in this book, including disaster mitigation, sustainability, marketing ethics, human resource management, policy frameworks, and technological implications, are multifaceted and often require multifaceted solutions. Research equips us with the data and insights needed to develop effective strategies and policies that address these challenges comprehensively.

As you embark on your journey through the chapters of this book, we encourage you to view research as your guiding light. Research is not just a means to an end; it is the compass that will help you navigate the intricate world of culture and heritage tourism. It empowers us to tackle the challenges, seize the opportunities, and create a future where culture and heritage are not just preserved but also celebrated through tourism.

In conclusion, research is the key that unlocks the potential of culture and heritage tourism to enrich our lives and connect us with the diverse tapestry of human history. We invite you to immerse yourself in the knowledge and insights presented in this volume, and we hope it inspires you to embark on your own research journey, contributing to the ever-evolving field of culture and heritage tourism.

CHAPTER OVERVIEW

Chapter 1: Proposed Methodological Framework for Assessing the Economic Value of Heritage Town and Local Economic Phenomena

This chapter by Varsha Vinod, Satyaki Sarkar, and Supriyo Roy explores the economic value of heritage towns and local economic phenomena. It highlights that heritage towns with their historic structures have a unique identity that can be economically beneficial if properly maintained. The authors discuss the role of entrepreneurship in the heritage economy and propose a methodological framework for assessing the economic value of heritage towns. Readers can expect to gain insights into the economic significance of heritage preservation and how to evaluate it.

Chapter 2: Augmented Reality and Virtual Reality in Cultural Heritage Tourism Enhancing Visitor Experiences

In this chapter by Shakeel Basheer, Sheezan Farooq, Viana Hassan, Younis Malik, and Murtaza Reshi, the focus is on the use of Augmented Reality (AR) and Virtual Reality (VR) in cultural heritage tourism. The authors explore how AR and VR enhance visitor experiences by providing immersive interactions with historical locations and artifacts. They discuss the potential of these technologies to transform the way tourists understand and appreciate heritage sites. However, they also address challenges like cost and authenticity. This chapter offers insights into the future of cultural heritage tourism through technology.

Chapter 3: Ancient Temples and Cultural Heritage of Odisha, India Issues and Challenges

Soumendra Biswas and Puspanjali Mohapatra delve into the ancient temples and cultural heritage of Odisha, India, in this chapter. They provide an overview of the unique ancient temples in coastal Odisha and their significance. The authors discuss how these temples contribute to the socio-economic development of local communities and protect their culture. This chapter sheds light on the role of ancient temples in promoting tourism and preserving cultural heritage in Odisha.

Chapter 4: Crafting an Experiencescape for Sustainable Cultural Tourism: A Case Study of Udaipur's Craft Village 'Shilpgram'

Saumya Kapil and Bindi Varghese examine sustainable cultural tourism development in Udaipur's craft village, 'Shilpgram,' in this chapter. The authors focus on the annual craft fair held in Shilpgram and assess its feasibility as an experiencescape model. Through case studies and interviews with stakeholders, they explore the challenges and best practices in balancing historical preservation with current expectations. This chapter contributes to our understanding of cultural tourism's responsibilities in cultural preservation and economic development.

Chapter 5: Motives of Cultural Heritage Tourism: The Case of World Heritage Sites in Oman

Mohammad Soliman, Samskrati Gulvady, Hamed Almuhrzi, and Musallam Al-Aamri investigate the motivations behind cultural heritage tourism in Oman, with a focus on UNESCO-listed World Heritage sites. This chapter seeks to understand why tourists are drawn to historical, cultural, and heritage-related sites. By addressing this gap in research, the authors offer insights into the factors influencing cultural heritage tourism in Oman. This conceptual work has practical implications for destination managers and policymakers.

Chapter 6: Community's Attitude towards the Development of Cultural and Heritage Tourism: An Exploratory Study

Amrik Singh and Supina Supina explore the attitudes of communities towards the development of cultural and heritage tourism in this chapter. They emphasize the significance of tourism, heritage, and culture in fostering economic development and preserving cultural identities. The authors discuss the emergence of cultural and heritage tourism, challenges, and future trends. This chapter provides valuable insights into the dynamics of community involvement in tourism and its impact on culture and heritage.

Chapter 7: Religious Tourism Areas and Cultural Conservation

Mokhlesur Rahman delves into the intersection of religious tourism and cultural conservation in this chapter. He highlights the role of religious attractions in cultural and economic activities. The chapter discusses how religion, as a part of culture, contributes to cultural conservation. By examining the connection between religious tourism and cultural preservation, this chapter sheds light on the complex dynamics of cultural heritage in religious contexts.

Chapter 8: Harmony and Holiness: Navigating the Challenges of Religious Tourism

Bhavna Taneja explores the challenges associated with religious tourism in this chapter. She examines the socio-economic impacts, environmental concerns, cultural clashes, and management challenges that arise from the influx of pilgrims and tourists visiting sacred sites. The chapter proposes sustainable strategies to navigate these challenges while preserving the sanctity and cultural integrity of sacred sites. Responsible tourism practices are emphasized for a positive impact on visitors and local communities.

Chapter 9: Institutionalizing Cultural Tourism for Heritage Preservation: Comprehending Shri Kashi Vishwanath Temple Corridor (SKVTC) Project

Atul Tripathi, Sujay Singh, and Rashmi Jha analyze the Shri Kashi Vishwanath Temple Corridor (SKVTC) Project in detail in this chapter. They aim to understand its impact on cultural tourism and heritage preservation. Through field visits, interviews, and document reviews, the chapter examines various factors related to cultural heritage tourism. The proposed chapter offers a unique perspective on using a combination of factors to improve cultural heritage tourism and provides important insights into heritage management.

Chapter 10: Animal Cruelty and Sacrifice in Religious and Mythological Tourism Perspective and Challenges in South Asia

Pratim Chatterjee, Debasish Batabyal, Namrata Chakrabortty, Yashwant Rawal, Sankar Mukherjee, and Soumyadeep Roy address the issue of animal cruelty and sacrifice in religious and mythological tourism in South Asia. They review the literature surrounding the ritualistic killing and cruelty of animals in religious tourism and highlight region-specific cruelty-conservation paradoxes. The chapter also discusses responsible and sustainable orientations in animal utilization in tourism. It concludes with recommendations for policy issues and business scopes in this context.

Chapter 11: Community Participation and Government Role in Religious Tourism Development in Kurukshetra

Lakhvinder Singh focuses on community participation and the government's role in the development of religious tourism in Kurukshetra in this chapter. Using survey data, the study assesses the potential of religious tourism in the area and evaluates the level of community participation in tourism development. The chapter concludes by emphasizing the need for a participatory approach in tourism planning to accelerate overall development.

Chapter 12: Hedonic Experience of Visitors at a Heritage Site: A Netnographic Study

Partho Seal and Debasish Batabyal explore the hedonic experiences of tourists at a UNESCO heritage site in India in this chapter. Using a netnographic approach, they identify three categories contributing to visitors' hedonic experiences: multisensory, imaginative, and emotional. The study provides valuable insights into improving tourist experiences and connecting with heritage monuments.

Chapter 13: Cultural Tourism and Community Development: Dhofar Governorate as a Case

Samskrati Gulvady, Mohammad Soliman, Hamed Almuhrzi, and Musallam Al-Aamri discuss the impact of cultural tourism on community development in Oman's Dhofar governorate. The chapter highlights the role of cultural heritage tourism, especially its intangible component, in the region's economic growth. It offers recommendations for maximizing the benefits of cultural heritage tourism while addressing challenges.

Chapter 14: Rhythms of Living Cultural Heritage: Co-Creating Memorable Tourism Experiences

Minna-Maarit Jaskari explores the role of folk music in creating memorable tourism experiences in this chapter. Focusing on the Kaustinen Folk Music Festival, the chapter discusses how folk music contributes to the preservation and promotion of living cultural heritage. It sheds light on the unique ways in which cultural heritage and tourism intersect through music.

Chapter 15: Unlocking the Power of Stakeholder Perception: Enhancing Competitive Heritage Planning and Place-Making

Aditi Nag and Smriti Mishra examine the Travel and Tourism Competitiveness Index (TTCI) as an evaluation tool for heritage sites in this chapter. They emphasize the impact of stakeholder perceptions on heritage planning strategies and competitiveness. The chapter highlights the importance of stakeholder engagement in enhancing a destination's appeal and proposes modifications to the TTCI to better serve underserved heritage sites.

Chapter 16: Nurturing the Human Capital - Essential Skills and Competencies for the Sustenance of Cultural and Heritage (ICH) Tourism: A Systematic Literature Review

In this chapter by Sarani Bhaumik, Sreeparna Bhattacharyya, and Sarbari Bhowmick, the authors focus on the critical role of human resources in the preservation and sustenance of Intangible Cultural Heritage (ICH) within the context of tourism. The chapter highlights that both tourism and ICH play significant roles in boosting a country's economy, with human resources being a vital component in managing and promoting ICH.

Chapter 17: Virtual Museums in the Conservation of Cultural Heritage and Satisfaction Level with Visits to Virtual Museums

Funda Varnaci Uzun emphasizes the importance of virtual museums in conserving cultural heritage in this chapter. The author discusses the concept of virtual museums and their significance in education and heritage conservation, especially during the Covid-19 pandemic. The chapter includes a study on student visits to virtual museums and their satisfaction levels, offering insights into the role of technology in heritage preservation.

Chapter 18: Sustaining Livelihoods and Culture through Tourism Development: The Case of Sriniketan in West Bengal, India

Priyakrushna Mohanty, Rupa Sinha, and Jyothi Kumar focus on the potential of culture-based tourism to generate sustainable livelihoods and preserve culture in Sriniketan, West Bengal, India. Using the DFID Sustainable Livelihood Framework, the chapter assesses the contributions of cultural tourism to local livelihoods and cultural preservation. It introduces a modified framework that incorporates cultural capital, providing guidance for practitioners.

Chapter 19: The Impact of Human Resource Challenges at Cultural Destinations and the Way Forward for Proper Destination Management

Udaidip Singh Chauhan explores the significance of human resource management in enhancing the tourist experience at cultural destinations. The chapter emphasizes the role of human interaction in creating a positive experience for tourists. It discusses how human resource management can be a differentiator in tourist experiences and leaves a positive impact on destination management.

In Summary

In concluding this preface for our edited reference book, "Exploring Culture and Heritage Through Experience Tourism," we would like to emphasize the profound importance of the themes and chapters presented in this comprehensive volume.

The exploration of culture and heritage through experience tourism is not merely an academic pursuit but a dynamic field that bridges the past with the present, and offers a vision for the future. As editors, our intention was to bring together a diverse range of scholarly perspectives and practical insights that collectively enrich the understanding of this multifaceted relationship.

Throughout the chapters, we have delved into various dimensions of cultural and heritage tourism, including economic assessments, technological advancements, community involvement, stakeholder perceptions, and the preservation of intangible cultural heritage. Each chapter contributes a unique perspective, offering readers a wealth of knowledge to draw upon.

It is evident that culture and heritage hold immense potential for not only driving economic growth but also fostering social cohesion, environmental sustainability, and the preservation of our shared human history. The chapters in this book underscore the need for careful planning, responsible management, and the engagement of diverse stakeholders to harness this potential.

As we move forward, it is imperative that academia, policymakers, practitioners, and communities work hand in hand to shape the trajectory of culture and heritage tourism. The challenges and opportunities discussed within these pages serve as a compass, guiding us toward a future where the richness of our cultural tapestry is celebrated, preserved, and shared with the world.

We extend our gratitude to the dedicated authors who have contributed their expertise and insights to this volume, as well as to the readers who will embark on a journey through these pages. We hope this book serves as a valuable resource for researchers, students, policymakers, and industry professionals, inspiring continued exploration and innovation in the realm of culture and heritage tourism.

In the ever-evolving landscape of global tourism, let us remain committed to the principles of responsible tourism and the preservation of our cultural and heritage treasures. May this book be a source of inspiration and knowledge that guides us in our efforts to explore, appreciate, and protect the diverse cultural and heritage assets that make our world so beautifully unique.

Warmest regards,

Yashwant Singh Rawal
Parul Institute of Hotel Management, Parul University, India

Rupa Sinha
Amity Institute of Travel and Tourism, Amity University, India

Sankar Kumar Mukherjee
Amity Institute of Travel and Tourism, Amity University, India

Debasish Batabyal
Amity Institute of Travel and Tourism, Amity University, India

Chapter 1
Proposed Methodological Framework for Assessing the Economic Value of Heritage Town and Local Economic Phenomena

Varsha Vinod
https://orcid.org/0000-0003-1153-2865
Birla Institute of Technology, Mesra, India

Satyaki Sarkar
https://orcid.org/0000-0002-5161-2344
Birla Institute of Technology, Mesra, India

Supriyo Roy
https://orcid.org/0000-0003-0600-0696
Birla Institute of Technology, Mesra, India

ABSTRACT

A historic town centers local economic activities and has a significant concentration of heritage structures. Such a place or historic core generally has a definite identity—a distinct townscape character worthy of preservation. It has been proven through several studies that cultural heritage has value and the ability to be a significant component of community development, provided it is properly maintained. This chapter aims to identify the prime movers of the heritage economy and utilize the potential of heritage towns to retain and enhance the role of heritage in its stakeholders' social and economic lives. Based on an extensive systematic literature review, the role of entrepreneurship in the heritage economy and various parameters that contribute to the economic assessment of heritage towns are explored. The proposed methodological framework contributes to identifying techniques for assessing the economic value of heritage towns for local economic development.

DOI: 10.4018/978-1-6684-9957-3.ch001

1. INTRODUCTION

A heritage town centers local economic activities and has a significant concentration of heritage structures. Such towns or historic cores generally have a definite identity - a distinct townscape character and image of the city worth preserving. Hence, in addition to its economic importance, the place gains a symbolic value that attributes common identities to its citizens. In their urge to develop, the inheritance from the past takes a backseat, allowing space for new developments (P.C. Guzmán, 2016). Such growth often results in a paradoxical situation. The need for a high return on economic values creates pressure for change, demolition, and renewal. The stress is further aggravated if any part of the historical fabric is marked by obsolescence.

On the other hand, the place's historical, townscape, or symbolic values urge its preservation. The idea of heritage has expanded over the past several decades, including both tangible and intangible cultural legacy, also increasing interest in extensive knowledge of the role heritage plays in our society (Babic, 2015). The major challenges historic towns face is to accommodate the ever-increasing pressure of economics and rejuvenating the existing fabric such that they become an integral part of the continuity of modern life. With appropriate policies and programs, it is possible to balance preservation and conservation issues with modern economic requirements. Heritage has unquestionably become one of the most critical drivers of global tourism.

The intrinsic aspect of small heritage towns is their significant historical value, and their economy is highly dependent on heritage, tourism, and cultural entrepreneurship. The heritage conservation aspect for small heritage towns has a significant linkage with the economic competitiveness of the area and their current standing in an economic value assessment framework. However, such a technique of heritage value assessment needs to be more present in small and more significant heritage towns (Florentina-Cristina Merciu, 2021). Hence it is difficult for such towns to understand their status and which sectors to invest in and improve on for enhanced economic development.

The significance of cultural components to the economic performance of towns has grown during the last several years. This tendency is strongly linked to urban cultural hotspots and tourism. Although cultural attractions, locations, and monuments dominated early cultural/heritage tourism efforts, it is evident that gradually performing arts, handicrafts, and other intangible cultural heritage has begun to attain great significance. The success of cultural destinations largely depends on their favorable economic impact and the long-term contribution they can provide toward regional economic growth (Alvaro Fierro, 2018).

There are plenty of entrepreneurship opportunities if the benefaction of resources is rightly executed in identified heritage areas. Since the economy is an integral part of any development process, there is an extreme need to identify a path for integrating entrepreneurship avenues with the local economy, thereby promoting the growth of local services and industry at the front end and having access to ample funds for maintenance and upkeep of heritage structures in the back end. Various studies and reports show that various small heritage towns in India must exhibit a frame to integrate all these aspects (Reher, 2021).

This highlights the need for economic value assessment. Heritage economic value assessments can play a vital role in realizing the level of acceptance of a particular heritage ecosystem to the stakeholders when compared with similar systems. There is a need for such systems to be modeled for heritage towns that will pave the way for self-assessment of such towns as well as help them to identify benchmarks to prosper and establish themselves economically in the national and global arena (Vegheș, 2018)

2. WHY ECONOMIC VALUE ASSESSMENT FOR HERITAGE TOWNS?

One of the biggest concerns is persuading society and the stakeholders that safeguarding cultural assets is vital. The importance given to the economic value of an asset can degrade other types of value, such as cultural, educational, or environmental values, that may be equally significant. "Worth" is often linked to a very econometric standpoint (Reher, 2021). The impact of heritage can be observed in areas like the production of direct economic activity, heritage preservation, creation of new cultural assets, development of a creative image of the city, etc. A heritage town faces various challenges as the consequences of importance towards the economic aspects of the town, as well as its institutional setup and the importance of the town at the national and global levels. A few of such challenges have been listed below -

- Tremendous economic pressure leads to ever-increasing issues of inadequate or obsolete infrastructure.
- Lack of community involvement and awareness negatively affects tourism and economic development.
- Lack of platform to assess the heritage initiative status and the resulting economic growth.
- Improper benefaction of heritage resources in terms of economic development
- Lack of direction to channel funds for the development of small heritage towns.
- Lack of adequate management policies about conservation and integrated development of the town.
- Lack of entrepreneurial activities at the desired level to boost the economy.

The significance of cultural components to a city's economic performance has escalated during the last several years. This pattern may be observed in urban cultural sites and tourism. The success of cultural destinations is largely attributed to their favorable economic effects and long-term contributions to local development (Alvaro Fierro, 2018).

The problems of heritage sites present prospects for future development to benefit the sites in economic, social, cultural, environmental, and integrated developmental terms.

- Better self-realization of current economic status,
- Enhanced entrepreneurship and community participation,
- Tourism opportunities and economic development,
- Better conservation and development policy inclusion

3. LITERATURE REVIEW

An extensive literature study was conducted to assess the existing research on heritage, local economic development, and the scope for entrepreneurship in heritage towns. The review contributes to understanding the role of urban heritage conservation in economic and socio-cultural development, identifying prime movers of the heritage economy, and inferring the various linkages and parameters that contribute to the economic assessment of heritage towns. A systematic review of 73 papers, comprising 18% (13 papers) on heritage, and conservation, 33% (24 papers) concerning heritage, society, and economy, 15%

(11 papers) exploring cultural entrepreneurship, and 34% (25 papers) focusing on various case studies and models was conducted to formulate an economic assessment model for heritage towns.

Heritage has been observed to have tremendous potential for fostering social innovation by fusing economic growth with sustainability. Several factors, ranging from social inclusion, volunteerism, employment opportunities, the environment (maintaining distinctiveness through conservation), sustainability (continued care and sustainable use of historic environment), laws, and regulations are considered when evaluating the significance of heritage and its effects on society (Reher, 2021). To maximize the impact and endurance of cultural heritage to empower local people, heritage management and research into its effects play a vital role. The present threats to heritage are found to be cultural globalization, unchecked development, exclusive tourist-oriented development, the commodification of culture, migration, and resettlement; and poverty, migration & social exclusion, poor or inadequate physical and social infrastructure, land tenure and ownership, gentrification, etc. as challenges to historic urban core (Bhavin Shukla, 2014). Empowering the travel and tourism industry to develop cultural tourism products & services and providing opportunities for cultural tourists to discover, explore and enjoy cultural experiences is extremely important. Revitalization of cultural heritage will enable communities to transform into alluring tourist destinations. (Vegheş, 2018).

In the quest to assess the status of heritage and management, understand the current challenges faced in heritage management in India, and provide a future roadmap for the development of national heritage, an initial survey of the archaeological heritage at the national level was conducted with an expert group discussion. It considered factors such as level of significance, use of technology for national heritage, conservation initiatives at the national level, evacuation & exploration of heritage sites, proposals for site museums, heritage tourism, revenue generation, heritage marketing, international collaborations, collaborations for skill development (Aayog, Improving Heritage Management in India, 2019).

Cultural involvement is said to have resulted in improved self-worth and leads to intellectual curiosity and the will to learn. The link between culture and education could be reinforced by digitization (Group, 2021). The culture-based development (CBD) model is significant for analyzing entrepreneurship behavior and indicates a substantial impact of culture on entrepreneurship activities (Dai, 2021). The characteristics such as cultural identity, support for consumption, stimulation of innovative tools, and in-situ experiences enhance local/small entrepreneurship development (Valentina Vasile, 2016).

Five major measurables of the economic impacts of heritage conservation were identified in research by (Rypkema, 2008), indicating (i) employment and income, (ii) revitalization of the city center, (iii) heritage tourism, (iv) real estate & property values (v) entrepreneurship development. A study was proposed to develop a systematic assessment methodology that correlates cultural heritage and sustainable development (P.C. Guzmán, 2016). It focused on various aspects such as trends in the urban monitoring of cultural heritage, strategic level: heritage conservation, a tool for development, operational level: best practices, integrating heritage conservation within urban planning, analysis of practices by urban development approach, analysis of cultural heritage-related indicators, comparative analysis on the coverage of sustainability dimensions. Parameters such as policy changes in the region, cultural heritage values, willingness to pay, attitudes and perceptions of residents, socioeconomic characters, resource allocation, and financing of the cultural heritage play a significant role in micro-level evaluation techniques.

The assessment of the value of architectural heritage using the weight of criteria and overall aggregation of collected data, research was formulated which has proven to be extremely useful to determine the level of importance of parameters using experts' opinions (Partha Sarathi Mishra, 2021). A study to explore the indicators currently used in cultural heritage research and practices for impact tool assessment highlighted that the parameters used to identify indicators ranged from geographic scale, usage, typology, address of sustainability dimension, etc. (Martina Bosone, 2021).

An evaluation system developed for World Cultural Landscape Heritage management comprises four general criteria - management foundation, management system, management measures, and management performance (Jiami Zhou, 2022). The primary aspects recommended for research are funding support, facilities, laws and regulations, management institutions, management mechanisms, heritage monitoring and maintenance, scientific research, heritage interpretation, and public education, management supervision, assessment and performance, and economic performance.

Table 1 highlights significant selective research that had been conducted, emphasizing the aim, research methodology, tools & techniques, and parameters considered. Understanding such aspects profoundly contributed to formulating study parameters for this research.

4. RESEARCH GAP

Inferring from the research and case studies referenced, certain drawbacks to the current approaches and assessment techniques require attention. The following are significant drawbacks and gaps that have been identified in the process –

- Lack of integration of geographical factors, environmental factors, local government potential, and heritage economics.
- The paucity of inclusion of stakeholders, their perceptions, and willingness to contribute and participate in the economy.
- Absence of a methodology interlinking social, economic, cultural, and socioeconomic aspects to determine the heritage economy.
- Lack of business models for converting enterprises to contribute to new heritage economy.
- Absence of an established relationship between cultural, entrepreneurial activities, local entrepreneurship development, and local economies promoting tourism.
- Lack of significant development in heritage marketing for small heritage towns.
- Lack of efficient model that considers heterogeneous industry, multiple users, etc.

The above observations highlight the dire need for an economic assessment model for heritage towns focused on integrating social, cultural, physical, economic, socioeconomic, institutional, environmental, and geographical factors. This lack of integration and interlinking is reflected in the heritage assessment models, business models, policymaking aspects, prioritization, fund allocations, etc. Hence, assessment is highly significant to highlight the sectors that require investments to uplift the condition of the respective cultural economy.

5. RESEARCH QUESTIONS

The literature studies lead one to ponder upon the following research questions:

i. How does economic value impact historic site preservation?

Table 1. Significant research relevant to the study

Aim	Methodology, Tools, and Techniques	Parameters
To determine the worth of architectural heritage for its management, using a weight of criteria and the overall aggregate of the data collected (Partha Sarathi Mishra, 2021).	i. Data Collection through primary and secondary sources to determine the parameters, ii. Identification of weights through the CRITIC method (correlation coefficient and standard deviation), iii. Identification of values of parameters through the GRA method, Ranking using the obtained index values.	• Architectural value - Aesthetic value, Architectural integrity, a rarity. • Economic value • Environmental and Ecological value - ecological threats, zoning compatibility, site integrity. • Historic Value - age value, associative value. • Socio-cultural value - cultural & religious value, functional value, public use value, service value, political value, etc.
To investigate how tangible and intangible cultural resources are related as a tool for development-focused preservation. (Bhavin Shukla, 2014).	i. Literature Review, ii. Case Study Method and Mapping	• Historic Built Form. • Characteristics of intangible heritage. • Evolution of protecting intangible cultural heritage. • Impact of ICH on tangible-built forms.
To highlight the role of microeconomic evaluation techniques in the cultural sector (Susana Mourato, 2002).	i. Description of economic valuation principles and tools, ii. Examination of a range of economic tools for the evaluation of cultural heritage iii. Analysis of advantages & disadvantages.	• Policy changes in the region, Cultural heritage values, Willingness to pay, • Attitudes and perceptions of residents, • Socioeconomic characteristics, resource allocation, • Financing the cultural heritage
To assess the impacts of heritage-led urban revitalization in George Town (David Throsby K. P., 2022).	i. Conduction of workshop. ii. Interview with stakeholders - business, residents, cultural institutions, visitors, and investors. iii. Summarization of surveys	• Impact on business, residents, cultural institutions, visitors, and investors concerning - economic aspects, social aspects, cultural aspects, • Willingness to pay, • Priorities for future development.
To simulate how external variables affect regional tourist potential (Nestor Shpak, 2021).	i. Identification of the elements and factors that contribute to the formation of regional tourist potential. ii. Development of a system for the direct external environmental factors. iii. Identification of indicators for evaluating the main direct external factors iv. Evaluation of the impact of the main direct environmental factors on the level of utilization of the regional tourist potential. v. Development of an appropriate methodology and testing based on a case study.	• Labor Resources –active population, registered unemployment. • Marketing Policy – number of tourist brands, prices of services, etc. • Tourism Management System • External Infrastructure – number of airports, railway stations, tourist centers, etc. • Quality of Environment – Capital Investments, Current Expenditure, Emission of pollutants, etc. • General Image of Region – Per capita income, crimes, etc.
To examine the impact of urban development pressure on CWHS To examine the function of the Heritage Impact evaluation as a tool for evaluation (Baharak Ashrafi M. K., 2020).	i. Recognizing the function of HIA in anticipating and analyzing possible risks and in minimizing disputes to protect heritage treasures, ii. Review and comparison of 4 case studies.	• Significance of the existing heritage, • Level of conservation works, role, and participation of stakeholders, • Role of international bodies in the urban heritage conservation process, timeline chart. • Perception of the community. • Possible challenges and threats.
To continually determine a particular set of indicators that may be applied to gauge the economic impact of historic preservation throughout time. (Donovan R. Rypkema, 2013).	i. Extensive Literature Review of preservation and economic links and measurements, ii. Interviews among the knowledgeable parties, iii. Conduction of an international symposium, iv. Briefing to the Advisory Board, Final Report.	• Jobs/Household Income, • Investments in the region, • Tourist Measurements (Demand side, Supply Side, Economic Measurements), Social Impacts • Property Values, Heritage Tourism, Environmental Measures, Downtown Revitalization,
To explore culture and creative industries' impact on the economy and urban resilience (Group, 2021).	i. Establishing that the Cultural and creative sectors are significant economic value and job sources. ii. Identify how culture can contribute to resilient recovery by driving innovation and entrepreneurship. iii. Formulation of future policies	• Share of Employment, the share of value-added, and a Comparison of the cultural sector's contribution. • Total consumption expenditure of households by type of expenditure. • Share of creative goods in total exports and imports.
To utilize the Cultural-Based Development paradigm to assess regional entrepreneurial behavior in China (Dai, 2021)	i. Literature Review - Driver of difference in regional entrepreneurship, ii. Application of CBD Model	• No. of private enterprises, • Immaterial living cultural capital, Material living cultural capital. • Human capital, Physical capital. • Material historic cultural capital. • Labor, immaterial historic cultural capital,

continued on following page

Table 1. Continued

Aim	Methodology, Tools, and Techniques	Parameters
To explore opportunities for cultural entrepreneurship to develop the local economy in Syria (M.A., 2019).	Descriptive method.	• Availability of Skilled workmanship for entrepreneurship development - promotion, management, creation of goods, cultural performers, etc. • Financial analysts, Planners, Number of tangible and intangible heritage present.
To design a framework that enhances eco-culture for small/ local heritage entrepreneurship development (Valentina Vasile, 2016).	i. Brief introduction and review of the eco-tourism and heritage value relationships. ii. Conduction of 2 surveys - scope survey and satisfaction survey	• Ease of accessibility of information. • Significance of business/ entrepreneurship development. • Significance of heritage marketing & inspiration for new business ideas.
To determine the necessary and suitable conditions to preserve historic places for cultural tourism while promoting local economic development (Edmunds, 2013).	Descriptive Case-study Approach.	• Demography. • Socioeconomic data. • The workforce in culture and creative industries, initiations by the government., • Community Engagement, • Presence of significant transit nodes.
To develop an evaluation system for the management of WCLH and provide suggestions for development (Jiami Zhou, 2022).	i. Selection of Evaluation Indicators. ii. Weight Assignment of Indicators. iii. Data Collection and Analysis from the case study area.	• Funding Support. • Facilities. • Laws and Regulations, • Management Institutions, Management Mechanisms. • Heritage Maintenance. • Management Supervision, assessment, and performance. • Economic Performance.
To create a methodological toolkit based on the cluster approach for the growth of cross-border tourism (Mukhambetov, 2021).	i. Selection of 19 indicators. ii. Grouping indicators into six groups. iii. Use Ward's Method and Euclidean square as objective functions and criteria for similarity and differences.	• Several cultural & historical monuments. • Rate of employment in the tourism industry, the proportion of income from the tourism sector, foreign tourism. • Several hotel accommodations. • Share of tourism investments in the region, the proportion of tourists visiting the facilities more than once, no. of tourists staying for more than three days. • The average annual generation of the number of jobs in the region as a whole and contribution from the tourism industry.
To ascertain the strategies for an effective management strategy at cultural heritage sites (Mahmoud Abdulkader Alghafri, 2020).	Data Collection through - primary surveys: questionnaires and interviews, secondary data for inputs.	• Stakeholders' attitude. • Cultural tourism and its benefits. • Significance of preserving the unused values of the CH in Syria. • Socioeconomic management. • Involvement of the local community.
To formulate a framework and tools for performing rapid evaluations of city economies to create city economic development plans (CEDPs) in Asian cities (Roberts, 2015).	i. Analyze, prepare and implement a CEDP and a variety of supporting plans. ii. Present a framework that guides the user through three processes to gather information. iii. Briefly describe the economic development challenges facing city planning and development.	• Economic data - qualitative and quantitative. • Changes in structural, socioeconomic, and economic geography activities. • Identify, map, and evaluate industry clusters, • Competitiveness of city economies. • Projections - population, economic growth, and demand for resources, • Futures risks to a city's economy.

Source: Author

ii. What key determinants govern the economic value assessment of small heritage towns?
iii. What is the contribution of each determinant in economic value assessment?
iv. How are entrepreneurship opportunities linked with economic values in the case of small heritage towns?
v. How can an entrepreneurship model be designed to boost heritage economics, the effects of which will multiply in the local economy?

Figure 1. Identified aspects of cultural heritage economy from literature studies

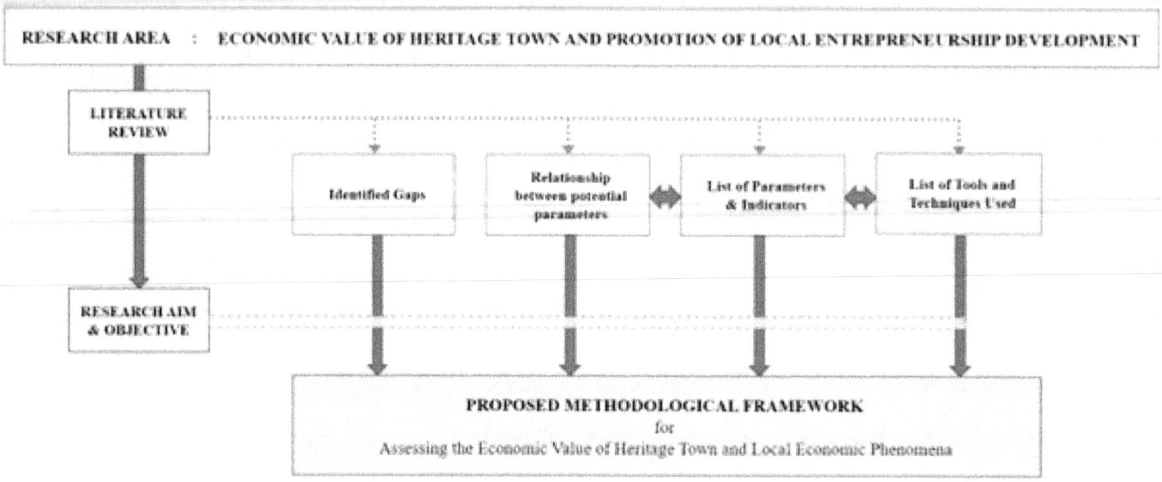

6. AIM AND OBJECTIVE

The paper aims to identify a methodological framework for the economic value assessment of heritage towns and utilize the potential of the protagonist to retain and enhance the role of heritage in its stakeholders' social and economic lives.

The objectives of this paper are enumerated below:

1. To identify the prime movers of the heritage economy and understand their interdependencies.
2. To create a platform for economic value assessment of small heritage towns by generating a model.
3. To establish the contribution of entrepreneurship activities to the overall economic value of small heritage towns.
4. To hypothesize heritage economy, assessment model.

The model that shall be formulated will contribute to the planning of long-term and short-term investments and promote local economic and entrepreneurship development. The problems of heritage towns and drawbacks of the current assessment approach act as prospects for future development to benefit the sites in economic, social, cultural, environmental, and integrated developmental terms such as follows:

- Developing an integrated assessment model that interlinks the social, cultural, economic, and physical aspects leads to identifying prime movers of the heritage economy for improved entrepreneurship activities, tourism opportunities, and economic development.
- Test the developed assessment model based on case studies.
- Formulation of better conservation and development policy for heritage towns to foster economic growth, promotion of entrepreneurship, and community participation.

7. METHODOLOGY

The research methodology to develop an assessment model to evaluate the economic value of heritage towns and thereby identify sectors for investment leading to local economic and entrepreneurship development has been phased into three stages, as illustrated in Figure 2.

In this research to formulate an assessment platform to evaluate the economic value of heritage towns, it is first necessary to understand the research and projects undertaken in the field of heritage, conservation, and local economic development. Thus, the first stage of this research is majorly based on identifying existing research gaps from an extensive literature review and formulating the research aim and objectives. This review shall also contribute to the identification of parameters and indicators that determine the heritage and economy in varied research. It also contributes to the knowledge pool of the tools, techniques and methodologies used for assessment of economic value of heritage and heritage towns.

Based on the aim and objectives of the research, the research then focuses on filtering the parameters and indicators relevant to this study and therefore tools and techniques associated with them. The parameters and indicators are shortlisted by the application of Delphi technique and Analytical Hierarchy Process (AHP), based on experts' surveys. The AHP method is one of the widely used methods that serve as a tool for choice, sorting, ranking and description.

Figure 2. Research methodology

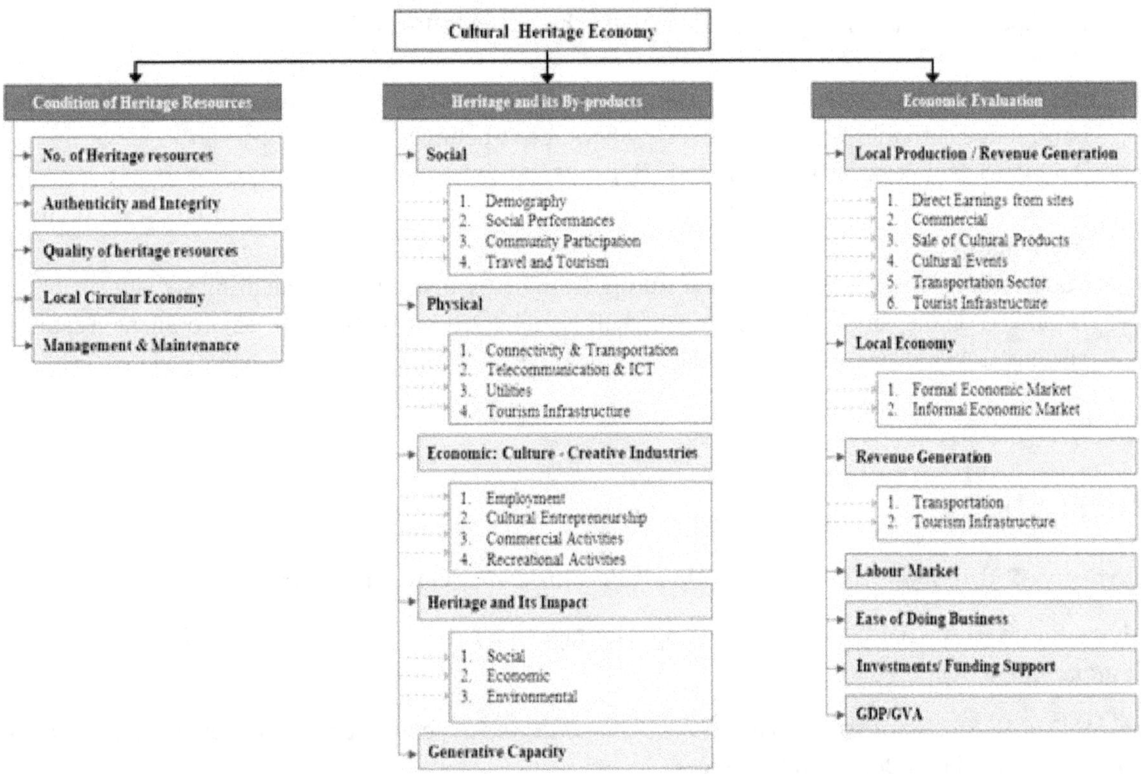

In the next stage, the association and interrelationship of the extracted parameters is assessed, which leads to the formulation of an economic assessment model for heritage towns. tools, and techniques will be formulated to generate an economic assessment model for heritage towns.

The model shall be hypothesized in the third stage of research. It shall consider the five selected study areas from the north, south, east, west, and central part of India to assess the specific scenario of small heritage towns in various parts of the country. Identifying specific parameters contributing to entrepreneurship development and strategies for investments and policymaking will henceforth be proposed.

8. CONCLUSION

The methodological framework mainly focuses on formulating an integrated heritage economic assessment model that includes heritage towns' social, cultural, physical, economic, geographical, and demographic aspects. The model shall be used for the economic assessment of identified heritage towns having similar constraints and problems. The model can also identify the sectors for investment for future entrepreneurship and policymaking aiming at Local Entrepreneurship Development (LED) by extracting primary parameters contributing to entrepreneurship activities.

1. Generation of Heritage Economy Assessment Model – A platform to assess the economic value of small heritage towns.
2. Identification of the investment sectors for future entrepreneurship and policymaking for Local Entrepreneurship Development.

The proposed methodological approach shall achieve the following:

- An economically vibrant society characterized by global outreach, primarily thriving from tourism-related activities being reinforced by heritage structures.
- An increase in per-capita income and GDP.
- An economically self-sustainable society will find its existence and growth based on demands from industry.
- Planned development of the town, resulting from better investments benefiting society.
- Improved quality of life for the citizens will be achieved by enabling an environment created by various stakeholders to make the place competitive.

REFERENCES

Aayog, N. (2019). *Improving Heritage Management in India*. Government of India.

Alvaro Fierro, I. A. (2018). Airbnb Branding: Heritage as a Branding Element in the Sharing Economy. *Sustainability*.

Babic, D. (2015). *Socially responsible heritage management - empowering citizens to act as heritage managers*. Social and Behavioral Sciences.

Baharak Ashrafi, M. K. (2020). Heritage Impact Assessment, beyond an Assessment Tool: A comparative analysis of urban development impact on visual integrity in Four UNESCO World Heritage Properties. *Journal of Cultural Heritage.*

Bhavin Shukla, K. B. (2014). *Development in the Context of Urban Conservation: Examining the Interrelationship between Architectural Heritage and Intangible Culture in Historic Settlements.* CEPT Archives.

Dai, Y. (2021). A Behavioral Cultural-Based Development Analysis of Entrepreneurship in China. *Sustainability.*

David Throsby, K. P. (2022). *Assessing the Impacts of Heritage-Led Urban Rehabilitation The Case of George Town, Penang, Malaysia.* Research Report - Getty Conservation Institute.

Donovan, R., & Rypkema, C. C. (2013). *Measuring Economic Impacts of Historic Preservation.* Advisory Council on Historic Preservation by Place Economics.

Edmunds, L. O. (2013). Maximizing the Local Economic Impact of Heritage Conservation for Cultural Tourism. *Asia Tourism Forum 2016 – The 12th Biennial Conference of Hospitality and Tourism Industry in Asia.*

Florentina-Cristina Merciu, A.-I. P.-L. (2021). *Economic Valuation of Cultural Heritage Using the Travel Cost Method: The Historical Centre of the Municipality of Bucharest as a Case Study.* Heritage.

Guzmán, A. P. (2016). Measuring links between cultural heritage management and sustainable urban development: An overview of global monitoring tools. *Cities (London, England).*

IG Group. (2021). *Economic and social impacts of cultural and creative sectors.* OECD.

M., A. (2019). The Effect of Cultural Heritage Sites in Local Economic Development - The Case Study of Syria. *IUPR (Russian Journal).*

Mahmoud Abdulkader Alghafri, V. I. (2020). Cultural Heritage Management as a Generator of Socio-economic Development (The Case Study of Syria). *SHS Web of Conferences.*

Martina Bosone, P. D. (2021). Indicators for Ex-Post Evaluation of Cultural Heritage Adaptive Reuse Impacts in the Perspective of the Circular Economy. *Sustainability.*

Mukhambetov, M. O. (2021). Cluster Approach. In *Cultural Heritage Tourism (Case Of The Central Asian Section Of Silk Road).* Central Asian Journal of Social Sciences and Humanities.

Nestor Shpak, O. M.-K. (2021). Simulation of the Influence of External Factors on the Level of Use of the Regional Tourism Potential: A Practical Aspect. *Administrative Sciences.*

Partha Sarathi Mishra, S. M. (2021). Value assessment of existing architectural heritage for future generation using criteria importance through inter-criteria correlation and grey relational analysis method: A case of Odisha temple architecture in India. *Current Science.*

Reher, G. (2021). Heritage as Action Search. *Sustainability, 14*(1), 282. doi:10.3390u14010282

Roberts, B. H. (2015). *Tool Kit for Rapid Economic Assessment, Planning, and Development of Cities in Asia.* Asian Development Bank.

Rypkema, D. D. (2008). *Heritage Conservation and the Local Economy*. Global Urban Development.

Susana Mourato, M. M. (2002). *Economic Valuation of Cultural Heritage: Evidence and Prospects*. The Getty Convention.

Valentina Vasile, M. R. (2016). *Heritage Entrepreneurship and Eco-tourism - A new vision on ecosystem protection and in-situ specific activities for cultural heritage consumption*. Economics and Finance.

Vegheş, C. (2018). Cultural Heritage, Sustainable Development and Inclusive Growth: Global Lessons for the Local Communities Under a Marketing Approach. *European Journal of Sustainable Development*, *7*(4). doi:10.14207/ejsd.2018.v7n4p349

Zhou, W. W. (2022). *Management effectiveness evaluation of world cultural landscape heritage: a case from China*. Heritage Science.

Chapter 2

Augmented Reality and Virtual Reality in Cultural Heritage Tourism Enhancing Visitor Experiences

Shakeel Basheer
https://orcid.org/0000-0001-7854-8285
Lovely Professional University, India

Sheezan Farooq
Islamic University of Science and Technology, India

Viana Hassan
https://orcid.org/0000-0002-7372-5059
University of Malta, Malta

Younis Mohammad Malik
https://orcid.org/0000-0001-7823-143X
Lovely Professional University, India

Murtaza Ahmad Reshi
Lovely Professional University, India

ABSTRACT

This book chapter explores the use of AR and VR in cultural heritage tourism, enhancing visitor experiences with historical locations and artefacts. AR and VR allow for immersive interactions, enabling visitors to comprehend and appreciate heritage places better. These technologies revive old structures and artefacts, offering personalized experiences and interactive storytelling. However, challenges like cost, hardware limitations, and maintaining authenticity need to be addressed. Future opportunities include wider accessibility, advancements in technology, and collaborations fostering innovation. AR and VR have the potential to revolutionize cultural heritage tourism by transforming visitor interactions and appreciation of historical sites.

DOI: 10.4018/978-1-6684-9957-3.ch002

1. INTRODUCTION

1.1 Background

Augmented Reality (AR) and Virtual Reality (VR) have emerged as game-changing technologies with the potential to transform how we experience and interact with cultural heritage tourism (González-Rodrguez et al., 2020). AR and VR technologies both produce immersive and interactive experiences that mix the physical and digital worlds, providing tourists with a new and improved approach to study historical locations, artefacts, and cultural monuments (Akpojaro & Bello, 2019). AR enriches tourist experiences in the context of cultural heritage tourism by superimposing digital information, such as historical data, 3D reconstructions, or multimedia content, onto the real-world environment (Papa et al., 2020). Tourists who utilise AR-enabled devices, such as smartphones or AR glasses, can see virtual features superimposed on the actual surroundings, which enhances their knowledge of the site's historical relevance and cultural context. The seamless integration of virtual content with the real environment provides visitors with a more engaging and instructive experience (Kamboj & Joshi, 2021).

As a way to promote historical and cultural places and support local economies, cultural heritage tourism has grown in popularity. The merging of augmented reality (AR) and virtual reality (VR) has changed how visitors interact with cultural heritage sites as technology develops. While VR develops completely immersive digital experiences, AR adds digital stuff over the actual world to enhance it. This essay explores the potential of augmented reality (AR) and virtual reality (VR) to improve visitor experiences in cultural heritage tourism and explores their consequences for engagement, preservation, and the tourist sector.

Augmented Reality (AR) and Virtual Reality (VR) have emerged as game-changing technologies with the potential to transform how we experience and interact with cultural heritage tourism (González-Rodríguez et al., 2020). AR and VR technologies both produce immersive and interactive experiences that mix the physical and digital worlds, providing tourists with a new and improved approach to study historical locations, artefacts, and cultural monuments (Bello et al., 2019). AR enriches tourist experiences in the context of cultural heritage tourism by superimposing digital information, such as historical data, 3D reconstructions, or multimedia content, onto the real-world environment (Papa et al., 2020). Tourists who utilise AR-enabled devices, such as smartphones or AR glasses, can see virtual features superimposed on the actual surroundings, which enhances their knowledge of the site's historical relevance and cultural context. The seamless integration of virtual content with the real environment provides visitors with a more engaging and instructive experience (Kamboj & Joshi, 2021).

By sending travellers to virtual copies of cultural heritage locations or historical periods, VR, on the other hand, gives a totally immersive experience (Jeong & Shin, 2019). Visitors can explore ancient ruins, historical buildings, or vanished civilizations as if they were physically present, even if the physical site is in ruins or no longer accessible, using VR headsets. This level of immersion not only aids in the preservation and promotion of cultural heritage, but also allows visitors to have a vivid and emotionally resonant experience of the past (Ayuningtyas & Ariwibowo, 2020). Destinations may attract a broader audience, especially younger generations who are more acclimated to digital encounters, by employing AR and VR technology in cultural heritage tourism. These technologies also help to preserve and conserve cultural sites by reducing the physical wear and tear caused by significant foot traffic (Tiwari et al., 2022). Finally, AR and VR technologies are revolutionising cultural heritage tourism by offering visitors engaging and immersive experiences that go beyond traditional methods. These technologies enhance visitor experiences and develop a better appreciation for our world's cultural heritage by blending the richness of history with the excitement of digital involvement.

Through the creation of immersive and compelling tourist experiences, augmented reality and virtual reality have proven their ability to alter cultural heritage tourism. Although obstacles still remain, they can be solved through technical improvements, teamwork, and ethical concerns. By combining AR and VR, we can provide travellers with fascinating and informative interactions with history, art, and culture while also assisting in the preservation of cultural heritage. In order to maximise the potential of the technology as it develops further and further for the benefit of cultural heritage tourism, stakeholders must collaborate.

2. LITERATURE REVIW

In order to give tourists a better understanding of the past and cultural significance, cultural heritage tourism involves exploring historical sites, artefacts, and traditions. Technologies like augmented reality (AR) and virtual reality (VR) have gained popularity recently, providing creative ways to improve visitor experiences in cultural heritage tourism. With a focus on how these technologies improve visitor experiences, this literature review aims to give a broad overview of the body of knowledge on the use of augmented reality and virtual reality in cultural heritage tourism.

The use of VR and AR in cultural heritage tourism has changed how tourists interact with historical monuments. When compared to virtual reality, augmented reality (AR) overlays digital material over the real world. The ability of these technologies to produce engaging and interactive experiences that make cultural heritage more approachable for a wide range of people has been highlighted by researchers (Guttentag, 2010). The involvement and education of visitors has been improved through the use of AR and VR technology. A richer knowledge of historical places and cultural relics may be achieved with the use of augmented reality (AR) smartphone applications, which can present 3D reconstructions, interactive components, and historical context in real time (Kersten et al., 2017). Virtual reality simulations let users virtually explore historical locations and events, providing a rich educational environment (Drot-Delange et al., 2018). In order to preserve and rebuild cultural heritage sites, AR and VR are essential. Through the use of these technologies, inaccessible or damaged sites can be virtually rebuilt, giving visitors the chance to experience historical settings that are no longer present (Subawa et al., 2021). This supports ongoing research efforts and helps to preserve cultural heritage.

Studies have discovered how AR and VR experiences affect our emotions and senses when we go to see cultural assets. Through the use of these technologies, visitors may be transported to other eras and fully immersed in historical narratives, which can elicit strong emotional responses (Xu et al., 2019). Particularly in the context of virtual reality, visitors might feel more emotionally connected to cultural items and venues (Gómez-Vega & J Picazo-Tadeo, 2019). There are difficulties to take into account, notwithstanding the obvious potential advantages of AR and VR in the tourism of cultural treasures. The user experience may be impacted by technical problems, such as connection and device compatibility (Hernández-Rojas et al., 2021). Another important factor is striking a balance between technical advancements and the preservation of genuine cultural experiences (De Luca et al., 2022). Applications for AR and VR in cultural heritage tourism must be well-liked by users and provide a positive user experience. Usability, content quality, and immersion are just a few of the elements that studies have looked into as they relate to user impression (Almeida et al., 2021). Improved visitor satisfaction and continuing participation in cultural heritage sites can result from good user experiences. The potential for augmented reality and virtual reality in cultural heritage tourism is still enormous as technology advances. The creation of tailored and adaptable experiences that use data analytics to customise information to users' tastes may be the main topic of future study (Braojos et al., 2020). Visitor experiences may also be improved by investigating how AR and VR might be combined with other cutting-edge technology like haptic feedback and artificial intelligence.

The implementation of augmented reality technology in a tourism-related exhibition significantly benefited visitors' attitudes, behavioural intentions, and experience values, with important elements including aesthetic impacts, usability, and a welcoming environment (S. T. Wu et al., 2020). In order to let users explore outdoor heritage sites throughout various eras, a new application was created using mobile augmented reality technology. A study (Boboc et al., 2022) revealed that this concept is not only true but also appealing to most people, suggesting that it may draw in more visitors while improving their experience while there. The suggested augmented reality tour system gives visitors the chance to have a more authentic experience by displaying virtual characters that replicate the way of living in the past at historical locations (Kim et al., 2009).

An empirical study (Olya et al., 2020) identifies the prerequisites for high festival visitor engagement through the use of augmented reality, including aesthetics, education, entertainment, and satisfaction. It also offers suggestions for festival organisers to raise visitor engagement and satisfaction in a memorable AR experience. A religious tourism experience model enables travellers to learn more about cultural and religious heritage, enhancing personal and spiritual experiences when visiting religious heritage sites linked to itineraries, using augmented reality and intelligent human-computer interactions displayed on personal mobile devices (Ramos et al., 1 C.E.). The psychological mechanism of mental imagery can be used to explain the effects of information type and augmenting immersive scenes on visitors' evaluations of the AR-facilitated museum experience and their subsequent purchase intentions. Augmented reality (AR) can improve visitor experiences and purchase intentions (He et al., 2018). The study (Cranmer et al., 2020) examined the perceived value of augmented reality for the tourism sector and identified five value dimensions, which have significant ramifications for developing strategies, implementing AR, and designing visitor experiences.

The findings of (Lee, 2022) demonstrate how information technology may be used to the travel and tourism sector, and they make recommendations on how and why virtual reality might be utilised to create travel experiences that will appeal to visitors. The findings (Vosinakis et al., 2020) suggest that mixed reality might be a potential medium for rich interactive experiences at museums that blend tangible and intangible heritage. Enhancing consumer value through the use of augmented reality at cultural heritage sites may provide businesses a competitive edge. This is a brand-new field of research with several prospects for further study (Tscheu & Buhalis, 2016). As suggested by (Huang et al., 2019) study's findings may be used to improve the intangible cultural heritage center's interactive methods and offer an efficient augmented reality interaction method for interacting with intangible cultural assets. In order to develop niche tourism, (X. Wu & Lai, 2021) offers recommendations for augmented reality tour app creators and governments of tourist destinations to improve their designs and formulate promotion strategies. The study looks at the factors that affect mainland Chinese film tourists' acceptance of an AR tour app.

Although augmented reality (AR) and virtual reality (VR) technology use and interest in cultural heritage tourism are growing, there is a glaring research gap in our knowledge of these technologies' long-term effects on the authenticity of visitor experiences and the preservation of cultural heritage sites. There has been little thorough research into how AR and VR may affect visitors' perceptions of authenticity and their emotional connections to cultural heritage sites, despite the fact that existing studies frequently concentrate on the immediate advantages of these technologies in enhancing engagement, interpretation, and learning for tourists.

Although AR and VR technologies allow for virtual trips, there isn't enough research to determine how they would affect actual visitors to culturally significant locations. What impact may the accessibility of immersive virtual experiences have on the quantity of visitors and their behaviour at physical locations? Furthermore, it is still largely unknown if these technologies may help promote sustainable tourism practises by lessening the effects of mass tourism on sensitive locations. Uncertainty exists around the extent to which AR and VR technologies serve a variety of consumers, including people with various cultural backgrounds and accessibility requirements. Research is needed to determine if these technologies can foster cross-cultural understanding and fill cultural gaps through immersive encounters.

Filling up these research gaps will offer insightful information on the complex influence of AR and VR technology on cultural heritage tourism. This knowledge is essential for guiding the creation of more efficient and long-lasting solutions that strike a balance between the advancement of technology and the maintenance of authenticity, emotional connection, and Utilizing Augmented Reality (AR) and Virtual Reality (VR) technology to Improve Tourist Experiences

2.1 Research Gap

Even though the abstract highlights the promise of augmented reality (AR) and virtual reality (VR) technologies in boosting cultural heritage tourist experiences, it does not specifically address the varied issues and solutions associated with applying AR and VR in a variety of different historical contexts. In the abstract, the benefits of these technologies are emphasized, but there is no in-depth analysis of the specific constraints and difficulties that may arise when merging AR and VR with a variety of historical places and objects. This is something that could be beneficial to readers. In addition, the abstract does not go into the potential disparities that may exist between the aspirations of individualized and immersive experiences and the actual restrictions that are imposed by both technological and cultural considerations.

Additional research might investigate the contextual complexities of merging augmented reality and virtual reality in a variety of situations associated with cultural heritage in order to help bridge this gap. This includes overcoming issues relating to the compatibility of hardware, the cost-effectiveness, and the adaptability to the historical and cultural relevance of locations. Researchers are able to provide a more thorough grasp of how augmented reality (AR) and virtual reality (VR) can be efficiently applied while keeping authenticity and cultural integrity if they examine a number of distinct case studies and scenarios. In addition, conducting research on user perceptions and reactions to augmented reality (AR) and virtual reality (VR) experiences within a variety of heritage contexts would help to a more comprehensive assessment of the influence of the technologies on tourist engagement and enjoyment. Such insights would offer significant assistance to experts and stakeholders in the cultural heritage tourism industry who are looking to harness the full potential of augmented reality and virtual reality.

3. METHODOLOGY

3.1 Objectives of the Study

1. The first objective is to improve tourist experiences by utilizing augmented reality (AR) and virtual reality (VR) technology. This entails developing fascinating and original ways for visitors to engage cultural heritage by way of immersive and interactive interactions with historical sites, artifacts, and tales.
2. The second objective is Fostering a More In-Depth Understanding and Appreciation of cultural Sites An additional goal is to encourage the development of a more in-depth understanding and appreciation of cultural sites. This can be accomplished by providing visitors with augmented reality (AR) and virtual reality (VR) experiences that enable them to have personal contacts with historical events and cultural locations.
3. The third objective Heritage Can Be Revived and Presented: Augmented and virtual reality can be utilized to bring back to life ancient buildings and artifacts that have been damaged, allowing them to be displayed in all of their former glory. The goal of this purpose is to display heritage sites and artifacts in such a way that visitors will be able to appreciate the historical significance as well as the aesthetic worth of these items.
4. This objective is Middle Ground Between Technological Advancements and Authenticity: The goal of this objective is to find a middle ground between the advancement of technology and the maintenance of authenticity and cultural integrity. It requires overcoming obstacles such as high costs, limitations of hardware, and complex development procedures, all while ensuring that the immersive experiences produced by AR and VR remain authentic to the historical and cultural context in which they are being used.

These aims illustrate the potential benefits and limitations of incorporating AR and VR technology into cultural heritage tourism. The overarching goal is to provide enriching and unforgettable experiences for tourists while preserving the originality of cultural heritage locations, cultural importance.

4. IN CULTURAL HERITAGE TOURISM, AUGMENTED REALITY

4.1 AR Definition and Overview

Augmented Reality (AR) is a cutting-edge technology that blends digital information, virtual objects, and computer-generated content into real-world environments, improving the user's perception and engagement with their surroundings. Unlike Virtual Reality (VR), which immerses users in a wholly computer-generated world, Augmented Reality (AR) overlays digital features onto the real world, allowing users to experience blended reality (Subasinghe et al., 2020). This can be accomplished via a variety of devices such as smartphones, tablets, smart glasses, and headsets. AR technology works by recognising the user's physical environment and superimposing virtual items onto it using sensors, cameras, and computer processing. 3D models, text, photos, animations, and other interactive information can all be included (Wu et al., 2016). Digital information is typically displayed in real-time, ensuring that it precisely aligns with the physical environment when the user moves or interacts with various items (S. Park & Kim, 2022).

AR applications are numerous and can be found in a wide range of industries, including entertainment, education, healthcare, marketing, retail, and gaming (Bolger, 2021). AR is employed in the entertainment industry to provide engaging and interactive experiences for users, such as mobile AR games like Pokémon GO, in which virtual animals are superimposed on the actual world (S. M. Park & Kim, 2022). AR enhances learning in education by providing interactive visualisations, allowing students to explore complex subjects in a more engaging way. AR also has practical applications in industries such as healthcare, where surgeons can use AR overlays during surgeries to display critical patient data without looking away from the operating area (Jovanovi & Milosavljevi, 2022). In retail and marketing, augmented reality (AR) allows buyers to try on virtual clothing or visualise furniture in their homes before purchasing, improving the buying experience. As augmented reality technology advances, its potential impact on numerous fields grows. Hardware advancements, such as more advanced and lightweight AR glasses, are making AR experiences more immersive and approachable. Furthermore, advancements in computer vision and AI algorithms are improving AR's ability to recognise and interact with the real environment (Damar, 2021). To summarise, Augmented Reality is a game-changing technology that blends the digital and physical worlds, increasing the user experience and offering up new opportunities

Figure 1. US virtual reality and augmented reality users, 2017-2021
(https://www.researchgate.net/figure/This-graph-shows-the-steady-increase-of-VR-and-AR-usage-from-2017-to-2021-in-the-United_fig2_341644383)

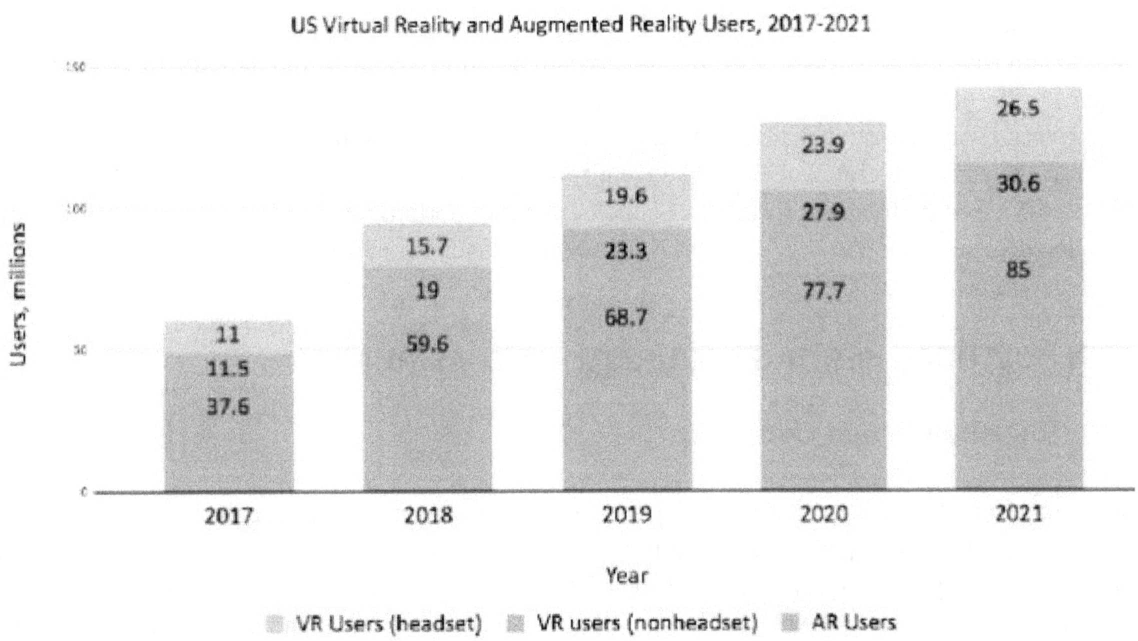

in a variety of industries. AR is poised to revolutionise how we see and interact with reality in our daily lives as its capabilities grow and become more widespread.

4.2 Applications of Augmented Reality in Cultural Heritage

Augmented Reality (AR) has transformed many industries, and one of its most intriguing applications is the preservation and promotion of cultural heritage. AR technology allows the merging of virtual content with the actual environment, making it a great tool for improving the way we experience and interact with historical locations, artefacts, and traditions (Kye et al., 2021). For starters, they provide tourists an immersive and fascinating experience by transporting them back in time to witness historical events or see ancient sites in their pristine magnificence. AR-guided tours can be built to provide interactive storytelling, allowing viewers to see and hear historical characters come to life, adding a personal touch to their historical journey (Suh & Ahn, 2022).

Museums and historical places can use augmented reality to bring ancient artefacts and artworks to life. Visitors can see 3D reconstructions, textual information, and animations layered on genuine artefacts by pointing their cell phones or tablets at displays (Young Lee Assistant Professor, 2021). This enhances learning, particularly for younger generations who are more acclimated to digital interactions. Furthermore, augmented reality apps can aid in the reconstruction and preservation of cultural content that has been destroyed or lost through time. AR simulations can reconstruct long-lost antique buildings, monuments, or archaeological sites, allowing researchers and the general public to appreciate their historical relevance and architectural brilliance.

AR also provides a welcoming environment for persons with physical limitations. AR allows everyone to participate in and appreciate cultural heritage by incorporating audio descriptions, sign language translations, and tactile interfaces. AR applications in cultural heritage can promote cultural exchange and tourism in addition to being instructive. Through AR-guided tours, language translations, and interactive experiences, tourists from all over the world can gain a more profound understanding and appreciation of a foreign land's culture. However, there are some considerations to make, such as the requirement for high-quality content development, maintaining accuracy and authenticity, and providing accessible for all users. Furthermore, technological hurdles such as the need for appropriate devices and consistent internet connectivity may hinder the mainstream adoption of AR applications in some areas (Chang et al., 2020).

Nonetheless, the potential of augmented reality in protecting and promoting cultural heritage is enormous. AR applications are likely to play an increasingly important role in connecting people with their rich and diverse cultural history as technology advances and becomes more accessible.

5. IN CULTURAL HERITAGE TOURISM, VIRTUAL REALITY

5.1 VR Definition and Overview

VR is a cutting-edge technology that immerses users in computer-generated, three-dimensional landscapes, duplicating real-world experiences or transporting them to wholly imaginary realms (Higgins-Desbiolles et al., 2022). VR gives a highly immersive and participatory experience by combining computer visuals, sensors, and interactive gadgets, blurring the lines between the physical and virtual worlds. This technology allows users to perceive and interact with digital content as if they were physically there in the simulated environment, providing a sense of presence and realism. A VR system's primary components include a headset or HMD (head-mounted display) that the user wears and encompasses their field of view, as well as input devices such as motion controllers or data gloves that track the user's motions and actions. The VR headgear displays stereoscopic images to each eye, creating the appearance of 3D space, while head-tracking technology guarantees that the presented material responds to the user's head motions, allowing them to gaze around and explore the virtual area from all perspectives.

VR is being used in a variety of areas, including gaming, entertainment, education, training, healthcare, and architecture. VR in gaming and entertainment provides users with a more immersive and realistic experience, allowing them to become totally immersed in the virtual world. VR is used in educational institutions and enterprises for training and simulations, allowing users to practise real-life scenarios without the hazards or costs connected with them. VR has been used in healthcare for therapeutic objectives such as pain treatment, exposure therapy, and rehabilitation. Architects and designers can use virtual reality to generate virtual prototypes and walkthroughs of buildings before they are built, speeding the design process and enhancing client communication.

Despite its wide applications and promise, VR confronts problems such as the requirement for expensive technology to display high-quality graphics, the risk of motion sickness or discomfort in some users, and concerns about the possibility of addiction or separation from reality (Lapointe, 2020). VR is predicted to become more accessible, inexpensive, and polished as technology advances, thereby expanding its reach and impact across numerous fields. Furthermore, the incorporation of other developing technologies, such as haptic feedback systems and eye-tracking, is anticipated to improve immersion and realism in virtual worlds, pushing the frontiers of what is feasible in the realm of Virtual Reality.

5.2 Virtual Reality Applications in Cultural Heritage

Virtual reality (VR) has developed as a disruptive technology with enormous potential across a wide range of industries, with one area that has seen tremendous progress being its application in cultural heritage protection and exploration (Kusuma et al., 2021). VR provides a one-of-a-kind opportunity to digitally reproduce historical sites, artefacts, and cultural events, allowing people to immerse themselves in history and get a better knowledge and appreciation for different civilizations.

The virtual recreation of ancient monuments and archaeological sites is one of the most significant applications of VR in cultural heritage. VR can realistically reproduce structures that have been lost to time, natural disasters, or human involvement using advanced 3D scanning techniques and historical study (Kusuma et al., 2021). This allows historians, archaeologists, and members of the general public to digitally explore and engage with these locations, bringing history to life in ways that traditional text and photos cannot. Museums and galleries have also embraced virtual reality to enhance their exhibitions and provide visitors with a more engaging experience. VR enables museums to display delicate or sensitive artefacts that would normally be stored for preservation. Visitors can inspect and engage with these artefacts in a virtual environment, which helps them understand cultural objects and historical contexts (Akhtar et al., 2021).

VR has also changed the way cultural stories are told. Historical narratives can be rebuilt and presented in an immersive and emotionally resonant manner using virtual reality. Users can become a part of key historical events, developing empathy and a stronger connection to the cultural stories offered. Furthermore, VR aids in the preservation of cultural heritage by capturing and archiving endangered cultural locations and customs. When sites are threatened by destruction due to conflict, climate change, or other circumstances, virtual reality can assist chronicle and preserve these priceless assets for future generations. VR applications in cultural heritage have also revolutionised education. Students can now take virtual field trips to famous historical locations or ancient civilizations, allowing them to immerse themselves in history and cultivate a love of study (Akhtar et al., 2021). Finally, virtual reality has created a new world of possibilities for the protection, exploration, and distribution of cultural material. VR has the potential to bridge the gap between the past and the present by digitally reproducing historical buildings, artefacts, and experiences, promoting a deeper understanding for other cultures and their contributions to human history. As technology advances, the potential for virtual reality in cultural heritage applications is predicted to expand, enriching our awareness of the world's common legacy (El-Said & Aziz, 2022).

5.3 VR's Advantages and Limitations

Virtual reality (VR) is an immersive technology that uses computer-generated pictures and sounds to create a simulated environment, allowing users to interact with and experience a virtual world. As this technology advances, it provides several benefits while also posing certain restrictions provides unmatched immersion, allowing users to feel as though they are physically present in the virtual environment. This realism improves a variety of applications, including training simulations, gaming, and virtual tourism. Training and Education: Virtual reality is extremely useful in training and educational contexts. It enables users to practise complex jobs and perilous events in a secure setting. This is especially useful for teaching professionals in industries such as medicine, aviation, and the military, where hands-on experience is essential has showed potential in treating a variety of mental health illnesses, including phobias, PTSD, and anxiety disorders. Exposure treatment in virtual reality environments allows patients to address their concerns in a safe and supportive atmosphere. Remote cooperation: Virtual reality allows for seamless cooperation among those who are geographically separated. Virtual meetings, conferences, and collaborative workplaces can boost productivity while also saving time and money. Entertainment and gaming: Virtual reality gaming offers a completely new degree of immersion and involvement. Players can physically interact with the virtual world, creating a very immersive gaming experience.

6. AUGMENTED AND VIRTUAL REALITY'S IMPACT ON CULTURAL HERITAGE TOURISM

In the field of cultural heritage tourism, combining augmented reality (AR) and virtual reality (VR) techniques has become essential for improving the visitor's experience. Even while cultural heritage places have inherent historical and cultural value, AR and VR technologies elevate these encounters by seamlessly fusing the past and present, providing visitors with immersive and engrossing experiences. The way tourists interact with cultural heritage locations, artifacts, and narratives has been revolutionized by AR and VR technologies. Visitors can interact with digital content that is projected on the real world by using AR overlays, which deepens their comprehension of historical events and cultural contexts. In contrast, virtual reality (VR) transports tourists to various historical periods, allowing them to virtually explore historical landscapes, monuments, and buildings. A greater understanding of the significance of heritage places is fostered by the juxtaposition of the real and virtual worlds.

The potential of AR and VR to recreate and portray historical sites and artifacts in their original magnificence is one of their primary advantages. These innovations bring decayed buildings and artifacts back to life, letting visitors experience them as they were in their heyday. Additionally, interactive narrative and gamification components provide individualized experiences catered to individual interests, transforming learning into an entertaining and memorable trip.However, integrating AR and VR into cultural heritage tourism is not without its difficulties. Wider use requires addressing technology barriers such high costs and hardware constraints. It is necessary to guarantee seamless user experiences while taking into account things like motion sickness and realistic visual representations. To establish a balance between the historical relevance of the place and the immersive experiences provided, it is also crucial to retain authenticity and cultural integrity while embracing technological innovations.

AR and VR have bright promises for the future of cultural heritage tourism. These experiences will become more approachable and user-friendly as technology develops, drawing in a larger audience. Collaboration between academics, business executives, and heritage organizations will spur invention and result in the creation of more advanced software and technologies. Modern innovations like haptic feedback and artificial intelligence will be included to improve these interactions' immersive quality. In conclusion, AR and VR technologies have become ground-breaking tools in the tourism of cultural resources. They alter how visitors interact with and value cultural heritage by offering intriguing and engaging encounters with historical locations and artifacts. Despite obstacles, the use of AR and VR in this industry has a promising future since it provides a way to both preserve the past and advance heritage tourism in the digital era.

7. AR AND VR INTEGRATION IN CULTURAL HERITAGE TOURISM

7.1 AR-VR Combination Experiences

Mixed Reality (MR), or combined AR-VR experiences, are a promising merging of augmented reality (AR) and virtual reality (VR) technologies (Toubes et al., 2021). This new type of immersive technology allows users to engage with both real-world and virtual aspects at the same time, blurring the line between the physical and digital worlds. These experiences, which combine the characteristics of AR and VR, bring up a wide range of possibilities in a variety of industries, including gaming, education, healthcare, architecture, and entertainment. Users in a mixed reality environment wear specialised headsets or glasses with cameras, sensors, and displays that overlay computer-generated content onto their real-world surroundings. Users may see and interact with digital objects as if they were physically present because to this mixing of the virtual and real worlds. Users can, for example, arrange virtual furniture in their actual living space, play games that interact with their surroundings, or receive step-by-step instructions overlaid on realistic things (Cai et al., 2021).

The combination of AR and VR technology creates a more immersive and participatory experience. Users can effortlessly move between the two modes, allowing them to explore virtual worlds while remaining aware of their real-world surroundings, resulting in a more interesting and safe experience. MR may take students on virtual field trips while also allowing them to interact with real-world artefacts and environments in educational settings. Surgeons can profit from MR during difficult procedures in the healthcare industry by visualising real-time patient data and virtual overlays of organs or surgical routes (Buhalis, 2020). This can improve precision and decision-making while also lowering hazards. AR-VR experiences that combine the two will also revolutionise gaming. Players can interact with virtual characters or things in their real-world surroundings, thereby transforming their living room or backyard into a gaming arena. This opens up new avenues for social involvement and cooperative gaming.

As technology advances, the potential for MR experiences grows, potentially altering the way we work, play, study, and communicate. However, issues such as technology limits, user interface design, and privacy problems remain. As these impediments are overcome, the adoption of mixed AR-VR experiences is expected to increase, changing the future of human-computer interaction and pushing the boundaries of what is possible in the digital environment (Chen et al., 2020).

7.2 Gamification and Interactive Storytelling

Gamification and interactive storytelling are effective methods for engaging audiences in novel and immersive ways, transforming passive consumers into active participants. These approaches generate captivating experiences that capture users' attention and promote meaningful interactions by mixing parts of narrative and gameplay.

Interactive storytelling elevates storytelling by allowing the audience to affect the narrative's path (Cai & McKenna, 2021). Instead of simply watching the story unfold, players take on the role of co-creators, making decisions that influence the plot's outcome. Individuals become emotionally attached to the characters and their choices as a result of this level of engagement, which develops a sense of agency and emotional investment. Interactive storytelling can be found in a variety of formats, including video games, virtual reality experiences, and interactive fiction, in which readers can change the plot by making choices. Gamification, on the other hand, refers to the incorporation of game-like aspects into non-game environments in order to increase engagement and motivation. Gamification motivates users to accomplish activities, learn new skills, or reach specific goals by appealing to the human urge for achievement, recognition, and competitiveness. Some popular gamification components that offer a sense of success and motivate continuing involvement are progress indicators, leaderboards, incentives, and badges.

When combined, interactive storytelling and gamification form a powerful synergy that engages people on several levels. The interactive aspect of the narrative draws people into the world of the story, making them emotionally invested, while gamification elements reinforce their desire to keep participating with the experience. Businesses and organisations have recognised the value of these ideas and have implemented them in a variety of settings. Employee training and development, for example, can be made more participatory and gamified, improving knowledge retention and skill-building. These approaches are used by educational institutions to make studying more interesting and successful for students, encouraging them to take an active role in their education.

Finally, the combination of interactive storytelling and gamification brings storytelling to life while raising engagement to new heights. These approaches, whether for entertainment, education, or training, have the potential to generate memorable and profound experiences that consumers remember long after their engagement has finished. As technology advances, the possibilities for interactive storytelling and gamification expand, providing even more fascinating and revolutionary experiences in the future.

7.3 Individualization and Customization

Personalization and customization are two essential concepts that have grown in importance across many industries, particularly in technology and consumer-centric firms. Personalization refers to the process of delivering tailored experiences to individuals based on their unique characteristics, behaviour, and preferences. They both aim to improve user experiences, increase engagement, and foster customer loyalty by tailoring products, services, or content to individual preferences and needs. This can be accomplished by collecting and analysing data, which may include demographic information, browser history, purchasing behaviour, and social media interactions. With this information, businesses may provide personalised recommendations, targeted advertising, and content that speaks to each user on a more personal way. Personalization fosters a sense of relevance, which can lead to improved customer satisfaction and conversion rates.

Customization, on the other hand, is letting users to actively participate in developing and configuring products or services to meet their individual needs. It enables customers to take a more active role in designing their own experiences. Customization might range from simple colour or size options to more advanced settings like personalised food programmes or tailored software solutions. Businesses can establish a deeper emotional connection and a sense of ownership by offering customers the chance to design something that matches with their tastes, leading to increased brand loyalty and advocacy.

Personalization and customisation are both strongly reliant on data analysis and technology. AI and machine learning are critical in analysing massive volumes of data in order to derive valuable insights, find trends, and make appropriate suggestions. While these practises have several advantages for both organisations and customers, they also raise issues about data privacy and security. To create and sustain consumer trust, the correct balance of personalisation and privacy must be struck. Finally, personalization and customization are strong methods that can revolutionise the way businesses interact with their customers. They result in more satisfied customers, increased brand loyalty, and improved corporate performance when implemented properly. Personalization and customization will remain at the forefront of customer-centric methods as technology advances, affecting the future of numerous businesses.

8. IMPROVING VISITOR EXPERIENCES

8.1 Reconstruction of Immersive Sites

Immersive site reconstructions have transformed our perception of and interaction with past and present built settings. These reconstructions, which make use of cutting-edge technologies such as virtual reality (VR) and augmented reality (AR), provide users with a captivating and informative experience that transports them to historical or existing locations in an engaging and interactive manner. One of the key benefits of immersive site reconstructions is their ability to bridge the gap between academic research, architectural visualisation, and public engagement. Scholars, historians, and architects can employ these technologies to bring historical ruins, forgotten civilizations, and architectural wonders from the past to life. These reconstructions provide vital insights into the spatial organisation, building processes, and cultural context of diverse structures and towns throughout history by integrating archaeological evidence, historical records, and precise computer modelling.

Aside from academia, immersive site reconstructions have significant benefits for tourism, education, and cultural preservation. Tourists can virtually explore faraway historical sites, gaining access to regions that might otherwise be difficult to visit due to geographical, political, or conservation constraints. This accessibility develops a greater respect for cultural assets and supports responsible travel. These reconstructions become excellent instruments in educational settings. Students can immerse themselves in historical events, see prominent landmarks as they once were, and obtain a deeper grasp of architectural and urban planning progress. Interactivity enhances learning by allowing students to engage with virtual items and experience various settings.

Furthermore, immersive site reconstructions can be extremely beneficial in terms of heritage preservation and urban planning. They aid in restoration efforts and inform decision-making processes for contemporary urban development initiatives by giving accurate and thorough depictions of historical sites. Stakeholders may see proposed alterations in the context of existing structures and assess potential cultural heritage consequences. As technology advances, immersive site reconstructions will become more realistic, interactive, and accessible. However, it is critical to find a balance between employing new technologies to promote understanding and appreciation of the past while ensuring that they complement rather than replace traditional preservation efforts. We can better connect with our shared history, cultural identity, and architectural legacy if we use immersive technologies ethically.

8.2 Interactions With Multiple Senses

The integration and synchronisation of information from several senses, including as sight, hearing, touch, taste, and smell, to generate a unified and full perceptual experience, is referred to as multi-sensory interactions. The human brain is remarkably adept at processing information from multiple sensory modalities at the same time, allowing us to interact with the world in a rich and meaningful way. Synesthesia is one of the most common examples of multi-sensory interactions, where stimulation of one sensory pathway leads to experiences in another. A person with synesthesia, for example, may sense colours when hearing certain sounds or link distinct tastes with specific textures. This condition shows the interdependence of our sensory experiences and underlines the human brain's complexity.

Multisensory interactions are essential in everyday activities. When we eat, for example, our perception of flavour is influenced by a combination of taste, smell, and even visual clues from the appearance of the meal. In social interactions, we rely on facial expressions, body language, tone of voice, and context to understand others' emotions and intentions. In recent years, researchers and designers have explored the concept of multi-sensory experiences to improve human-computer interactions and virtual reality environments. Developers may create more immersive and engaging experiences by adding several sensory modalities like as haptic feedback, spatialized audio, and visual signals. This has significant implications for a variety of sectors, including gaming, education, and therapy. Furthermore, multi-sensory interactions can help people who have sensory impairments. Individuals with vision impairments, for example, may benefit from aural input or tactile signals to properly navigate their environment. Similarly, tactile and auditory signals could improve learning experiences for people who have hearing loss.

Understanding and utilising the power of multisensory interactions can lead to better human-machine interfaces, communication techniques, and more inclusive designs for a variety of applications. By recognising the interplay of our senses, we can gain a better understanding of how we perceive the world around us and open up new avenues for invention and creativity.

8.3 Content for Education and Interpretation

Educational and interpretative content is critical to effective learning and communication in a variety of settings, including universities, museums, science centres, online platforms, and more. This sort of content seeks to inform, explain, and convey complicated material in a clear and understandable manner, increasing comprehension and engagement across a wide range of people.

At its core, instructional content is intended to provide learners of any age or background with knowledge, skills, or insights. It frequently takes an organised approach, breaking down complicated ideas into manageable chunks and utilising numerous media formats to improve comprehension and retention, including as text, photos, videos, infographics, and interactive features. Educational content is found in textbooks, e-learning platforms, educational websites, and digital courses, and it caters to both formal and informal learning environments. Interpretive content, on the other hand, goes beyond mere factual information to provide context, analysis, and personal connections to the subject matter. This content delves into the underlying meanings and relevance of topics, encouraging audiences to engage on a deeper level and connect the knowledge to their own experiences and opinions. In museums, art galleries, historical sites, and heritage centres, interpretive content is frequently used to provide immersive and rewarding experiences that inspire curiosity and critical thought.

The ability of instructional and interpretative information to achieve a balance between being informed and engaging is important to its efficacy. Well-crafted content considers the audience's past knowledge and interests, tailoring the presentation to their specific requirements and preferences. Furthermore, the provision of educational and interpretative content requires careful research, fact-checking, and expert evaluation to assure correctness and credibility. In conclusion, educational and interpretative material is critical in sharing knowledge and generating meaningful connections with information. Such content, whether in traditional educational settings or on new digital platforms, provides learners with useful insights, fosters curiosity, and promotes lifelong learning. We can respond to changing needs by constantly refining and inventing content creation strategies.

8.4 Improving Accessibility and Inclusion

Improving access and inclusivity is critical for creating a more equal and caring society. It entails creating an environment in which everyone, regardless of ability or background, is able to actively participate and interact in all parts of life. We can establish a more cohesive and sympathetic community that celebrates diversity and empowers individuals by tearing down barriers and fostering an inclusive mindset. Physical accessibility is one significant area of focus. It is critical to ensure that public places, buildings, and transit systems are constructed to accommodate those with impairments or mobility issues. Ramps, lifts and broad doorways, for example, make it easier for everyone to move around independently. Additionally, creating tactile pathways and signage can help persons with visual impairments navigate more easily.

Another critical factor is digital accessibility. As technology becomes more prevalent in our lives, it is critical that websites, applications, and digital material are created with accessibility in mind. This includes offering alternative language for images, captions for films, and compatibility with screen readers, allowing people with disabilities to easily access and interact with online information. Education is also important in developing inclusion. Schools should develop inclusive teaching practises that cater to different learning styles and give students with disabilities with the required accommodations. Emphasising the importance of empathy and understanding differences can assist establish an accepting and respectful learning environment for all kids. Promoting diversity and inclusion in the workplace should be a top focus. Companies can create rules that accommodate disabled personnel, provide flexible working arrangements, and assure equitable opportunities for professional growth and promotion. In addition to improving the workforce, emphasising diversity generates a sense of connection and loyalty among employees.

To prevent prejudice and discrimination, public awareness and sensitivity training are vital. Through media, advertising, and community efforts that challenge prejudices and celebrate variety, society should actively promote inclusiveness. In conclusion, improving accessibility and inclusivity is a multifaceted endeavour that requires a collaborative effort from individuals, communities, businesses, and politicians. We can build a more equitable and inclusive society where everyone's rights and dignity are respected, and all individuals can thrive and fulfil their potential by creating accessible physical spaces, embracing digital inclusivity, promoting inclusive education and workplaces, and fostering empathy and awareness.

9. FUTURE OPPORTUNITIES AND DIRECTIONS

8.1 Long-Term Maintenance and Sustainability

Long-term maintenance and sustainability are key components of every successful project, initiative, or organisation. They include the ability to meet current requirements without jeopardising future generations' ability to meet their own needs, as well as the ability to maintain and enhance systems, processes, or products through time. Sustainability and long-term maintenance are critical to ensuring a prosperous and secure future for both human society and the planet, whether applied to environmental conservation, infrastructure development, or business operations. In the context of environmental sustainability, it is critical to protect and conserve natural resources, biodiversity, and ecosystems. Implementing practises that reduce carbon emissions, promote renewable energy sources, and encourage responsible consumption and waste management are all part of this. Long-term sustainability of these initiatives necessitates continuous monitoring, adaptive management, and persistent innovation in response to changing environmental circumstances and developing threats.

In the case of infrastructure projects, sustainability entails designing and building structures that are long-lasting, energy-efficient, and socially beneficial. It necessitates a life-cycle strategy in which the environmental and social implications are taken into account from planning through decommissioning. Long-term maintenance includes routine inspections, repairs, and upgrades to ensure that infrastructure remains safe, reliable, and meets evolving standards and regulations. In business, sustainability entails ethical and responsible practises that respect the well-being of employees, communities, and the environment. Adopting sustainable practises can result in cost savings, a better reputation, and more staff engagement. Long-term maintenance entails regularly reviewing the company's impact, identifying areas for improvement, and executing long-term positive change methods.

Collaboration and shared responsibility among governments, businesses, communities, and individuals are required to accomplish sustainability and long-term maintenance. Policymakers must pass favourable legislation and incentives, while corporations must adopt environmentally friendly practises and technologies. Communities may help by lobbying for responsible development and holding stakeholders accountable. Education and awareness are also important in supporting long-term sustainability and upkeep. In conclusion, sustainability and long-term maintenance are fundamental principles that guide us towards a more resilient, equitable, and prosperous future. Empowering individuals with knowledge about the importance of these concepts can drive behavioural change and foster a culture of responsibility and stewardship. We may create a better world for future generations by incorporating these concepts into environmental conservation, infrastructure development, and economic practises.

10. CONCLUSION

In order to improve visitor experiences, this study explores the application of augmented reality (AR) and virtual reality (VR) technology in cultural heritage tourism. The study looks into how these immersive technologies affect how engaged, educated, and emotionally connected visitors are to cultural heritage places. The study focuses on key elements that promote the successful integration of augmented reality and virtual reality into historical attractions. It also discusses the advantages and difficulties of implementing them in cultural tourism. Particularly, AR and VR have the capacity to imaginatively improve tourists' experiences before and after visits in cultural heritage areas1. Additionally, the tourist industry has made extensive use of these immersive technologies, which show potential for improving the entire visitor experience2. The use of these technologies in cultural tourism has the potential to alter how visitors view and engage with historical monuments, providing them with richer and more impactful emotional experiences.

11. FINDINGS

Enhanced Visitor Engagement: AR and VR technologies create a dynamic and interactive experience that allows tourists to actively engage with cultural heritage sites. AR overlays historical contexts and information on real-world environments, allowing visitors to explore complex narratives and visualisations. VR, on the other hand, immerses tourists in realistic 3D environments, transporting them to bygone eras and significantly increasing engagement levels. These technologies have the potential to turn traditional learning into an exciting and memorable experience, particularly for younger visitors who may find traditional methods less engaging. Emotional Connection: Immersive experiences enabled by AR and VR technologies foster a deeper emotional connection between tourists and cultural heritage sites. The technologies evoke empathy and emotional responses by allowing visitors to virtually walk in the footsteps of historical figures or experience significant events, making heritage sites more meaningful and memorable. Accessibility and Inclusivity: AR and VR technologies can improve accessibility for differently-abled tourists, making cultural heritage sites more inclusive. Virtual tours and experiences help visitors with physical limitations, allowing a broader audience to participate in the cultural adventure.

The report cites numerous issues that must be addressed in order to properly integrate AR and VR in cultural heritage tourism. Some of the main challenges that stakeholders must overcome include technological limits, cost consequences, and the necessity for adequate content curation and interpretation.

Sustainability and conservation: While augmented reality and virtual reality can enhance tourist experiences, they should supplement rather than replace physical visits to heritage sites. To preserve the authenticity and conservation of these sites, a balance between digital and physical aspects is required. Finally, the integration of AR and VR in cultural heritage tourism has the potential to revolutionise visitor experiences by improving engagement, education, and emotional connections. To reap the full benefits of new immersive technologies while protecting the authenticity and value of cultural heritage places, rigorous planning, content moderation, and a balanced approach are required.

12. IMPLICATIONS

The practise of travelling to experience and appreciate the cultural and historical significance of a destination's heritage sites, artefacts, traditions, and customs is referred to as cultural heritage tourism. This type of tourism has far-reaching consequences for both host communities and guests. Economic Impact: Cultural heritage tourism can significantly enhance the local economy. Tourist spending on lodging, meals, transportation, and souvenirs can produce jobs and revenue for the city. Furthermore, investments in cultural heritage preservation and promotion can lead to infrastructure development and improved business opportunities. Tourist influx frequently necessitates the preservation and conservation of cultural heritage places, traditions, and crafts. This may result in increased awareness and attempts to maintain and conserve local culture, ensuring its survival for future generations.

Cultural Exchange and knowledge: Cultural heritage tourism encourages cross-cultural exchanges, allowing tourists to get a better knowledge and appreciation for various cultures. This communication can foster tolerance, mutual respect, and a sense of global interconnectedness, eventually leading to peace and harmony. Traditions comeback: Some traditional practises and arts may have a comeback as a result of tourism. Local communities may benefit by preserving and showcasing their distinctive cultural components, so strengthening cultural pride and identity. Cultural heritage tourism, while its benefits, can offer social issues. Overcrowding, cultural commodification, and deterioration of authenticity may result from over tourism in particular regions. Local communities may struggle to strike a balance between sharing their culture and preserving its integrity.

Environmental Impact: If not handled effectively, the inflow of tourists can place a pressure on natural resources and contribute to environmental degradation. To offset these effects, proper planning and responsible tourism practises are required. Educational Opportunities: Cultural heritage tourism can be used to educate people about history, art, and architecture. It allows tourists to learn about a destination's cultural evolution and gain a deeper respect for its legacy. Infrastructure creation: Promoting cultural heritage tourism frequently necessitates the creation of infrastructure, such as museums, visitor centres, and enhanced transportation networks. These developments have the potential to help local communities by improving their quality of life. Finally, cultural heritage tourism has both beneficial and negative consequences for both host communities and visitors. Proper planning, sustainable practises, and community involvement are required to ensure that cultural heritage tourism benefits all stakeholders.

REFERENCES

Akhtar, N., Khan, N., Mahroof Khan, M., Ashraf, S., Hashmi, M. S., Khan, M. M., & Hishan, S. S. (2021). Post-covid 19 tourism: Will digital tourism replace mass tourism? *Sustainability (Basel)*, *13*(10), 5352. doi:10.3390u13105352

Akpojaro, J., & Bello, R.-W. (2019). Smart Healthcare System: A Primer. In *Peer-Reviewed, Refereed, Indexed Journal with IC*, (5). https://www.kernelsphere.com/four-internet-things-communications-models/

Almeida, F., Buzady, Z., & Ferro, A. (2021). Exploring the role of a serious game in developing competencies in higher tourism education. *Journal of Hospitality, Leisure, Sport and Tourism Education*, *29*, 100347. doi:10.1016/j.jhlste.2021.100347

Ayuningtyas, D., & Ariwibowo, D. A. (2020). The strategic role of information communication technology in succeeding medical tourism. *Enfermeria Clinica*, *30*, 170–173. doi:10.1016/j.enfcli.2020.06.039

Ayuningtyas, D., & Ariwibowo, D. A. (2020). The strategic role of information communication technology in succeeding medical tourism. *Enfermeria Clinica*, *30*, 170–173. doi:10.1016/j.enfcli.2020.06.039

Bello, R.-W., Akpojaro, J., & Olubummo, D. A. (2019). Impact of Information Communication Technology in Enhancing the Management of the Natural and Cultural Heritage Resources in Idanre Hills for E-tourism Attraction Design of a Liquid Level Detector with a Digital Readout View project. *Machine Learning and Cloud Computing View project.* Research Gate. https://www.researchgate.net/publication/336020387

Boboc, R. G., Băutu, E., Gîrbacia, F., Popovici, N., & Popovici, D. M. (2022). Augmented Reality in Cultural Heritage: An Overview of the Last Decade of Applications. *Applied Sciences, 12*,(19), 9859. doi:10.3390/app12199859

Bolger, R. K. (2021). Finding wholes in the metaverse: Posthuman mystics as agents of evolutionary contextualization. *Religions*, *12*(9), 768. doi:10.3390/rel12090768

Braojos, J., Benitez, J., Llorens, J., & Ruiz, L. (2020). Impact of IT integration on the firm's knowledge absorption and desorption. *Information & Management*, *57*(7), 103290. doi:10.1016/j.im.2020.103290

Buhalis, D. (2020). Technology in tourism-from information communication technologies to eTourism and smart tourism towards ambient intelligence tourism: A perspective article. *Tourism Review*, *75*(1), 267–272. doi:10.1108/TR-06-2019-0258

Cai, W., & McKenna, B. (2021). Knowledge Creation in Information Technology and Tourism Research. In Journal of Travel Research, 60(4). doi:10.1177/0047287520903142

Cai, W., McKenna, B., Wassler, P., & Williams, N. (2021). Rethinking Knowledge Creation in Information Technology and Tourism. In Journal of Travel Research, 60(6). doi:10.1177/0047287520946100

Chang, C. L., McAleer, M., & Ramos, V. (2020). A charter for sustainable tourism after COVID-19. *Sustainability (Basel)*, *12*(9), 3671. doi:10.3390u12093671

Chen, S., Law, R., Xu, S., & Zhang, M. (2020). Bibliometric and visualized analysis of mobile technology in tourism. In Sustainability (Switzerland), 12(19). https://doi.org/ doi:10.3390u12197975

Cranmer, E. E., tom Dieck, M. C., & Fountoulaki, P. (2020). Exploring the value of augmented reality for tourism. *Tourism Management Perspectives*, *35*, 100672. doi:10.1016/j.tmp.2020.100672

Damar, M. (2021). *Metaverse Shape of Your Life for Future: A bibliometric snapshot.*

De Luca, V., Marcantonio, G., Barba, M. C., & De Paolis, L. T. (2022). A Virtual Tour for the Promotion of Tourism of the City of Bari. *Information, 13*(7), 339. doi:10.3390/info13070339

Drot-Delange, B., Fleury, S., Lafourcade, P., & More, M. (2018). *Un tour de magie pour introduire la représentation binaire des nombres*. CS Unplugged. https://csunplugged.org

El-Said, O., & Aziz, H. (2022). Virtual Tours a Means to an End: An Analysis of Virtual Tours' Role in Tourism Recovery Post COVID-19. *Journal of Travel Research*, *61*(3), 528–548. doi:10.1177/0047287521997567

Gómez-Vega, M., & Picazo-Tadeo, J. (2019). Ranking world tourist destinations with a composite indicator of competitiveness: To weigh or not to weigh? *Tourism Management*, *72*, 281–291. doi:10.1016/j.tourman.2018.11.006

González-Rodríguez, M. R., Díaz-Fernández, M. C., & Pino-Mejías, M. Á. (2020). The impact of virtual reality technology on tourists' experience: A textual data analysis. *Soft Computing*, *24*(18), 13879–13892. doi:10.100700500-020-04883-y

González-Rodríguez, M. R., Díaz-Fernández, M. C., & Pino-Mejías, M. Á. (2020). The impact of virtual reality technology on tourists' experience: A textual data analysis. *Soft Computing*, *24*(18), 13879–13892. doi:10.100700500-020-04883-y

Guttentag, D. A. (2010). Virtual reality: Applications and implications for tourism. *Tourism Management*, *31*(5), 637–651. doi:10.1016/j.tourman.2009.07.003

He, Z., Wu, L., & Li, X. (2018). When art meets tech: The role of augmented reality in enhancing museum experiences and purchase intentions. *Tourism Management*, *68*, 127–139. doi:10.1016/j.tourman.2018.03.003

Hernández-Rojas, R. D., del Río, J. A. J., Fernández, A. I., & Vergara-Romero, A. (2021). The cultural and heritage tourist, SEM analysis: The case of The Citadel of the Catholic King. *Heritage Science*, *9*(1), 1–19. doi:10.118640494-021-00525-0

Higgins-Desbiolles, F., Bigby, B. C., & Doering, A. (2022). Socialising tourism after COVID-19: Reclaiming tourism as a social force? *Journal of Tourism Futures*, *8*(2), 208–219. doi:10.1108/JTF-03-2021-0058

Huang, W., Xiang, H., & Li, S. (2019). The application of augmented reality and unity 3D in interaction with intangible cultural heritage. *Evolutionary Intelligence*, *1*, 1–1. doi:10.100712065-019-00314-6

Jeong, M., & Shin, H. H. (2019). Tourists' Experiences with Smart Tourism Technology at Smart Destinations and Their Behavior Intentions. *Sage*, *59*(8), 1464–1477. https://doi.org/ doi:10.1177/0047287519883034

Jeong, M., & Shin, H. H. (2020). Tourists' Experiences with Smart Tourism Technology at Smart Destinations and Their Behavior Intentions. *Journal of Travel Research*, *59*(8), 1464–1477. doi:10.1177/0047287519883034

Jovanović, A., & Milosavljević, A. (2022). VoRtex Metaverse Platform for Gamified Collaborative Learning. *Electronics (Basel)*, *11*(3), 317. doi:10.3390/electronics11030317

Kamboj, S., & Joshi, R. (2021). Examining the factors influencing smartphone apps use at tourism destinations: A UTAUT model perspective. *International Journal of Tourism Cities*, *7*(1), 135–157. doi:10.1108/IJTC-05-2020-0094

Kamboj, S., & Joshi, R. (2021). Examining the factors influencing smartphone apps use at tourism destinations: A UTAUT model perspective. *International Journal of Tourism Cities*, *7*(1), 135–157. doi:10.1108/IJTC-05-2020-0094

Kersten, T. P., Tschirschwitz, F., & Deggim, S. (2017). DEVELOPMENT OF A VIRTUAL MUSEUM INCLUDING A 4D PRESENTATION OF BUILDING HISTORY IN VIRTUAL REALITY. The *International Archives of the Photogrammetry, Remote Sensing and Spatial Information Sciences, XLII-2-W3*(2W3), 361–367. doi:10.5194/isprs-archives-XLII-2-W3-361-2017

Kim, K., Seo, B. K., Han, J. H., & Park, J. Il. (2009). Augmented reality tour system for immersive experience of cultural heritage. *Proceedings - VRCAI 2009: 8th International Conference on Virtual Reality Continuum and Its Applications in Industry*, (pp. 323–324). IEEE. 10.1145/1670252.1670325

Kusuma, P. A., Mutiarin, D., & Damanik, J. (2021). STRATEGI PEMULIHAN DAMPAK WABAH COVID PADA SEKTOR PARIWISATA DI DAERAH ISTIMEWA YOGYAKARTA. *Journal of Tourism and Economic*, *4*(1), 47–59. doi:10.36594/jtec.v4i1.110

Kye, B., Han, N., Kim, E., Park, Y., & Jo, S. (2021). Educational applications of metaverse: Possibilities and limitations. In Journal of Educational Evaluation for Health Professions, 18. doi:10.3352/jeehp.2021.18.32

Lapointe, D. (2020). Reconnecting tourism after COVID-19: The paradox of alterity in tourism areas. *Tourism Geographies*, *22*(3), 633–638. doi:10.1080/14616688.2020.1762115

Lee, U. K. (2022). Tourism Using Virtual Reality: Media Richness and Information System Successes. *Sustainability, 14*(7), 3975. doi:10.3390/su14073975

Olya, H., Jung, T. H., Tom Dieck, M. C., & Ryu, K. (2020). Engaging visitors of science festivals using augmented reality: Asymmetrical modelling. *International Journal of Contemporary Hospitality Management*, *32*(2), 769–796. doi:10.1108/IJCHM-10-2018-0820

Papa, A., Mital, M., Pisano, P., & Del Giudice, M. (2020). E-health and wellbeing monitoring using smart healthcare devices: An empirical investigation. *Technological Forecasting and Social Change*, *153*, 119226. doi:10.1016/j.techfore.2018.02.018

Papa, A., Mital, M., Pisano, P., & Del Giudice, M. (2020). E-health and wellbeing monitoring using smart healthcare devices: An empirical investigation. *Technological Forecasting and Social Change*, *153*, 119226. doi:10.1016/j.techfore.2018.02.018

Park, S., & Kim, S. (2022). Identifying World Types to Deliver Gameful Experiences for Sustainable Learning in the Metaverse. *Sustainability (Basel)*, *14*(3), 1361. doi:10.3390u14031361

Park, S. M., & Kim, Y. G. (2022). A Metaverse: Taxonomy, Components, Applications, and Open Challenges. *IEEE Access : Practical Innovations, Open Solutions*, *10*, 4209–4251. doi:10.1109/ACCESS.2021.3140175

Ramos, C. M. Q., Henriques, C. H. N., & Lanquar, R. (2016). *Augmented Reality for Smart Tourism in Religious Heritage Itineraries: Tourism Experiences in the Technological Age*. IGI Global. Https://Services.Igi-Global.Com/Resolvedoi/Resolve.Aspx?Doi=10.4018/978-1-5225-0435-1.Ch010, 245–272. https://doi.org/ doi:10.4018/978-1-5225-0435-1.ch010

Subasinghe, M., Magalage, D., Amadoru, N., Amarathunga, L., Bhanupriya, N., & Wijekoon, J. L. (2020). Effectiveness of artificial intelligence, decentralized and distributed systems for prediction and secure channelling for Medical Tourism. *11th Annual IEEE Information Technology, Electronics and Mobile Communication Conference, IEMCON 2020*, (pp. 314–319). IEEE. https://doi.org/10.1109/IEMCON51383.2020.9284898

Subawa, N. S., Widhiasthini, N. W., Astawa, I. P., Dwiatmadja, C., & Permatasari, N. P. I. (2021). The practices of virtual reality marketing in the tourism sector, a case study of Bali, Indonesia. *24*(23), 3284–3295. doi:10.1080/13683500.2020.1870940

Suh, W., & Ahn, S. (2022). Utilizing the Metaverse for Learner-Centered Constructivist Education in the Post-Pandemic Era: An Analysis of Elementary School Students. *Journal of Intelligence*, *10*(1), 17. doi:10.3390/jintelligence10010017 PMID:35324573

Tiwari, A., Dhiman, V., Iesa, M. A. M., Alsarhan, H., Mehbodniya, A., & Shabaz, M. (2021). Patient Behavioral Analysis with Smart Healthcare and IoT. *Behavioural Neurology*, *2021*, 1–9. doi:10.1155/2021/4028761 PMID:34900023

Tiwari, S., Rosak-Szyrocka, J., & Żywiołek, J. (2022). Internet of Things as a Sustainable Energy Management Solution at Tourism Destinations in India. *Energies, 15*(7), 2433. doi:10.3390/en15072433

Toubes, D. R., Vila, N. A., & Fraiz Brea, J. A. (2021). Changes in consumption patterns and tourist promotion after the covid-19 pandemic. *Journal of Theoretical and Applied Electronic Commerce Research, 16*(5), 1332–1352. doi:10.3390/jtaer16050075

Tscheu, F., & Buhalis, D. (2016). Augmented Reality at Cultural Heritage sites. *Information and Communication Technologies in Tourism, 2016*, 607–619. doi:10.1007/978-3-319-28231-2_44

Vosinakis, S., Nikolakopoulou, V., Stavrakis, M., Fragkedis, L., Chatzigrigoriou, P., & Koutsabasis, P. (2020). Co-Design of a Playful Mixed Reality Installation: An Interactive Crane in the Museum of Marble Crafts. *Heritage, 3*(4), 1496–1519. https://doi.org/ doi:10.3390/heritage3040083

Wu, H. C., Li, T., & Li, M. Y. (2016). A Study of Behavioral Intentions, Patient Satisfaction, Perceived Value, Patient Trust and Experiential Quality for Medical Tourists. *Journal of Quality Assurance in Hospitality & Tourism, 17*(2), 114–150. doi:10.1080/1528008X.2015.1042621

Wu, S. T., Chiu, C. H., & Chen, Y. S. (2020). *The influences of innovative technological introduction on interpretive experiences of exhibition:a discussion on the intention to use augmented reality*. Taylor & Francis. doi:10.1080/10941665.2020.1752754

Wu, X., & Lai, I. K. W. (2021). The acceptance of augmented reality tour app for promoting film-induced tourism: The effect of celebrity involvement and personal innovativeness. *Journal of Hospitality and Tourism Technology, 12*(3), 454–470. doi:10.1108/JHTT-03-2020-0054

Xu, C., Jones, C., & Munday, M. (2019). Tourism inward investment and regional economic development effects: perspectives from tourism satellite accounts. *Regional Studies, 54*(9), 1226–1237. doi:10.1080/00343404.2019.1696954

Young, L. (2021). A Study on Metaverse Hype for Sustainable Growth. *International Journal of Advanced Smart Convergence, 10*(3).

Chapter 3
Ancient Temples and Cultural Heritage of Odisha, India:
Issues and Challenges

Soumendra Nath Biswas
https://orcid.org/0000-0002-6664-252X
Assam University (A Central University), India

Puspanjali Mohapatra
Utkal University of Culture, India

ABSTRACT

India represents one of the oldest extant civilizations in the world. Odisha, the eastern state of India has rich source of unique ancient temples and ethnic culture. The ancient city Bhubaneswar is well known as the 'Temple City.' Every year, good numbers of domestic and international tourists visit Odisha to experience this marvelous temple architecture of ancient period. These are the sources of knowledge for the scholars of ancient history, archaeology, heritage, Tourism studies and others. This article begins with the understanding of ancient temples existing till today in some spiritual places of coastal Odisha, mainly in the golden triangle of eastern India. Socio-economic development of the local community and protection of their culture are also discussed here. This chapter analyzes the role and importance of ancient temples and cultural heritage of Odisha for the development and promotion of tourism business in the state.

INTRODUCTION

Life has important station in the temple for religious people of India. India is the place of birth for four religions, viz. Hinduism, Buddhism, Jainism and Sikhism respectively. Odisha (or Orissa) is the state of eastern India. The land is known for ancient temples, caves, cultures and heritage. Maximum numbers of people of Odisha people especially from coastal Odisha are religious in nature and they give priority to their ethnic culture. People of Odisha, often visit religious places for performing rituals and praying

DOI: 10.4018/978-1-6684-9957-3.ch003

to God for attaining *moksha*. *Tirtha Kshetra* or religious place are the sacred place with full of active power where moksha or final release may be obtained as per the religious belief. The capital of Odisha, Bhubaneswar is recognizedas a temple city for having numbers of ancient temples;*Puri Dham*is well known as one of the C*hardham* of India. Konark Sun temple is an international Heritage site. Throughout the world, every year huge numbers of people visit these sites and experience the unique ancient Kalingan style Temple Architecture. Not only Hindu temple, Odisha also has numbers of ancient Buddhist and Jain temples or their ruins. Attracting Tourists in these destinations influence the socio-economic factor of the local community.

The holy state of Odisha occupies a unique position in the tourist map of India. It has rich tourism potential to attract a large number of inbound tourists, both foreign and domestic. This chapter highlights on the ancient temples and cultural heritage of Odisha. Ethnic Temple structures came to Odisha after the Gupta period. The Hindu dynasties Somavamsi, Bhaumakaras, Gangas had inspired the Odishan architects to create their own style of temple architecture representations without any influence. The period of that unique style of Orissan architecture was from 6th century A.D. to 16th century A.D (Chand, 2005). In Odishan temple sculpture, erotic arts appear from earliest form of temple architecture. Except the Mukteswara temple, erotic arts appear in almost all other temples of Orissa even in the temples of poorest in the decoration (Donaldson, 1975). Indian Silpashastras have categorized Indian Temples into three as per their styles and architecture. These are: the Nagara style of Architecture, the Dravida style of Architecture, and the Vesara style of Architecture (Saraswati, 1989). But the unique features of Odishan temples have also mentioned in the inscription on the Mukhamandapa of Amartesvara temple at Holal dated 1235 A.D. Kalingan type as one of the four categories of the temples in India along with Nagara, Vesara and Dravida (Das, 1994). In Indian context, the Odishan temples that is known as Kalingan style of temple architecture received a special attention by scholars to focus on architecture of temples of Odisha and also analyzed the purpose behind the decorative images of temple of Odisha (Panda, 2016). The word 'temple' is originated from the Latin word '*Templum*' which means a square or a rectangular place for the purpose of worship. From the ancient time, this word represents to a place forworshipingGod.It is used to denote various religious associations except Christianity and Islam. Temples are the meeting points of culture, arts and religions. Arts and cultures are closely associated with religious festival. For performing ritual festivals, people need a place of worship. Therefore temple in Hindu religion has immense important for performing ritual practices and flourishing ethnic culture. The architectural structure is mainly based on the myth. Hindu mythological stories and characters had influence on sculptures of those ancient temples.

People are living in the physical environment that includes landform, water bodies, air, flora, fauna etc. Therefore environment plays an important role to influence the culture and ritual practice of the people. Orissan temples have influence of these environmental phenomena. Along with god and goddesses, many other sculptures have been made of animals, plants, landforms etc. In the temple architecture, political, social, economical and cultural lives of ancient peoples are also revealed. Social structures have huge impact on construction of temples and its architecture. It designates the quality or status gained in the society. It is also symbol of prestige and wealth of the founder and ruling dynasty of temple. Temples are the showcase of artistic and technical knowledge of the craftsmen in the period of specific dynasty. Temples also served as the meeting place of the village folk irrespective of caste and creed. The temples with their rituals and celebrations are closely associated with gatherings during festivals. Being a cultural hub, temples witness the evaluation of different school of art, architecture, sculpture, painting, music and dance forms. Temples are the source of inspirations to many artistic people like painters, poets, dancers, composers etc.

The Orissan art and architecture is the gift of her rich historical period. Ashoka's rule in ancient Orissa was great start to Orissan (Odian) art and architecture. Odisha is bestowed with numerous archaeological and religious monuments, which attract pilgrims and tourists from the other states of India and different parts of the world. To study the ancient ethnic culture of a society, these monuments and heritage sites have immense value. Conservation and beautification of these monuments are essential for maintaining sustainability.

OBJECTIVES

The main objective of this study is to find out the issues and challenges are facing by the ancient Temples and Cultural Heritage of Odisha. However, the objectives in particular are:

1. To understand the importance of ancient temples of Odisha.
2. To find out the significance of cultural heritage for sustainable development of Tourism in the study area.
3. To analyze the issues and challenges are faced by these Tourism destinations.

METHODS

To have in depth knowledge of some of the issues of cultural heritage related with ancient temples, extended conversations were made with the potential stakeholders of these destinations.This study was conducted using self-administered questionnaires. Along with observation, interview methods were also followed for the above said purpose. Secondary data were collected from books, literature, journals, websites etc. Collected data were analyzed and accordingly findings and suggestions have been given.

ANCIENT TEMPLES OF ORISSA (ODISHA)

Temples of Odisha are incredible for plenty of sculptures. This sculpture includes various forms of human beings, erotic arts, animals, plants, and various decorative designs. The Orissan temple style was flourished with the *Sailodbhava* kings ruled during middle of 6th century A.D. to first quarter of 8th century A.D. in ancient Orissa (or Odisha). Those temple structure was got in pick moment under the *Bhaumakaras*ruler (736-950A.D.) and the *Somavamsis* (950-1112 A. D) dynasty and reached the pick form throughout the Ganga period (1112-1435 A.D.) till under the rule of *Suryavamsi-Gajapatis* (1435-1542A.D.). Examples of these ancient temples are: *parasuramesvara, lingaraja, Muktesvara, Vaital, Puri Jagannath* temple and the *Konark* temple etc.

Table 1. Ancient temples and heritage sites of Odisha

Name of the temple	Location
Parsurameswar Temple, Mukteswar Temple, Rajarani Temple, Lingaraj Temple, Brahmeswar Temple, Bhaskareswar Temple. Vaital Deul (Vesera style temple), Megheswar Temple, Swarnjaleswar Temple, Ananta Basudev temple	Bhubaneswar
Jaganath Temple	Puri (60 Kms from Bhubaneswar)
The Sun Temple	Konark (65 Kms from Bhubaneswar)
Ruined Buddhist Monasteries (Diamond Triangle of Odisha)	Lalitgiri, Ratnagiri and Udaygiri (105 kms from Bhubaneswar)
Sakhigopal Temple	Sakhigopal (20 Kms from Puri)
Ashokan Rock Edit Forepart of Elephant Sculpture Modern Buddhist Stupa	Dhauli (Outskirts of Bhubaneswar)
Goddess Mangala temple	Kakatpur (65Kms from Bhubaneswar)
64 Yogini Temple-1	Hirapur (15 Kms from Bhubaneswar)
64 Yogini Temple-2	Ranipur Jharial (110 Kms from Bolangir)
Khichekeswari Temple	Khichiing (270Kms from Bhubaneswar)
Nilamadav temple	Kantilo (100Kms from Bhubaneswar)
Varahi temple	Chaurasi (50Kms from Bhubaneswar
Panchayatan Vishnu temple	Chhatia (40Kms from Cuttack)
Bakreswar Temple	Huma (35 Kms from Sambalpur)
Samaleswari Temple	Sambalpur
Simhanath Temple	Baideswar (65 Kms from Bhubaneswar)

Source: https://odishatourism.gov.in/content/tourism/en/discover/attractions/temples-monuments.html

DISTINCTIVE FEATURES OF THE TEMPLES OF ORISHA (ODISHA)

Ancient India observed many articulations at the hands of *Kalingan* craftsmen especially in the post-Gupta period. The ancient Odisha was known as *kalinga*. Many stylistic designs had been constructed in the form of Temples in ancient period. Most important architectural features and art forms were started in *Kalinga*during Gupta dynasty. The unique innovation '*Nagar sikhara*'type of architecture was made by the Imperial Guptas slowly became the symbol of kalingan architectural activity. It was started from 6th-7th century A.D. and followed till 13th century A.D. Easy availability of suitable raw materials, conquest and cultural interaction were also the motivational factor for that.

The famous *Lingaraja* temple in Bhubaneswar became complete in eleventh century A.D. The magnificent sun temple at *Konark* was constructed during the middle of thirteenth century A.D.After that the innovation in temple architecture was gradually declined and came to an end in the sixteenth century

A.D. with the termination of Hindu rule.The demonstration of that school of temple architecture is known as *Kalingan* Style of temple architecture in Odisha. This unique style of *kalingan* architecture has been mentioned in the literary references in *Silpasastras*. To study the *Kalingan* architecture, one may refer textual tradition that is preserved in *Bhuvanpradip* and *Slipa Sarini* and *Silparatnakosa*.

The *Kalingan* Style of temple architecture, as mentioned in the*Silpasastras*, there are three types of temples, *Rekha, Pidha,* and *Khakhara*. Two components *Rekha* and *Pidha* from part of one architectural system, the *Rekha* was indicated by sanctum and the *Pidha* represented by the porch of pyramidal roof made of The *Pidhas*. The *khakhara* system temples have semi-cylindrical roof, and mainly for *sakti* worship. In the earlier phase, there were only flat roof; there were no *Pidha deul* and the frontal hall *Jagamohana*. In the later stage, during the *Ganga* period, two more structures namely *natamandira* and *bhogamandapa* were added.

MAJOR ANCIENT TEMPLES OF ORISSA (ODISHA)

Parasuramesvara Temple

The *Parasuramesvara* temple is situated at Bhubaneswar was constructed during c.7th century AD. It represents early stages of Orissan temple. It is a small temple but fully decorated shrine of Shiva and contains many sculptures of romantic couples, animals and floral design. The temple is 12.8 meter high and consisting of a *deula* and *jagamohana*. The *deula* is *triratha* on plan. The *parsvadevata* are *Ganesa* and *Kartikeya* in the south and east side respectively. The *chaitya* window side are inflated with *Ravana* shaking mount *Kailasa, Nataraja, Bhiksatanamurti* of Siva and *Mahisamardini*. The southern side comprises with Vishnu, *Siva-Parvati, Ardhanarisvara* and *dikpals*.

Muktesvara Temple

The *Mukteswar* Templewas constructed in the 10th century and considered as one of the most beautiful *Kalingan* styles of architecture and considered as 'Gem of *Orissan* architecture'. The temple architecture and iconography of *Muktesvara* temple is the combination of early phase and the next phase of experimentation of temple architecture. Beautiful, elegant carvings and the harmonious blending of decorative elements make this temple marvellous though its height is 10.5 meter consisting of *vimana* and *jagamohana*. The *deula* is *pancharatha* in figure. The central projection has *parsvadevata* arranged between four pilasters, of which two inner ones carry representations of *Ganga* and Yamuna. The ceiling part of the *jagamohana* is delicately carved with full blown lotus and contains the representation of *Virabhadra* and the *Saptamatrika* (with babies in their laps).

Rajarani Temple

The *Rajarani* temple is situated in the heart of Bhubaneswar and the temple is ancient to 11th century A.D. It consists of a deul and a *jagamohana*. The temple is now without a deity, but it is considered as a Shiva Temple because the presence of *Saivite* doorkeepers, the scenes of *linga* worship and *Nataraj* and *Parvati* on the platform. The temple is *pancharatha* type and stands on a platform consisting of three mouldings. In the lower and upper *jangha*, there are several facets with carvings of floral designs. The *jagamohana* is a pyramidal structure.

Lingaraj Temple

The most admired and tallest temple of Bhubaneswar is *Lingaraj* Temple. It was constructed in the 11th century AD. The great King*JajatiKeshari* who had placed the foundation of the *Lingaraj* temple and his great grandson *LalatenduKeshari*had completed the work. The temple dominates in the Bhubaneswar city. The temple complex has520 feet long wall and the main gate is safeguarded by two lions and ceiling was made by a pyramidal roof. This temple complex includes 65 secondary temples. Lord *Lingaraj*, is the God in the temple is characterized as a black stone. The *Lingaraj*in this temple is called '*Swayambhu*' and considered as a natural *linga*. The height of the main *deul* rises to a height of 45.11 meter.

Vaital Temple

The Vaital temple of Bhubaneswar is among the earliest type of the *Khakara*style of Temple Architecture preserved today. The temple has*deula* and *jagamohana*. The uniqueness of this Temple is that the *Deul* is rectangular in nature unlike curvilinear in other temple. This temple isdedicated to the Goddess *Chamundaand* was considered as a Tantric worship place. The deula is 11.5 meter in height andhas semi-cylindrical roof with three *kalasa* fixed on it. Therefore, it is popularly known as *Tinmundiya deula*.The *jagamohana* is rectangular in shape. The sculptural and decorative feature of the temple is considered to the period of 8th century A.D.

Anata-Vasudeva Temple

The *Anata-Vasudeva* Templewas built on the east Bank of *Bindu-Sarovara* (lake) in Bhubaneswar. It was built in1278 AD by *Chandradevi*, the daughter of *Anangabhima*-III. Facing towards west, it consists of a *deul*, a *jagamohana*, a *nata mandapa* and a *bhoga mandapa*. This temple is dedicated to Lord Vishnu. The *parsva-devatas* are three of the ten inscriptions of Vishnu, of which the four-armed *Varaha* and *Trivikrama*, exist in the south and north niches respectively. The pyramidal *jagamohana* has two windows, northern windows of which are with figures of *Rama*, *Lakshmana*, *Sita* and *Hanuman*.

Sun Temple, Konark

The Sun Temple of Konark is recognised as a World Heritage site is situated in Konark on the coastal area of Bay of Bengal. In the *kapila Samhita*, it was called as *Arka-Kshetra* or *Padma-Kshetra*. It is a part of the Golden triangle of eastern India along with Bhubaneswar and Puri. At present, the temple has ruined but the *jagamohana* is still intact. It was built in thirteenth century A.D. by *Narasimhadeva*-I of the *Ganga* dynasty. It was also named as 'Black Pagoda' by European mariners. The temple has been listed as World Heritage Monuments by UNESCO in 1984.

The temple is looks like a massive chariot with 12 pairs of wheels with full of sculptures and pulled by seven spirited horses. The temple is full of sculptures like: lions, musicians, dancers and erotic couple. The temple represents the culmination of art and architecture of medieval Odisha. Laterite stone was mainly used for the foundation whereas the sculptures have made with Khondalite and Chlorite stone.

Sculptures include many activities like agriculture to war and struggle from men, women to enormous variety of animals and Trees. The walls of the temple contain finest architecture of divine, semi divine, human, amorous couple, various animal, and plants figures.

Jagannatha Temple, Puri

The super architectural temple of Lord Jagannatha was built by the *Chodaganga Deva* of the *Ganga* dynasty during second half of the 12th century A.D. The temple is facing towards east, and consists of *deula, jagamohana, natamandira* and *bhogamandapa*. Height of the sikhara of the *deul* is 65.2 meter. The main deities are Lord *Jagannath, Balavadra* and *Subhadra*. The temple complex is also including numbers of small temples of shrines like: *Narasimha*, Surya, *Vimala, Lakshmi, Ganesa* and others. The temple has four gateways. The main gate on the east is called Simhadvara, south gate is called asvadvara, west gate is called *vyaghradvara* and north gate is known as *hastidvra* In front of the *simhadvara* there is *Arunastambha*. The deul is *pancharatha* style.The carvings on the exterior walls of the main temple include fine sculptures of various forms of *Vishnu; dikpalas, naga-nagi, Vyalas* and other *Brahmanical* divinities etc.

The Jagannath temple is considered as a *dham* of eastern India among the *chardham* of India. Every year during the annual festival of *Rathajatrya*, millions of people used to visit here. Puri is the religious place and a part of the golden triangle of eastern India. Along with this temple, Puri sea beach has also potential to attract millions of tourists. Therefore, it has become a mass Tourism destination.

MAJOR ISSUES AND CHALLENGES

Ancient Temples are the heritage site of any country. In India, studying ancient temples means going back to the history. These ancient temples are unique attractions for tourists visiting in the destinations. Unplanned developments near to the temple complex, lack of proper master plan are the major issues now. Lack of conservation, environmental catastrophe, safety and security, social issues, management issues with local communities, Changing Demographics and Religious Practices etc. are some of the issues related with tourism development in the heritage sites. Some of the important major issues and challenges are given in the following chart.

FINDINGS AND SUGGESTIONS

While studying various issues and challenges of ancient temples and cultural heritage of Odisha, it has been founded that ancient temples being ethnic resources are not always protected. They are also facing various challenges, both external and internal. Major issues for protecting ancient temples are financial constraints. Huge ancient superstructures need funds for maintenance. Some famous temples like Puri Jagannath temple, Konark Sun Temple and Lingaraj Temple in Bhubaneswar generate revenue through donations and offerings. But for smaller and lesser-known temples are still struggling for finance. In many ancient sites, encroachment by the local people and disputes of lands are major issues. Violation of law of Archaeological Survey of India and other regulatory bodies are also happening here. Many of these ancient temples are attracting not only tourists; huge numbers of local devotees are also visiting every day

Figure 1. Issues and challenges of ancient temples and cultural heritage of Odisha

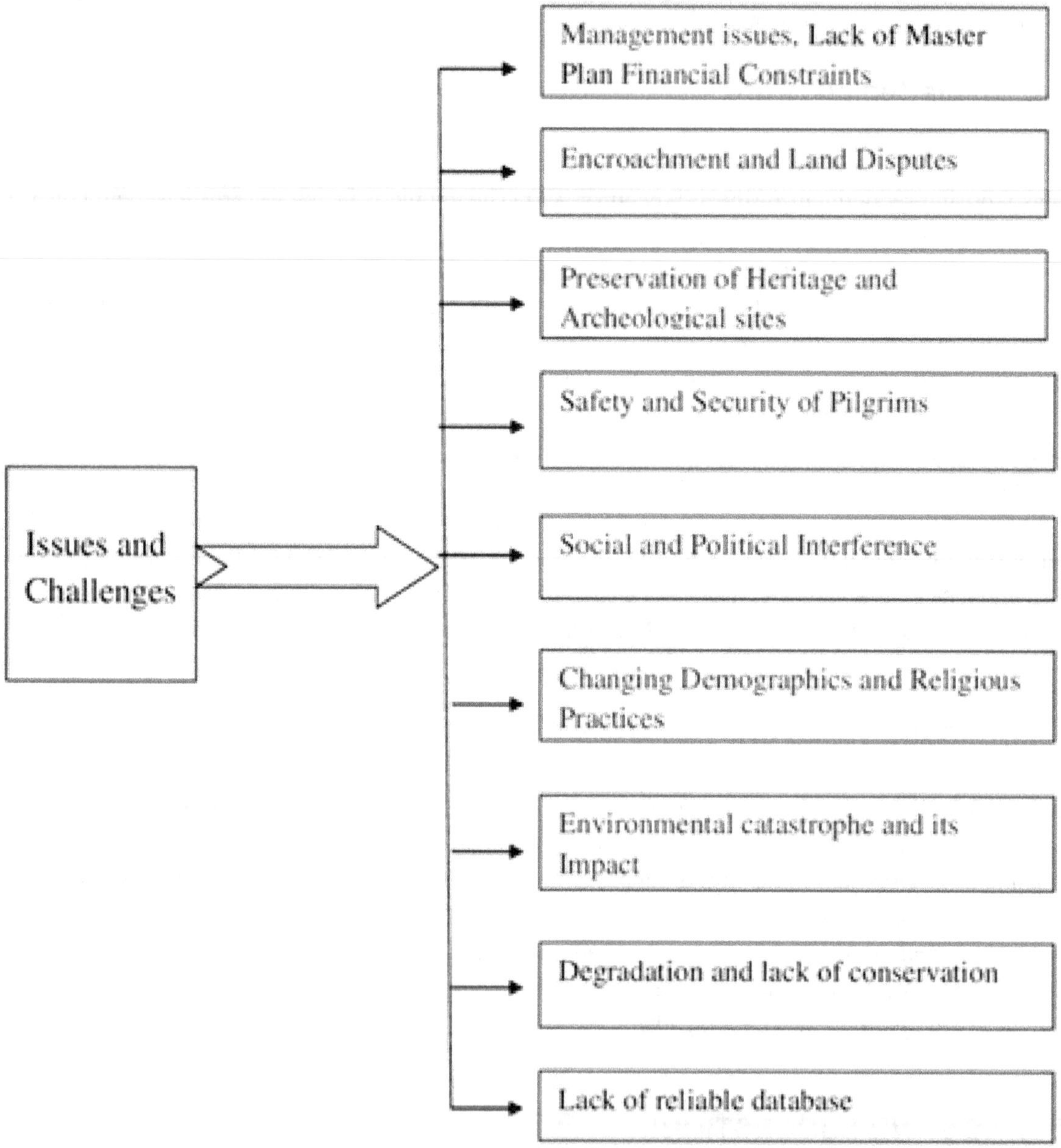

especially in religious celebrations. Managing and controlling crowds, avoiding stampedes, protecting assets of temples and other security measures are also challenges ((Batabyal et al., 2023a; Batabyal et al., 2023b) . Preservations of ancient temples are important for future generations. Some of these ancient temples suffer from negligence, lack of proper renovation, and not availability of techniques of ancient craftsmanship are also significant challenges. Many of these ancient temples are facing unnecessary intervention from local political or social groups. Various disputes related with religious practices also can create to maintain sanctity of the temples (Batabyal et al., 2022). Many of these temples are situated in environmentally sensitive areas like near to the river, lake, forest etc. More gatherings in the temple premises can deteriorate the carrying capacity, pollution, deforestation, and hamper environmental sustainability. Urbanization and changing demographic profiles are also major challenges. People perception, changes of religious practices and beliefs have direct impact on temple activities.

CONCLUSION

Ancient temples are important resources for any civilisation. These are the jewels of historical references. Cultural heritage of any society is directly related with ancient religious places. It is the responsibility of everyone to protect these for the use of future generations. Ancient Temples are considered not only the place of worship but also the warehouse of our culture and heritage. Odisha is home to some of the most beautiful presently existing ancient temples that attract tourists from every corner of the world. Bhubaneswar, the capital of Odisha is well known as Temple City. Odisha has numbers of ancient temples. Each temple is extraordinary and has its own unique history. The unique architecture of Odishan temple is known as Kalingan style of temple architecture. Temple architecture in Odisha plays a crucial role in preserving and promoting cultural heritage of the state. Local materials were used for construction of temple structure. Temples were mainly constructed by kings, rulers, and wealthy individuals with the help of super artisans, craftsmen of the locality. Kalinga, the ancient Odisha was well known for cultural heritage. Unique handicrafts and handlooms are continuing its glory. Government has setup many craft villages for production of traditional arts with ethnic setting. There are many issues for preserving and protecting ancient temples in the state. For addressing these challenges combined efforts are require between temple authorities, government agencies, religious organizations, and the local community. Sustainability is the core issue here. By improving governance, maintaining law and order, having financial sustainability, and promoting cultural heritage, these ancient temples can be protected for future generations.

REFERENCES

Allen, L. R., Long, P. T., Perdue, R. R., & Dieselbach, S. (1988). The impact of tourism development on resident's perception of community life. *Journal of Travel Research*, *27*(1), 16–21. doi:10.1177/004728758802700104

Bachleitner, R., & Zins, A. H. (1999). Cultural tourism in rural communities: The Residents perspective. *Journal of Business Research*, *44*(3), 199–209. doi:10.1016/S0148-2963(97)00201-4

Backman, S., Uysal, M., & Backman, K. (1991). Regional Analysis of Tourism Resources. *Annals of Tourism Research*, *8*(1), 323–327. doi:10.1016/0160-7383(91)90016-5

Bagri, S. C. (1999). Status of Travel & Tourism Industry in India, A Futuristic View of Growth. *Journal of Tourism, 41*.

Baloglu, S., & Brinberg, D. (1997). Affective images of tourism destinations. *Journal of Travel Research*, *35*(4), 11–15. doi:10.1177/004728759703500402

Batabyal, D., Das, D. K., Sen, S. K., Porya, A., Halam, H., & Chatterjee, P. (2022). Evolution of Shakti Peetham for religious, mythological and dark tourism in Indian sub-continent. *Revista Turismo & Desenvolvimento*, *39*, 291–300.

Batabyal, D., Rawal, Y. S., Chatterjee, P., & Goswami, S. (2023b). Temple service quality model for future pilgrimages in Indian sub-continent through user-generated contents (UGC) in TripAdvisor. *International Journal of Tourism Anthropology*, *9*(2), 113–135. doi:10.1504/IJTA.2023.130849

Batabyal, D., Ray, N., Bag, S., & Nag, K. (2023a). Tourists' Harassment During Pilgrimage: A Case Study of TripAdvisor's Review for the Hindu Pilgrim Centers in India. In Resilient and Sustainable Destinations After Disaster: Challenges and Strategies (pp. 139-156). Emerald Publishing Limited.

Biswas, S., & Mohapatra, P. (2022). Analysis of Impact of Covid-19 on Religious Tourism Destinations of Odisha, India. IMPACT: International Journal of Research in Applied. *Natural and Social Sciences*, *10*(9), 1–12.

Biswas, S. N. (2020). *Odisha: The Land of Chromatic Tourism Destinations*. Mousumi Prokashani.

Chand, S. (2005), Orissa Review Vol. LXII No. 4, Pp- 45-47

Cohen, E. (1992). Pilgrimage and tourism: convergence and divergence. In *Sacred journeys: The anthropology of pilgrimage* (pp. 47–61). Greenwood Press.

Das, R. P. (1994). *Terminology of Orissan Temple Architecture*. Lalita Kala Academy, New Delhi, December.

Dash, P. K. (2000). *Management of Tourism Sector in Orissa, Problems and Future*. [Thesis, Utakal University].

Dhar, P., Sarkar, A. K. (1998). *Indian Tourism*. Kanishka Publication.

Donaldson, T. (1975). Propitious Apotropaic Eroticism in the Art of Orissa. *Artibus Asiae*, *37*(1), 75–100. doi:10.2307/3250213

Hall, M. C., & Zeppel, H. (1990). History, Architecture, Environment: Cultural Heritage and Tourism. *Journal of Travel Research*, *29*(2), 54–55. doi:10.1177/004728759002900212

Murphy, P., Pritchard, M. P., & Smith, B. (2000). The destination product and its impact on traveller perceptions. *Tourism Management*, *21*(1), 43–52. doi:10.1016/S0261-5177(99)00080-1

OTDC. (2023). *Temples*. OTDC. https://odishatourism.gov.in/content/tourism/en/discover/attractions/temples-monuments.html https://www.orissatourism.org/odisha-destinations/ https://magazines.odisha.gov.in/orissaannualreference/ORA-2004/pdf/listof_temples_of_orissa.pdf

Padmanabhan Rao, R. A. (1999). Tourism; Medium of Social and Cultural Development. *Yojana*, *43*(8), 41.

Panda B. (2016). Temples of Orissa. *Global journal of Indian history and culture*, *1*, 11-12.

Panigrahi, K. C. (1993). History of Orissa. *Kitab Mahal, Cuttack*, *1993*, 1–9.

Polácek, M., & Aroch, R. (1984). Analysis of cultural sights attractiveness for tourism. *Tourism Review*, *39*(4), 17–18.

Rinschede, G. (1992). Forms of Religious Tourism. *Annals of Tourism Research, 19*(1), 51–67. doi:10.1016/0160-7383(92)90106-Y

Ryan, C., & Montgomery, D. (1994). The Attitudes of Bakewell Residents to Tourism and Issues in Community Responsive Tourism. *Tourism Management, 15*(5), 358–369. doi:10.1016/0261-5177(94)90090-6

Schroeder, T. (1996). The Relationship of Residents' Image of their State as a Tourist Destination and their Support for Tourism. *Journal of Travel Research, 34*(4), 71–73. doi:10.1177/004728759603400411

Seth, R. (1999). Tourism: Problems and Prospects. *Yojana*, 21–24.

Singh, P.K (1998). *Fifty years of Indian Tourism*. Kanishka Publication.

Singh, R. (1998) *Infrastructure of Tourism in India*. Kanishka Publication.

Timothy, D., & Olsen, D. (Eds.). (2006). *Tourism, religion and spiritual journeys*. Routledge. doi:10.4324/9780203001073

Tripathy, M. (1998). *Patta painting society and economy of the chitrakaras of Orissa a study of change and development.* [Thesis, Utkal University]. https://shodhganga.inflibnet.ac.in/handle/10603/189442

Chapter 4
Crafting an Experiencescape for Sustainable Cultural Tourism:
A Case Study of Udaipur's Craft Village "Shilpgram"

Saumya Kapil
Christ University, India

Bindi Varghese
Christ University, India

ABSTRACT

India, known for its rich and varied cultural heritage, hosts numerous fairs and festivals as vibrant platforms to showcase its diversity. Events like the Kala Ghoda Arts Festival, Ellora-Ajanta International Festival, Surajkund Crafts Mela, Pushkar Fair, and many others celebrate the country's artistic and cultural wealth. This book chapter explores sustainable culture and heritage development in Udaipur's craft village, 'Shilpgram.' The village hosts an annual craft fair during the peak tourist season, attracting artists across India to showcase their unique crafts, folk dances, and paintings. This chapter assesses Shilpgram's feasibility as an experiencescape model using a case study method. Interviews with key stakeholders provided insight into the destination's difficulties, strengths, and best practices. The study adds to our understanding of cultural locations' responsibilities in cultural tourism by balancing historical preservation with current expectations.

1. INTRODUCTION

Festivals and fairs in India, such as Diwali and the Kumbh Mela, provide considerable economic possibilities while reinforcing cultural values. Despite India's long history of handicrafts dating back to the Indus Valley civilisation, the industry has failed due to a lack of exposure and marketing. Artist Villages are a possible option, providing workshops and design studios to foster creativity while linking artisans to national and worldwide forums. They give artisans a global platform to present their work, bridging

DOI: 10.4018/978-1-6684-9957-3.ch004

the gap between tradition and modernity and helping preserve and prosper India's distinct handmade legacy. According to Shastri (2017), an artist village is aptly characterized as

A space for recreation and interaction, where the general masses can come, see the displayed art form, watch the performing Art and interact with the artists. Artisans can reside and enlighten others about their skills. (Shastri, 2017 p. 25)

Shilpgram, Udaipur, is an exceptional illustration of an artist community that is self-sustaining. The art and craft of India, preserved in the hands of rural artisans and explored for potential connections to the modern world today. Moreover, with "Shilpgram - a Craft Village" in Udaipur, the West Zone Cultural Centre (WZCC) embodies this aim by displaying the different cultural traditions of Western Zone states. Shilpgram also holds the annual "Shilpgram Utsav," which attracts over 500,000 visitors and hundreds of artists and entertainers. WZCC's headquarters, 'Bagore-Ki-Haveli,' also has a museum dedicated to conserving royal legacy (Mehmood & Aazem, 2002). These facilities are critical for preserving and appreciating India's diverse cultural mix.

According to UNESCO, oral traditions and expressions, performing arts, social practices, rituals and festive events, knowledge and practices, and traditional handicrafts are all areas of ICH (UNESCO, 1989). In reality, another implicit feature strongly related to the previous domains produces a liveable place for appreciation (i.e., an experiencescape). The significance of ICH lies not only in its cultural form but also in the social and economic worth of culture and tradition transmission for communities or specific social groupings. This chapter aims to apply a case study technique to determine the viability of Shilpgram as an experiencescape model. This evaluation will mostly depend on qualitative interviews with significant Shilpgram stakeholders. Finally, this chapter aims to contribute to a broader understanding of how cultural destinations such as Shilpgram can effectively serve as experiencescape models, preserving and promoting cultural heritage while addressing contemporary challenges and opportunities in the field of cultural tourism.

2. LITERATURE REVIEW

2.1. Understanding Cultural and Heritage Tourism

'Heritage' is defined as "our legacy from the past, what we live with today, and what we pass on to future generations," as enshrined in an international treaty known as the 'Convention for the Protection of the World Cultural and Natural Heritage,' which was ratified by UNESCO in 1972. 'Cultural heritage' is defined as "the irreplaceable sources of life and inspiration that include monuments such as architectural works of monumental sculpture and painting, elements of archaeological structures, inscriptions, cave-dwelling and combinations of features; groups of buildings, groups of separate or connected buildings; and sites including archaeological sites of outstanding historical, aesthetic, ethnological, or anthropological value" Articles 1 and 2 of the *World Heritage Convention*. As a result, more than half of the proportion of 'cultural heritage' in the tourist sector is concerned with intangible cultural heritage, which (a sub-type of tourism) is immeasurable (Halder & Sarda, 2021). Intangible cultural heritage, on the other hand, refers to "the practices, representations, expressions, knowledge, and skills—as well as the instruments, objects, artefacts, and cultural spaces associated in addition to that—that communities, groups, and, in some cases, individuals recognize as part of their cultural heritage" (UNESCO, 1989, Art. 2, para 1).

The United Nations Educational, Scientific, and Cultural Organization (UNESCO) adopted the first specific international legal instrument on ICH in 1989, namely the Recommendation on the Safeguarding of Traditional Culture and Folklore (UNESCO, 1989), with a minimal scope because the concept of "folklore" is more restrictive than ICH (Lenzerini, 2011). Later, in 2003, UNESCO declared five domains of intangible cultural heritage: oral traditions and expressions, including language as a vehicle of intangible cultural heritage; performing arts; social practices, rituals, and festive events; knowledge and practices concerning nature and the universe; and traditional craftsmanship (Bedjaoui, 2004). Following this, governments and researchers in many nations began to promote this new type of "immaterial" or "living" culture, and governments all over the globe are now competing to get their countries' ICH recognized by UNESCO (Loulanski & Loulanski, 2011; Qiu et al., 2022).

Su (2019) goes so far as to argue that even the commercialization of ICH is desirable as long as the local bearers of ICH are allowed to actively engage in the process by associating their traditional meanings and values with its commodified form. As a result, the influence of tourism on ICH must be examined on two levels: the cultural artefact, form, or practice itself and its bearers or practitioners. Tourism may have both beneficial and harmful effects on intangible heritage and its bearers. According to existing research, the adverse effects of tourism on intangible heritage include homogenization, commodification, altering traditional meanings and values associated with it, marginalization of local practitioners involving their replacement by more professional performers, and even their disempowerment or exploitation. Tourism's beneficial impact includes the rebirth of cultural traditions, enhanced visibility, intercultural interchange, job prospects for local populations, and enhanced pride and self-esteem (Ranwa, 2022).

2.2. The Concept of "Experiencescape"

Cultural resource consumption has shifted from permanent material legacy to movable intangible forms of culture, such as events, festivals, dance, and music. The interconnected tangible and intangible tourism resources co-create a destination's spatial experience, allowing tourists to form physical and emotional attachments to the location (Cetin & Bilgihan, 2016). Cultural festivals featuring live performative components are increasingly appealing to tourists seeking to experience a unique cultural aspect of a site, mainly traditional festivals still performed by local populations (Chen, 2022). The creation of ICH for tourism is viewed as a strategy for economic revitalization. Furthermore, preserving such ICH assets may assure a genuine presentation to the audience (Choi et al., 2020); hence, a quality tourist experience can satisfy both heritage management aims and tourism development demands.

Tourists are more interested in destinations that provide a variety of entertainment options. Tourists envisage various alternatives based on location attractiveness before commencing their journey. The landscape of experience, or experiencescape, refers to the settings where experience is designed and received (O'Dell, 2005). For example, many ancient industrial city facilities have been turned into art galleries, exhibitions and amusement centres, with new aesthetic components and original purposes replaced by new ones. Tourism, leisure, labour, and play have all merged in modern civilizations, blurring divisions between them. The interaction between economic activity and socio-cultural development has shaped consumption and its values.

Pine and Gilmore (1999, 2011) introduced the experience economy, which characterizes the visitor experience along two axes. Tourist activities and experiences are classified into four categories based on consumer immersion and engagement: entertainment, educational, escapism, and esthetic (Pine & Gilmore, 1999). The entertaining and esthetic experiences need passive involvement, but the instructional and escape experiences require active participation. Customers are engrossed by events or offerings when they engage in an entertainment or educational experience (Zhang et al., 2021). Customers must be fully immersed in a setting for both esthetic and escapism experiences.

2.3. The Government Initiatives and Cultural Centres in India

Rajasthan is rich in cultural heritage and there have been various government intitiatives to revive ICH that are struggling. With the collapse of snake charmer as rural artists' traditional employment owing to the restriction put on snake capturing by the Indian government's Wildlife Protection Act, 1972, a tiny part of Kalbeliya women switched to their dance as an occupation during the 1980s for survival (Ranwa, 2022). This was followed by the Indian state's nomination of the Kalbeliya dance and songs for Intangible Cultural Heritage (ICH) designation by UNESCO, which resulted in its inclusion on the ICH Representative List in 2010. Further, the implementation of cultural centres as a part of an initiative for entire India play an important role in preserving and promoting the country's unique cultural history.

The concept of Zonal Cultural Centers (ZCCs), first proposed in 1985, was enlarged under the 7th Five Year Plan to include seven centres spanning geographic and linguistic boundaries: Patiala, Nagpur, Udaipur, Prayagraj, Kolkata, Dimapur, and Thanjavur (About West Zone Cultural Center – WZCC, n.d.). These places, frequently outside state capitals, are chosen for their profound cultural relevance and accessibility. Table 1 shows the financial support provided to ZCCs throughout the last three years and the current fiscal year. These institutions run various programs, including the "Young Talented Artists" initiative, which provides a one-time financial prize of Rs. 10,000 to young artists aged 18 to 30 specialising in uncommon art forms. The "Guru Shishya Parampara" plan aims to pass down ancient art forms by teaching pupils under established artists, whilst "Theatre Rejuvenation" encourages theatrical activities with honoraria of up to Rs. 30,000 per show. The "Research & Documentation" project works with state Cultural Departments to conserve fading visual and performing arts. Furthermore, "Shilpgram" promotes folk and tribal arts, "Octave" promotes the cultural legacy of the North East area, and the "National Cultural Exchange Programme (NCEP)" supports festivals and cultural exchanges, strengthening India's cultural variety and history.

The Western Zonal Cultural Centre (WZCC) in India, which is credited with pioneering the establishment of Shilpgram, a craft village in Udaipur, serves as a multifaceted platform for artistic expres-

Table 1. Grant-in-aid provided to ZCCs

Sl. No.	Year	Amount (Rs in Lakhs)
i.	2016-17	6085.07
ii.	2017-18	4689.71
iii.	2018-19	5952.69
iv.	2019-20 (as of date)	5655.75

Source: PIB dated 02/03/2020

sion, encompassing both performance and visual arts, as well as the promotion of individual and tribal craftsmanship in the country's western region. This one-of-a-kind crafts community gets its inspiration and participation from a broad socio-ethnic terrain that stretches from Rajasthan's medieval deserts to the coastal state of Goa, which has vestiges of Portuguese colonial influence (About West Zone Cultural Center – WZCC, n.d.). WZCC is a registered society led by the Governor of Rajasthan, with participation from governmental and non-governmental organizations, mainly from the Western Zone. The overarching mission of the Zonal Cultural Centres is to foster a profound societal appreciation for the region's and the nation's rich cultural heritage, with a particular emphasis on nurturing individual artistic expressions and cultural artefacts to revitalize endangered art and craft forms and ensure their continuity.

2.4. Successful Cultural Tourism Destinations in India

There are various case studies to understand the scenario of cultural heritage destinations in India. Sharma and Sharma (2015) stress the dual function of fairs and festivals in Pithoragarh, both the obstacles and economic advantages they provide, in sustaining cultural traditions while injecting vigour into the local economy. The study of the "Pandrandu Vilakku" festival in Kerala by Babu and Munjal (2015) highlights the issues faced by commercialization and the necessity to preserve the interests of local traders to minimize economic leakage. Tiwari's (2015) research on Shilpgram in Rajasthan exemplifies a successful public-private collaboration that supports traditional crafts, ensuring cultural heritage continuity while delivering economic advantages to rural and urban artists.

Munjal (2015) underlines the need to incorporate cultural resources into fairs and festivals to stimulate urban development in Sohna. It emphasizes the significance of discussion between traditional and emerging stakeholders in the sustainability of major cultural events. Durgamohan's (2015) appraisal of the Hampi festival highlights the importance of local community support and the necessity for a collaborative master plan encompassing all critical stakeholders to fully achieve such events' cultural and economic potential. Gaur and Chapnerkar (2015) investigate the statewide "Ganapati festival," highlighting the necessity of a well-coordinated strategy and environmentally sustainable methods. They also emphasize the need for cultural preservation to preserve the festival's originality. Similarly, Sharma and Sharma (2015) emphasize the importance of infrastructure development, such as roads and municipal services, in their study on Pithoragarh.

Artist villages in India, such as Belapur Artist Village, Sanskriti Kendra in Delhi, and Shilpgram in Udaipur, contribute significantly to the country's cultural history. Ar. Charles Correa's Belapur Artist Village presents a unique housing plan that fosters communal engagement among artists while protecting privacy (Perry, 2008). Sanskriti Kendra, an organic artist community, provides a temporary home for various artists and craftspeople while blending in with the natural surroundings. Shilpgram, a government-sponsored initiative, acts as a vibrant hub for artisans, conserving tribal culture through its rural market, vocational training, and traditional huts (Shastri, 2017). These artist villages preserve and promote India's rich cultural traditions and represent the country's architectural variety, highlighting the value of social spaces, sustainability, and cultural authenticity in the context of artist communities (Munjal & Jauhari, 2015). In conclusion, these case studies show that successful cultural festivals and heritage sites in India need a careful mix of cultural preservation, economic advantages, stakeholder engagement, infrastructural development, and cultural authenticity.

3. STUDY SETTING

Shilpgram, located in the gorgeous Aravali range, offers an example of a self-sufficient artist community that honours India's vast creative legacy and forges current links (About West Zone Cultural Center – WZCC, n.d.). This 70-acre rural arts and crafts complex in Udaipur meticulously depicts the different ways of life and traditions of society and tribal tribes from India's Western Zone, which includes Rajasthan, Gujarat, Maharashtra, and Goa. Shilpgram's 31 hut typologies truly depict the architectural and cultural forms of the Western Zone, representing various regional traditions. These cottages, created by local artists using indigenous materials, reflect the distinct characteristics of their multiple locations.

For example, the Ramaroa huts in Rajasthan are meant to accommodate large families and have privacy walls, kitchens, grain storage, and weaver's workshops. South Gujarat's Dang huts are mostly made of dang wood and thatch grass for insulation, and they have wonderfully painted Pithoda-style walls. Huts erected on a flat clay plain in West Gujarat sparkle with granaries and beautiful mud-mirror work by household ladies (Shastri, 2017). The coastal Koli community's huts in Maharashtra are made of poal wood and coconut leaves, giving the walls a characteristic mat-like look. Goa's hutments are distinguished by two styles: one in Vasco, characterised by rich iron and red laterite stone, and another in Kundi Kora, distinguished by clay construction and thatched, sloping roofs strengthened with date leaves and tree trunks.

The complex also includes four museums: the Chaupal Museum, which displays domestic goods; the Gol Museum, which displays masks and musical instruments; the Tribal Museum, which immerses visitors in tribal life; and the Kothi Museum, which displays various home styles. An open-air amphitheatre with seating for 8,000 people, which can be observed in the site map of Shilpgram in *Figure 1,* emerges as a cultural focal point, staging dramatic events that fascinate tourists and foster respect for traditional folk arts. The craft bazaar is designed to seem like an ancient "haat" or traditional market. It allows visitors to immerse themselves in the colourful culture while purchasing magnificent art and craft objects from the region, maintaining history and supporting local craftsmen (Shenoy, 2022).

Shilpdarshan, a recurring event, welcomes traditional artisans and skilled workers to demonstrate their abilities and artwork, giving them a platform to sell their wares while fostering acknowledgement of their brilliance. This event deepens the bond between makers and customers, broadening respect for local handicrafts (Tnn, 2016). Shilpgram, in essence, is a tribute to India's creative variety and cultural legacy, providing a platform for artists and tourists to connect with and celebrate the country's rich traditions while contributing to local people's economic well-being (Tiwari, 2015).

4. METHODOLOGY

This chapter investigated Shilpgram as an experiencescape model using a case study technique. Shilpgram, a former famine site, was chosen as the research location owing to the state government's transformation of the area into an artist village, implying the establishment of a one-of-a-kind landscape of experience integrating economic activity with socio-cultural growth. This study's data came from both primary and secondary sources. Preliminary data was gathered through anthropological research, which includes personal encounters, meetings with authorities, organizers, artists, tourists, and the Shilpgram local population. A total of 20 persons aged 18 and up were questioned, including three organizers, five

Figure 1. Sitemap of Shilpgram
(Source: Shastri, 2017)

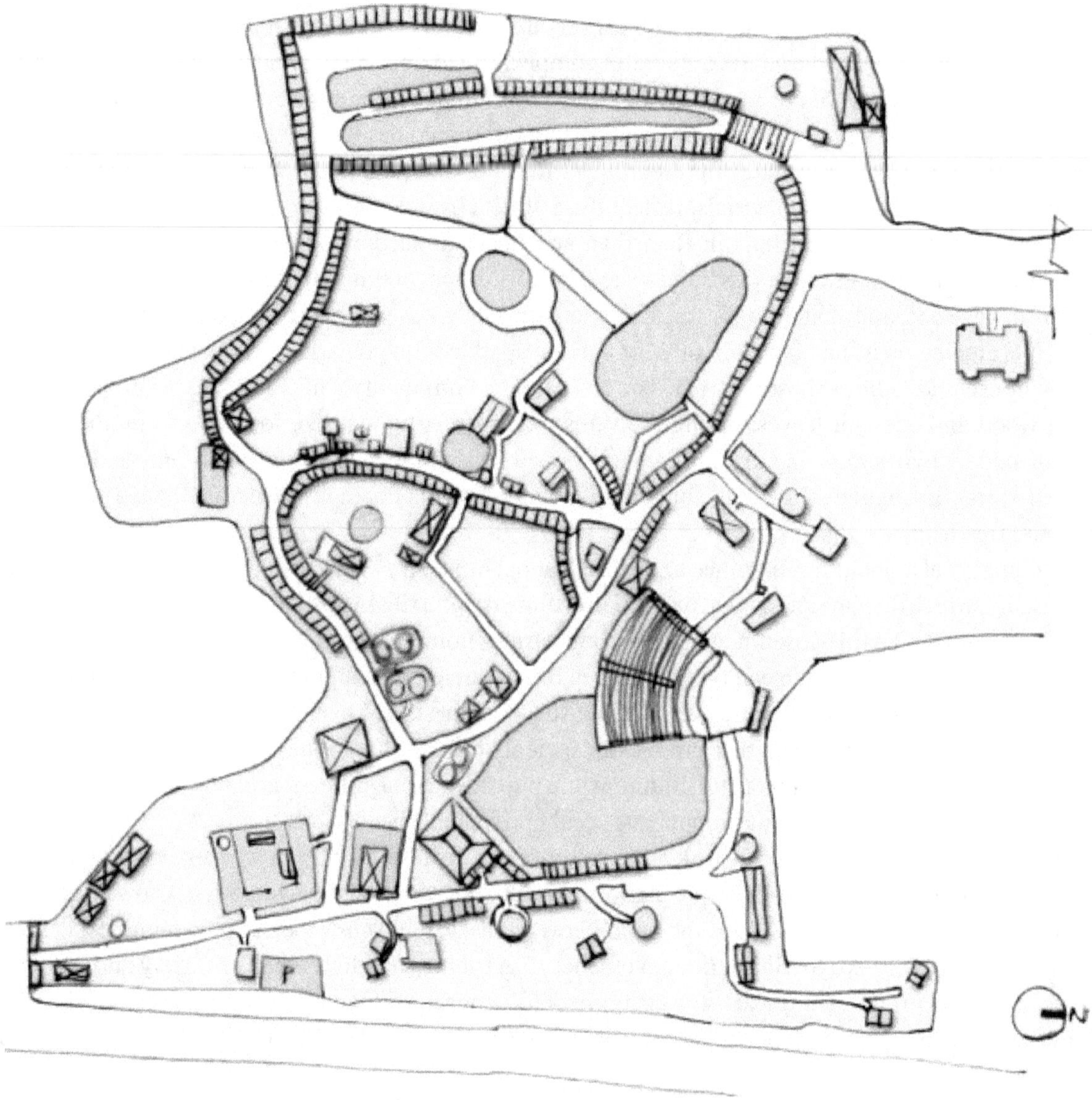

artists, five crafters, two locals, and five tourists. Based on the availability of respondents, convenient sampling was used.

The information gathered allowed for an examination of stakeholder involvement in the development of Shilpgram's Experiencescape (Zhang et al., 2021), which was divided into four themes: entertainment (sub-categorized as actor performance and theatre service), educational (culture and Art; and social connection), escapism (relax and escape; and interactive participation), and esthetic experiences (theatre design; and props and facilities). The survey data were used to assess Shilpgram's efficacy as an experiencescape and highlight the concerns and challenges that must be solved to thrive as a cultural heritage destination. This study approach enabled a thorough examination of Shilpgram's function and influence within the cultural and economic environment, revealing insight into its potential to stimulate socio-cultural development and promote the preservation of traditional art forms.

5. SHILPGRAM'S ALIGNMENT WITH THE EXPERIENCESCAPE FRAMEWORK

This section delves into the applicability of Pine and Gilmore's (1990, 2011) Experiencescape Model to the setting of Shilpgram, using insights from interviews with various stakeholders. Su (2019) argues that commercializing ICH is beneficial as long as local carriers actively participate in the process, linking traditional meanings and values with its commodified form. Shilpgram, a platform for creative expression and craftsmanship developed by India's Western Zonal Cultural Centre (WZCC), is a fulfils the requirements of commodification of ICH as a diversified platform for artistic expression and craftsmanship from Rajasthan to Goa, representing the multifarious character of ICH participation (About West Zone Cultural Center - WZCC, n.d.). Pine and Gilmore's Experiencescape Model divides visitor encounters into four categories: entertainment, education, escapism, and esthetic, with different levels of passive or active interaction (Zhang et al., 2021). The application of this model to Shilpgram underscores the importance of active participation in fostering immersive experiences, aligning with the diverse cultural offerings and practices showcased in this unique craft village.

5.1. Educational Experience

Shilpgram, nestled in the vibrant city of Udaipur, stands as a bastion of educational experiences for tourists seeking cultural enrichment and artistic insight. Through a series of illuminating interviews, it becomes evident that Shilpgram's commitment to providing an educational voyage is unwavering. A volunteer at the tribal museum highlights the cultural and artistic significance of the art exhibition, connecting it to the historical context of paintings of Mavji Maharaj of Beneshwar, Dungarpur, from 1784 to 1789. Talking about the paintings he mentioned,

These paintings are inspired by the prophecies [also called "Bhakdi"] of Mavji Maharaj, [total 1,766,000 in number] converted into four scriptures. Now, [i.e., in the 20th century] the prophecies of the second scripture are proving to be true., ... he predicted that there would be telephone communication, water would be sold, and people would go into space...

Further, they also mentioned the presence of 30 paid volunteers in varied areas such as Cultural/ Art events, folk art management or graphic designing. It demonstrates the collaborative nature of the event. A zari artisan from Bareilly, Uttar Pradesh, discusses the government's role in promoting craft and providing training from fashion societies before presenting their Art to exhibitions such as Shilpgram. These insights suggest that Shilpgram is an educational platform fostering a deeper understanding of culture, Art, and crafts.

5.2. Esthetic Experience

Shilpgram, a vibrant tapestry of India's diverse cultural heritage, offers visitors an extraordinary visual experience that transcends the conventional boundaries of Art and craft. We discover a tapestry woven with echoes of tradition, authenticity and exceptional visual appeal from the heart of this artistic haven. The relevance of village-like architecture, as can be observed in *Figure 2: The Gujarat Hut,* is elaborated further by the organizer,

"The materials for the huts are sourced from these villages, and the artists themselves come from these villages. Even the decorations used are brought from their homes and adorned here in the same traditional manner. The significance of this craft fair lies in the fact that artists feel like they are in their own homes, showcasing their Art and culture to visitors. The perspective is to establish a connection between people and the process of crafting these items. Unlike with ready-made products, here, you can engage directly with the artisans and witness the crafting process."

The architecture of this craft fair is designed so that every craft from different aspects of India has a space. An artisan from the Bunkar (weaver) community of Latur, Maharashtra, exclaimed,

"People here are dressed in quite different attire, and the Shilpgram festival is quite unique. Every aspect has a separate section or area."

A blue pottery artist from Bagru village with beautifully coloured pottery, as can be observed in *Figure 3*, explained the making of this GI-tagged tangible heritage of India,

"Blue pottery, which is world-famous, also has a GI tag; Raja Man Singh [Prince of Jaipur] got this Art from Iran since then, it is only made in blue colour. A researcher went again to research in Iran and introduced all the colours. It is made from a slurry of the pawad stone and not ceramic."

These narratives highlight the esthetic dimension of Shilpgram, where visual appeal and craftsmanship play a pivotal role in shaping the overall experience. However, a performing artist from Manipur highlights the low payment aspect and the significance of participation.

5.3. Escapism Experience

Shilpgram, nestled amidst the rustic landscapes of Udaipur, unfolds a captivating narrative of escapist experiences that transport visitors into a realm of forgotten crafts, age-old skills, and cherished traditions. A tourist from the USA came to Shilpgram through the Recommendation of locals; he mentioned,

"It is amazing how they are supporting different crafts, forgotten skills and trade... while me and my wife are immersing ourselves in it... my four kids here... are enjoying the swings."

A group of college students, still in their college IDs mentioned,

Figure 2. Gujarat Hut
(Source: Photograph taken by the author)

Figure 3: Blue pottery artisan discussing the origin of the craft
(Source: Photograph taken by the author)

"We visit Shilpgram yearly with our friend's group... it is like a ritual. We usually come in the initial days when there are fewer tourists... We visit all the stalls, click photos and then go to Fateh Sagar."

An artisan of musical instruments from Delhi was thrilled to see the reaction of visitors while putting up a stall for the first time in Shilpgram. He said,

"These are traditional musical instruments that are part of our heritage. We have miniatures as well as actual-sized instruments for professional use. We mainly cater to kids..., aiming to divert their attention from mobile phones and encourage their interest in learning how to play musical instruments."

Further, a freelance artist in the tribal museum mentioned their concern for the behaviour of visitors in the museum,

It's a selfie point for the general public, [although] artists who are interested inquire [about the paintings]. The development of the gallery is needed... They touch the paintings; they don't know how to behave in galleries.

The paintings lacked descriptive information boxes that could provide details about the profound prophecies of Saint Mavji Maharaj, as shown in *Figure 4*. Moreover, various artisans with the stalls criticized the bathroom facilities and the dust, which is quite pervasive, occasionally causing breathing issues.

5.4. Entertainment Experience

In essence, Shilpgram is more than a cultural spectacle; it is an enduring source of entertainment that captivates diverse audiences, inviting them to immerse themselves in the enchanting rhythms and narratives of India's rich artistic traditions. Siddi Dhamal, a folk dance artist, praises the cultural appeal of the event and mentions the significance of folk dance performances.

"We have been coming here for ten years now... we get to see the original culture of India here. The African folk dance that we perform is more than 800 years old; it is an expressive dance form inspired by the sounds and actions of animals... I feel very proud to perform that on such a huge stage."

Another dance group from Bhuj, Kutch, expressed how proud they were to represent their dance form and be a part of the "closing event," i.e., on 30 December. The performances in the amphitheatre are not

Figure 4: The painting exhibited in the Tribal Museum
(Source: Photograph taken by the author)

pre-recorded, and they do not use modern music recorders. They use authentic instruments originally used with folk dances from several parts of India. The artists can be seen practising for the closing event in *Figure 5.* The organizer passionately exclaimed,

"Shilpgram operates year-round, with artists coming to stay and perform here for 15 days, honing their craft through continuous practice and specialized training for high-profile performances. Folk artists nationwide participate, with 700 performing artists and 900 artisans. It's not like a reality show; one must continuously strive and practice for these performances. Artists undergo intensive training to master their craft."

These accounts emphasize that Shilpgram serves not only as a cultural experience but also as a source of entertainment, attracting diverse audiences. In context to Ranwa, 2022, tourism exerts both advantageous and detrimental influences on ICH. Detrimental effects include homogenization, commodification, altering traditional meanings, marginalization of local practitioners, and potential disempowerment. Shilpgram has continued to maintained the authenticity of west zone culture for decades. It includes invitations to artisans & performing artists from their villages and the use of folk instruments. Conversely, Ranwa (2022) also argues that tourism can lead to the revitalization of cultural traditions, increased visibility, intercultural exchange, job opportunities, and heightened pride and self-esteem which is effective in the case of Shilpgram as reviewed by stakeholders in their interviews.

6. CHALLENGES AND FUTURE PROSPECTS

Shilpgram is well-known for spreading Indigenous knowledge. Indigenous knowledge is therefore linked to environmental problems and the possibility of local place sustainability (Veldpaus, 2015). On the one

Figure 5: Artists rehearsing for the closing event in the amphitheatre
(Source: Photograph taken by the author)

hand, indigenous knowledge is utilized to preserve natural resources, adapt to disasters, and conserve species and biodiversity. It will be feasible to seek some remedies from the wisdom of ancient peoples, that is, to tackle environmental problems using indigenous knowledge. Indigenous knowledge, on the other hand, is linked to ecotourism (Wang et al., 2021; Pal & Rawal, 2020; Mehta et al., 2022), following global trends and boosting tourist destinations' competitiveness. However, several concerns with forming esthetic experiences in Shilpgram demand immediate attention. There were salary concerns for the artists, bathroom facilities and the plantation's demand to regulate dust in the atmosphere.

Another issue that has to be addressed is the proper commercialization of ICH through Shilpgram. ICH was overlooked during place-making studies in comparison to tangible cultural assets. ICH plays an essential role in branding and marketing during this process. When museums, theme parks, monuments, events, and souvenirs are changed from ICH, they become a tool and scale in place creation, combining tangible and intangible aspects. As part of the cultural landscape and touring circuit, ICH may help travellers realize their desire to learn about and experience local culture (Qiu et al., 2022). Online communities can be used to brand these experiencescapes more ethically. Online communities developed with distinct cultural practices or grouped by a common theme based on historic sites or other cultural heritage have evolved lately, coinciding with the beginning of the Web 2.0 era (Ginzarly et al., 2019).

In contrast to offline community involvement, online community contact is entirely geo-free and may encourage mutual understanding amongst people from diverse cultural backgrounds. Others can exchange memories or sensations that are part of their "sense of place" with others who live nearby or with persons on the opposite side of the world (Dhonju et al., 2018). Furthermore, it expands the potential for engagement between local communities and experts while breaking down occupational barriers (Li et al., 2020; Liang et al., 2021). For example, Gomez-Oliva et al. (2019) developed a unique and co-created "Be Memories" application for tourists to spread the ICH during local events in a Spanish community with significant cultural value. According to this viewpoint, social media material provides a low-cost and long-term supplement to official information, documentaries, and local tours. There is a need for Shilpgram to garner this co-creation of cultural assets to stand as a successful model of Experiencescape.

7. CONCLUSION

Shilpgram, a renowned craft village in Udaipur, offers diverse experiences to tourists, categorized into entertainment, educational, escapist, and esthetic domains. Tourists from all walks of life are drawn to its cultural richness and immersive activities. The entertainment experiences include vibrant folk dances, highlighting India's diverse heritage. Shilpgram fosters educational experiences by connecting visitors with art exhibitions and craft demonstrations. As an escapist haven, it allows visitors to escape the mundane and engage with forgotten skills. Finally, the esthetic dimension is showcased through its architecture and craftsmanship, providing visual appeal. These facets collectively make Shilpgram a multifaceted cultural destination.

Shilpgram is an example of an experiencescape, providing a complex trip through India's rich cultural fabric. Its dedication to conserving indigenous knowledge meshes with worldwide ecotourism trends, strengthening its attractiveness as a sustainable tourist destination. However, it has urgent issues such as addressing artist remuneration, improving restroom facilities, and regulating dust levels. Shilpgram's cultural environment and tourism circuits can benefit from the ICH's role in branding and marketing in conjunction with tangible cultural assets. Integrating online communities and social media provides a viable channel for co-creating cultural assets, increasing engagement with a global audience and experts, and bridging cultural divides. Shilpgram may secure its place as a successful and culturally enriching experiencescape model by tackling these difficulties and embracing co-creation, conserving India's legacy for future generations.

REFERENCES

About West Zone Cultural Center. (n.d.). WZCC. https://wzccindia.com/about-west-zone-cultural-center-wzcc/

Babu, V., & Munjal, S. (2015). Oachira Panthrandu Vilakku: A study of culturally embedded festival aligned with economic benefits. *Worldwide Hospitality and Tourism Themes*, *7*(4), 403–416. doi:10.1108/WHATT-03-2015-0009

Bedjaoui, M. (2004). The Convention for the Safeguarding of the Intangible Cultural Heritage: The legal framework and universally recognized principles. *Museum International*, *56*(1-2), 150–155. doi:10.1111/j.1350-0775.2004.00469.x

Cetin, G., & Bilgihan, A. (2016). Components of cultural tourists' experiences in destinations. *Current Issues in Tourism*, *19*(2), 137–154. doi:10.1080/13683500.2014.994595

Chen, Z. (2022). Visualizing experiencescape–from the Art of intangible cultural heritage. *Current Issues in Tourism*, *25*(4), 559–578. doi:10.1080/13683500.2021.1892040

Choi, S., Imon, S. S., & Couto, U. (2020). Negotiating with authenticity to ensure sustainability of traditional cultural festivals: Residents' perspectives about Macao's Drunken Dragon festival. *Sustainability (Basel)*, *12*(3), 885. doi:10.3390u12030885

Dhonju, H. K., Xiao, W., Mills, J. P., & Sarhosis, V. (2018). Share Our Cultural Heritage (SOCH): Worldwide 3D heritage reconstruction and visualization via web and mobile GIS. *ISPRS International Journal of Geo-Information*, *7*(9), 360. doi:10.3390/ijgi7090360

Durgamohan, M. (2015). Cultural and economic significance of Hampi festival. *Worldwide Hospitality and Tourism Themes*, *7*(4), 377–392. doi:10.1108/WHATT-03-2015-0013

Gaur, S. S., & Chapnerkar, S. M. (2015). Indian festival's contribution to the cultural and economic well-being: A case study of Ganapati festival. *Worldwide Hospitality and Tourism Themes*, *7*(4). doi:10.1108/WHATT-03-2015-0017

Ginzarly, M., Roders, A. P., & Teller, J. (2019). Mapping historic urban landscape values through social media. *Journal of Cultural Heritage*, *36*, 1–11. doi:10.1016/j.culher.2018.10.002

Gomez-Oliva, A., Alvarado-Uribe, J., Parra-Meroño, M. C., & Jara, A. J. (2019). Transforming communication channels to the co-creation and diffusion of intangible heritage in smart tourism destination: Creation and testing in ceutí (spain). *Sustainability (Basel)*, *11*(14), 3848. doi:10.3390u11143848

Halder, S., & Sarda, R. (2021). Promoting intangible cultural heritage (ICH) tourism: Strategy for socio-economic development of snake charmers (India) through geoeducation, geotourism and geoconservation. *International Journal of Geoheritage and Parks*, *9*(2), 212–232. doi:10.1016/j.ijgeop.2021.02.008

Kolay, S. (2016). Cultural heritage preservation of traditional Indian Art through virtual new-media. *Procedia: Social and Behavioral Sciences*, *225*, 309–320. doi:10.1016/j.sbspro.2016.06.030

Lenzerini, F. (2011). Intangible cultural heritage: The living culture of peoples. *European Journal of International Law*, *22*(1), 101–120. doi:10.1093/ejil/chr006

Li, J., Krishnamurthy, S., Roders, A. P., & Van Wesemael, P. (2020). State-of-the-practice: Assessing community participation within Chinese cultural World Heritage properties. *Habitat International*, *96*, 102107. doi:10.1016/j.habitatint.2019.102107

Liang, X., Lu, Y., & Martin, J. (2021). A review of the role of social media for the cultural heritage sustainability. *Sustainability (Basel)*, *13*(3), 1055. doi:10.3390u13031055

Loulanski, T., & Loulanski, V. (2011). The sustainable integration of cultural heritage and tourism: A meta-study. *Journal of Sustainable Tourism*, *19*(7), 837–862. doi:10.1080/09669582.2011.553286

Luke, W. (2008). *Correa's Artist's Village Today - The incremental house*. Blogpost. http://incrementalhouse.blogspot.co.nz/2008/04/correas-artists-village-today.html

Mehmood, F., & Aazem, G. (2002). Shilpgram in Udaipur: Organising craftspeople. *Economic and Political Weekly*, *37*. https://architexturez.net/doc/10-2307/4412740

Mehta, K., Rawal, Y. S., & Pal, S. (2022). The Role of Smart City Mission in Tourism Development and Sustainability in Udaipur. *Atithya: A Journal of Hospitality*, *8*(2).

Munjal, P. (2015). Potential of participatory approach in sustaining fairs and festivals of small towns: The case of Sohna, Haryana. *Worldwide Hospitality and Tourism Themes*, *7*(4), 347–366. doi:10.1108/WHATT-03-2015-0008

Munjal, S., & Jauhari, V. (2015). Leveraging the cultural and economic value of fairs and festivals in India. *Worldwide Hospitality and Tourism Themes*, *7*(4), 417–423. doi:10.1108/WHATT-03-2015-0006

O'Dell, T. (2005). Experiencescapes: Blurring borders and testing connections. In T. O'Dell & P. Billing (Eds.), *Experiencescapes – Tourism, culture, and economy* (pp. 11–33). Copenhagen Business School Press.

Pal, S., & Rawal, Y. S. (2020). Perception and attitude of the host community on socio-economic development through tourism-a case study on Udaipur Rajasthan. *International Journal of Advanced Science and Technology*, *29*(3), 9202–9217.

Pine, B. J., & Gilmore, J. H. (1999). *The experience economy: Work is theatre & every business a stage*. Harvard Business Press.

Pine, B. J., & Gilmore, J. H. (2011). *The experience economy*. Harvard Business Press.

Preservation and promotion of languages, folk dance, Art and culture of tribals. (n.d.). PIB. https://pib.gov.in/PressReleaseIframePage.aspx?PRID=1604892

Qiu, Q., Zuo, Y., & Zhang, M. (2022). Intangible cultural heritage in tourism: Research review and investigation of future agenda. *Land (Basel)*, *11*(1), 139. doi:10.3390/land11010139

Ranwa, R. (2022). Impact of tourism on intangible cultural heritage: Case of Kalbeliyas from Rajasthan, India. *Journal of Tourism and Cultural Change*, *20*(1-2), 20–36. doi:10.1080/14766825.2021.1900208

Sharma, S., & Sharma, S. (2015). Understanding the socio-economic impact of Pithoragarh's cultural jewels: Its fairs and festivals. *Worldwide Hospitality and Tourism Themes*, *7*(4). doi:10.1108/WHATT-03-2015-0007

Shastri, J. (2017). *Artist Village.* [Master's thesis, Unitec Institute of Technology].

Shenoy, S. (2022). Rajasthani Paintings To Jewellery, Shilpgram Is Udaipur's Best Spot For Budget Shopping. *Curly Tales*. https://curlytales.com/rajasthani-paintings-to-jewellery-shilpgram-is-udaipurs-best-spot-for-budget-shopping/

Su, J. (2019). Understanding the changing intangible cultural heritage in tourism commodification: The music players' perspective from Lijiang, China. *Journal of Tourism and Cultural Change*, *17*(3), 247–268. doi:10.1080/14766825.2018.1427102

Tiwari, S. (2015). Culture based development: Empirical evidence for Shilpgram Fair, Rajasthan. *Worldwide Hospitality and Tourism Themes*, *7*(4), 393–402. doi:10.1108/WHATT-03-2015-0005

Tnn. (2016, December 27). Shilpgram Utsav: Real festival of craftsmen, folk artists from far off villages. *The Times of India*. https://timesofindia.indiatimes.com/city/jaipur/shilpgram-utsav-real-festival-of-craftsmen-folk-artists-from-far-off-villages/articleshow/56190759.cms

UNESCO. (1989). *Recommendation on the Safeguarding of Traditional Culture and Folklore*. In Proceedings of the General Conference of the United Nations Educational, Scientific and Cultural Organization, Paris, France. http://portal.unesco.org/en/ev.php-URL_ID=13141&URL_DO=DO_TOPIC&URL_SECTION=201.html

UNESCO. (2003). *International convention for the safeguarding of the intangible cultural heritage.* UNESCO.

Veldpaus, L. (2015). Reconnecting the city: The historic urban landscape approach and the future of urban heritage. *Cultural Trends*, *24*(4), 340–342. doi:10.1080/09548963.2015.1106036

Wang, N., Fang, M., Beauchamp, M., Jia, Z., & Zhou, Z. (2021). An indigenous knowledge-based sustainable landscape for mountain villages: The Jiabang rice terraces of Guizhou, China. *Habitat International*, *111*, 102360. doi:10.1016/j.habitatint.2021.102360

Zhang, Q., Liu, X., Li, Z., & Tan, Z. (2021). Multi-experiences in the art performance tourism: Integrating experience economy model with flow theory. *Journal of Travel & Tourism Marketing*, *38*(5), 491–510. doi:10.1080/10548408.2021.1952148

Chapter 5
Motives of Cultural Heritage Tourism:
The Case of World Heritage Sites in Oman

Mohammad Soliman
https://orcid.org/0000-0002-9359-763X
University of Technology and Applied Sciences, Salalah, Oman & Fayoum University, Fayoum, Egypt

Samskrati Gulvady
University of Technology and Applied Sciences, Salalah, Oman

Hamed Almuhrzi
Oman Tourism College, Muscat, Oman & Sultan Qaboos University, Muscat, Oman

Musallam S. Hawas Al-Aamri
University of Technology and Applied Sciences, Salalah, Oman

ABSTRACT

Cultural heritage tourism (CHT) is one of the most significant tourist patterns that has grown substantially in recent years. This is because the majority of visitors have motives for seeking to learn more about the historical, cultural, and heritage-related sites in popular tourist destinations. Given the relative paucity of studies that addressed the key motivations for the CHT pattern in many countries throughout the world, including Oman, this highlights the necessity for further research into the motives that contribute to this pattern of tourism. As a result, this study seeks to focus on the motivations for CHT in Oman, with an emphasis on the UNESCO-listed World Heritage sites. This conceptual work helps scholars better understand the factors that influence the pattern of CHT while also giving them a number of perspectives for future research on the subject. The current work also offers destination managers and decision-makers in the tourism industry a set of practical contributions.

INTRODUCTION

Numerous patterns of tourism have developed as a result of the significant economic, social, and cultural impact that tourism has had in many countries, including those of the Middle East countries

DOI: 10.4018/978-1-6684-9957-3.ch005

(Selmi et al., 2021; Soliman, 2015). These include cultural heritage tourism (CHT), which has gained popularity among both domestic and international tourists as an intriguing type of alternative tourism (Ismail et al., 2014). Cultural tourism, of which CHT is part (du Cros, 2001) is one of the most popular and quickly growing tourist subsectors globally. It is a type of economic activity that is seen as giving people access to culture and attracting tourists. Furthermore, tourism flows generated by culture and heritage have an uneven and irregular spatial distribution, with tourist destinations relying on cultural amenities, architectural treasures, museums, etc. (Secondi et al., 2011). According to Matteucci et al. (2022), a thorough study of the various socioeconomic changes and worldwide influences that have influenced how people live is necessary for any attempt to predict the future of cultural tourism. CHT could contribute to economic development and provides sustainable benefits for the future, promoting and preserving cultural and natural heritage. With its vast range of sub-sectors (including archaeology, architecture, art, historic tours, museums, music festivals, and opera), CHT will continue to expand in popularity over the upcoming years (Surugiu & Surugiu, 2015). Cultural heritage is thus made up of both tangible and intangible assets. This work concentrates on tangible cultural heritage (ICH) which is a valued asset among the resources and attractions of tourism destinations (Hsu et al., 2022).

According to Seyfi et al. (2019), a sizeable portion of the worldwide tourism market is CHT. This includes the Middle East, where Oman is one of the most popular travel destinations thanks to its unrivaled natural and cultural attractions (Alkathiri & Soliman, 2022; Elbaz et al., 2022; Mathew & Soliman, 2021). Oman is distinguished in this sense by its cultural heritage sites. The nation has a number of sites that are part of the United Nations Educational, Scientific, and Cultural Organization (UNESCO) world heritage list as well as other tentative lists of sites that are up for nomination. The UNESCO world heritage sites in Oman are the ancient city of Qalhat; Bahla Fort; Archaeological sites of Bat, Al-Khutm, and Al-Ayn; Aflaj Irrigation Systems of Oman; and Land of Frankincense- are listed in UNESCO (2023). In addition to these, there are seven sites on the tentative list: the forts of Rostaq and Al-Hazm; Al Hallaniyat Islands proposed nature reserve; Bar Al Hakman proposed nature reserve; Smahan's mountain nature reserve; Al Dimaniyat Islands nature reserve; Ras Al Hadd turtle reserve and the Heritage site of Ras Al Jinz; the cultural landscape of Bisya and Salut and its archaeological remains.

To understand how cultural tourism, including CHT, is being performed and consumed, it is vital to explore the relationship between the concept of culture and heritage, way of life, local values, beliefs (Matteucci et al., 2022). Indeed, this necessitates concentrating on the motives of CHT. In this regarding, CHT, like many other leisure and tourism practices, is a form of experiential consumption because the experiences have the potential to become a significant component of tourists' memorability (Rasoolimanesh et al., 2021). In this vein, the vast proliferation of knowledge, increasing mobility, and improved accessibility of travel have created a tremendous need for people to travel and experience the cultural heritage of other nations firsthand. CHT serves to draw the attention of the modern traveler looking for chances for immersive tourism and engaging experiences (Hughes & Carlsen, 2010). Thus, CHT primarily necessitates satisfying a market that needs to wonderfully relive the past (Hughes & Carlsen, 2010). This emphasizes the need to draw attention to the most significant motivators for tourists to visit cultural heritage sites. In this respect, a variety of motivational theories are used to explain why people act in specific ways and what motivates them, including the push-pull model (Crompton, 1979; Dann, 1977), Iso-Ahola's motivation theory, (Iso-Ahola, 1982), the travel career ladder (TCL) model (Pearce & Lee, 2005), and others. According to Nguyen and Cheung (2016), heritage motivation describes the drives to learn more about the site. Higher motivational travelers are more interested in the topic of accurate information and experiences of heritage visits. As a result, it can be argued that CHT is motivated by tourists' interest in historical, artistic, scientific, or heritage offerings by a community, region, group, or institution (du Cros, 2001). Tourists have needs and want to learn about other cultures, getting a unique experience from CHT. Tourists of CHT could be also motivated by lifestyle offerings of a community, region, group, or some institutions. Hence, CHT may take many forms, according to the motivation behind tourists' impetus to travel to a certain destination. Cultural heritage and historical sites, dance, music and theatre performances, art galleries, museums and exhibitions, and religious and ethnic traditions are the main attractions for tourists motivated by knowledge seeking, self-development, and the desire to experience other cultures. In addition, a major motivation behind travel is cultural heritage, which is produced through travel as well as by cultural sights, attractions, and events (Richards, 2018). Nostalgia is another key motivator. of CHT (Hughes & Carlsen, 2010). Irrespective of the motivations, travel offers a chance for self-development because it exposes people to various cultural traditions, ways of life, and worldviews. In fact, tourism is a cultural experience in and of itself (De Ascaniis et al., 2018).

Although there have been several studies on visitor motivations, there have been relatively few studies on the motivating factors behind cultural and heritage tourism (e.g., Chhabra et al., 2003; Secondi et al., 2011). More specifically, no prior research has been done to highlight the motivations for CHT, particularly with regard to the UNESCO World Heritage Sites in Middle Eastern countries, including Oman. Given the preceding discussion, the current chapter seeks to fill this gap by highlighting the motives of CHT, concentrating on the UNESCO World Heritage Sites in Oman. By identifying and filling in these research gaps, researchers can gain a deeper knowledge of why people engage in CHT, which could lead to more insightful conclusions and possibly inform approaches to improving the visitor experience and fostering cultural preservation. Additionally, this chapter contributes to the implementation of policy and future orientations by taking into account the perspectives of both tourists and policymakers in Oman by comprehending the motivations of CHT in Oman.

CULTURAL HERITAGE TOURISM (CHT)

Tourism offers a means for tourists to gain knowledge about various cultural customs, lifestyles, and the globe as seen from a distance. It fosters interaction between people from various social classes and encourages mutual aid, which strengthens cultural traditionalism (Cravidão et al., 2018). In addition, tourist destinations are primarily motivated by their culture, along with their natural attractions. According to Seyfi et al. (2019), heritage tourism forms a bigger segment of cultural tourism and can be presently considered as one of the significant and rapidly advancing spheres in the travel industry, and the oldest practices for travel. Historical evidence shows that ancient Egyptians, Romans, and nobility of the medieval times undertook to travel to savour places of historical and cultural importance (Timothy & Boyd, 2006).

Cultural heritage tourism (CHT) signifies visitors' engagement with activities, artefacts, and places reflecting realistic past and present representations of tales and individuals. Visitors seek exceptional touristic understandings containing a fusion of ethos, enjoyment, learning, and realism, making CHT a considerable avenue for international tourism (Rasoolimanesh et al., 2021). Considering cultural landscapes, Gravari-Barbas et al. (2018) demonstrated developing standards for cultural and natural heritage sites. A common approach recognizes that World Heritage Sites are dynamically living and applies to cultural landscapes (such as agrarian and rural areas, gardens, parks, and industry-based areas). Heritage is defined as "monuments, groups of buildings, and sites" in the World Heritage Resource Manual of UNESCO. These sites may comprise several important cultural heritage locations, including urban centres, industrial heritage routes, archaeological sites, cultural landscapes, and cultural routes. This necessitates the right, skill-based management of such areas.

As national and international tourism has grown over the past few decades, Visitors now recognize the significance of this "good" form of tourism, which benefits the economy and aids in the preservation of culture (Timothy & Boyd, 2006). Therefore, CHT can be viewed as a notable tourism element that offers visitors a variety of advantages. Travelers have a need and a desire to learn about different cultures and gain special experiences through CHT. The popularity of this type of tourism will increase over the coming years, and its growth rates are quicker than the average global rate in part because it includes a number of sub-sectors (such as archaeology, architecture, art, historic tours, museums, music festivals, and opera) (Surugiu & Surugiu, 2015). CHT motivation is one of the persuasive forces that make people embark on their cultural journey. In addition, CHT is one of the significant global tourism resources. Timothy (2014) looks into how cultural heritage contributes to economic and social advancement while observing tourists' emerging travel preferences, such as the heritage tourism market, the heritagization of everyday life, the value of authenticity, branding heritage locations, the growth of cultural routes, intersections between heritage and other forms of tourism, and the heritagization of non-traditional heritage spaces, which significantly impact destination development. According to Lussetyowati (2015), CHT is a method for preserving traditional culture and modern ideals while restoring historic buildings and public spaces for avant-garde purposes. Additionally, there are numerous other goals, such as encouraging revenue from cultural heritage and safeguarding cultural resources. Other goals include providing visitors with an accurate understanding of resources, actual encounters with resources, and promoting tourism.

Matlegiewicz (2020) classifies tourism related to cultural heritage into three broad categories, namely: a) intellectual culture (e.g., cultural heritage, museum); b) education (language seminar tours, study, specialist); and c) general culture (e.g., culinary, ethnic, exotic, industrial, military, pilgrimage, regional, urban). The following are listed by Ismagilova et al. (2015) as the key components of culture-related tourism: a) Tourist involvement is increased as a result of exposure to cultural heritage, which is a significant source of motivation; b) Cultural heritage sites and objects become vital assets of cities, boosting profits and significantly promoting economic progress; c) Helpful on a social level as well to facilitate seasonal tourist intake and provide a uniform distribution of visitors throughout the area; d) A positive brand image of cultural heritage as a useful instrument for leadership adoption in the tourist destination; and e) Generates cash for the area, as well as a sense of pride for the locals and helps them share it with tourists. It has been discovered that culinary, cuisine, or gastronomy tourism—also known by a variety of names—is a significant factor in determining the choice of destination, particularly CHT, which cannot be disregarded (Lima et al., 2023).

CHT is challenged by threats that can hamper the heritage sites. There may be conflicting growth nearby in addition to direct threats to the cultural place's components. The wellbeing of the cultural heritage site will require that appropriate decisions be made in relation to more general economic or social issues (UNESCO, 2013). CHT will benefit the local economy, but more questions are being raised regarding the long-term viability of cultural heritage as tangible and intangible heritage are being integrated more deeply into the tourism industry. A few tactics for creating CHT are to "discover the fit; concentrate on quality and authenticity; preserve as well as protect (Lussetyowati, 2015). Furthermore, digital technologies are causing the line between the actual and virtual worlds to become more and more blurry, which could result in more immersive tourist experiences. Travelers now have more varied experiences, which increases their interest in distinctive cultures and heritage (Jiang et al., 2023).

Along with digital innovations, cultural innovations are also necessary at CHT destinations, as it revives cultural heritage and keep pace with tourists' requirements, while adapting to global changes (Wang et al., 2023). Utilisation of innovative 3D scanning and modelling could be beneficial for the visitors to engage themselves in the virtual version of CHT (Little et al., 2020). In this vein, CHT can resourcefully target the digital native generation. Wu et al. (2023) indicate that the positive attitude of youth (e.g., college students) towards CHT positively influences their objective to partake in CHT. Thus, it can be claimed that CHT serves as a bridge between the past, present, and future, helping interact with people from all walks of life, including youth, and from different cultural and social backgrounds.

UNESCO WORLD HERITAGE SITES: THE CASE OF OMAN

The Paris-based cultural committees of UNESCO publish lists of the items that are a part of World Heritage Sites (WHS) and Intangible Cultural Heritage (ICH) each year. Every inscribed WHS property is also categorized as a cultural, natural, or mixed aspect. Each WHS property or ICH component may be categorized as being at risk. The designation of a location as a UNESCO WHS is essential for both preserving its historical and aesthetic legacy and promoting tourism there (Santa-Cruz & López-Guzmán, 2017).

In this vein, Oman is one of the world's earliest human settlements. It is a nation with an abundant history, culture, and heritage. As a result, tourists from all over the world travel to the country. In Oman, there are many different kinds of tourism, with CHT possibly being the most well-known form. As previously stated, there are five sites in Oman that are on the UNESCO World Heritage List, including the Bahla Fort, the Archaeological Sites of Bat, the Land of Frankincense, the Aflaj Irrigation Systems of Oman, and the Ancient City of Qalhat (UNESCO, 2023). UNISCO's preliminary list also includes seven additional sites.

First, the Bahla Fort, located inside the city walls on an earth mound, is surrounded by agricultural fields, date palms, and historic sites like the ancient market. The fort is a triangle-walled complex with its main constructions located on the high terrain in its south-eastern and south-western portions. It previously had an area of around 200 m2. Bahla Fort, which was included on the list in 1987, is the only fort that has been classified by UNESCO. The formal repair programs, which got underway in the 1990s, received a $9 million investment from the Omani government (al-Busaidi, 2004).

Second, the 1966-founded *Archaeological Site of Bat* is situated inside Oman close to a palm grove at the boundary between the Rub al-Khali Desert and the Hajar Mountains (Kondo et al., 2016). Together with the adjacent sites (i.e., Al-Khutm and Al-Ayn), it constitutes the world's greatest collection of settlements and necropolises from the third millennium B.C. The site was included on the list of the UNESCO World Heritage in 1988 (Benkari, 2018; Ermertz et al., 2019).

The *Land of Frankincense* is the third place to be included on the 2000 list of UNESCO World Heritage Sites. The finest frankincense in the world used to be produced in the Land of Frankincense. Its aromatic resin is referred to as "the king of scents" and is commonly used in incense and perfume. The frankincense trade, which was one of the most significant trading activities of the ancient and medieval world, was illustrated vividly by the frankincense trees of Wadi Dawkah, the remnants of Shisr/Wubar, the caravan oasis, and the associated ports of Khor Rori and Al-Baleed (UNESCO, 2023).

The *Aflaj Irrigation System* of Oman was recognized by UNESCO as a World Heritage Site in 2006. The five irrigation systems that make up this are Falaj Al-Khatmeen, Falaj Al-Malki, Falaj Daris, Falaj

Al-Jeela, and Falaj Al-Muyassar. This is a representation of the approximately 3,000 similar systems still in use in Oman. Water is channeled for home and agricultural usage using gravity from underground sources or springs. The site has several watchtowers constructed to protect the water systems, illustrating the historical reliance of people on the Aflaj system. The Aflaj is an extremely well-preserved land use type that is now in danger due to the declining level of the subsurface water table (Al-Ghafri, 2008; Al-Marshoudi et al., 2021; UNESCO, 2023).

Moreover, on Oman's eastern coast, the historic town of *Qalhat* was added to the UNESCO World Heritage List in 2018. The city, which is encircled by internal and external walls, first became well-known as a significant port in Arabia between the 11th and the 15th centuries. The city was a significant commerce center, as evidenced by the surface ceramics. Numerous shards reveal trade routes between Asia and Africa (Ermertz et al., 2019).

Besides the aforementioned five sites, according to the UNESCO (2023), there are seven sites in Oman that are on the tentative list: The forts of Rostaq and al-Hazm (1988); Al Hallaniyyat Islands Proposed Nature Reserve (2013); Bar al Hakman Proposed Nature Reserve (2013); Smahan's Mountain Nature Reserve (2013); al Dimaniyyat Islands Nature Reserve (2013); Ras al Had Turtle Reserve and the Heritage Site of Ras al Jinz (2013); and Cultural Landscape of Bisya & Salut and its Archaeological Remains (2014). A Tentative List is a list of the properties that each State Party plans to consider nominating.

MOTIVES OF CULTURAL HERITAGE TOURISM

Understanding why visitors undertake a tourist trip is very central to the overall understanding of tourism phenomena. Motivation theory is one of the theoretical underpinnings and guidelines that enable tourism researchers to understand why visitors engage in each activity. Tourism literature, in general, tries to answer two main questions related to motivation: 1) what motivates tourists to undertake travel, and 2) what motivates tourists to undertake a certain trip, activities, and/or visit a certain attraction. There are several theories and frameworks put forward to unpack visitors' motivation and engagement in travel behaviour. The push-pull model (Crompton, 1979; Dann, 1977) is widely employed by tourism researchers to understand what motivates people to travel in relation to a given destination. Recently, Iso-Ahola's escaping-seeking dichotomy (Iso-Ahola, 1982) is getting used by several researchers to explain visitors' motivation. The dichotomy postulates that one is constantly subjected to two forces which direct their motivational directions: escaping from personal and/or interpersonal situations, and/or seeking personal and/or interpersonal rewards (Huang, 2021). Another well-cited tourism motivation model is the travel career ladder (TCL) (Pearce & Lee, 2005). Drawing upon Maslow's hierarchy of needs theory, it is postulated that visitors with more travel experience will tend to exhibit higher needs (such as self-actualisation). In comparison, people with less travel experience will tend to exhibit basic needs identified by Maslow. Visiting cultural heritage attractions is a core tourism activity. The report by the UNWTO (2018) on the synergy between tourism and culture (culture is being used as a wide concept that incorporates different tangible and intangible cultural heritage resources) indicates that different established and emerging tourism destination representatives believe that cultural heritage resources are core activities for international travellers. In the same line of thought, the tourism literature also suggests that cultural heritage tourism resources are core motivations for people to travel and/or in their travel decision-making process (López-Guzmán et al., 2019).

Understanding what motivates people to visit and/or engage with CHT activities and experiences becomes crucial because it is a significant tourist form. The literature offers different models to explore visitors' motivation at cultural heritage attractions. Packer and Ballantyne (2002) correctly, drawing from psychology literature, employed three core questions to guide to understanding visitors' motives at cultural heritage attractions: 1) where the visitors invest time and energy, 2) how much time and energy visitors invest, and 3) how long visitors are willing to keep on investing. By deploying these three questions, they provide five core motives driving visitors to visit cultural heritage sites and engage in cultural heritage experiences: learning and discovery, passive enjoyment, restoration, social interaction, and self-fulfilling. Theoretically, the literature offers two main models that were developed specifically to explore visitors' motives for visiting cultural heritage attractions. One of the earliest models is McKercher's (2002) motivation-experience model. The model departs from two dimensions: *first*, the importance of cultural heritage attractions in one's travel decision and destination selection; *second*, the depth of one's experiences during the visit. According, to the nexus between these two dimensions there are five typologies of cultural heritage tourists. In these five cultural visitors' motivation will stretch from a high level of motivation, where cultural heritage attractions and experiences are central to one's travel/destination decision-making, to low motivation where cultural heritage sites and experiences have little contribution to one's travel/destination selection decision-making.

Falk (2006) developed a new model to explore visitors' motivation to visit different cultural heritage attractions. The central notion of Falk's model is that visitors' identity and motivation are indivisibly entwined, i.e., the identity of a visitor on the day of visiting cultural heritage sites will contribute significantly to defining their driving motives. Falk's model starts by offering five identities of visitors that contribute to understanding the motives of their visit. The first identity is the 'explorer', as an explorer visitor, they tend to be motivated by learning and exploration reasons. The second identity is the 'facilitator', facilitators are driven to visit the cultural heritage sites by social motives, i.e., to facilitate the visit of the site for a family member, friends, guests, or others. The professional/hobbyist is the third identity. This type of visitor is the professional ones who visit the site for a specific purpose related to their profession, study, hobby, and/or work. The fourth identity, according to Falk is the experience seeker. This experience seeker is the most common category according to Falk. Experience seeker visitors tend to be motivated by enjoyment, and entertainment. The last identity is spiritual pilgrim/recharger, which is associated with spiritual and tranquillity motives. Falk's model was revisited and further developed to include additional contextual-related identities striving to explain visitors' motives for visiting cultural heritage attractions (Dawson & Jensen, 2011; Falk & Storksdieck, 2010).

In the last decade, a new focus emerges when it comes to understanding visitors' motivation for visiting CHT. Several studies suggest that the cultural background of visitors and their level of association/closeness with the sites' heritage and history is a defining factor of their motive to visit. Fountain et al. (2010) suggested that Chinese visitors were more geared toward natural-cultural motivation. For example, Chinese visitors visiting New Zealand were more motivated to see the natural landscapes, learn about the Māori culture and learn about nature-based stories. Almuhrzi et al. (2020) suggested that insider visitors were more motivated by national-pride and familial motives compared to international visitors. The international visitors, as outsiders, were motivated to learn and discover a new culture and nation.

Several recent studies start to offer new dimensions on visitors' motivation for visiting cultural heritage attractions. While traditionally, visiting museums and cultural heritage sites were among the most cited cultural heritage tourism motives, recent studies indicated that intangible tourism activities such as gastronomy were considered the most attractive cultural experiences in Budapest (Smith et al.,

2022). Zhao and Agyeiwaah (2023) explored Chinese visitors' value-based motivations for visiting The Five Great Avenues in Tianjin, China. Their study shows that visitors prioritize self-improvement and socialisation as the most important value associated with visiting the site. Other motivational values such as hedonism, differentiation, existentialism, escapism, and freedom are important.

CONCLUSION

The current work seeks to study the most important motives of cultural heritage tourism (CHT), highlighting the World Heritage Sites in the Sultanate of Oman. Due to their reputation as branding tools, World Heritage Sites have a desired position in niche tourism, increasing their visibility to potential tourists. Although protecting and maintaining these sites is the major objective of UNESCO recognition, in many cases they have seen a significant increase in tourists, particularly from foreign countries, leading to a strong association between these lists and tourism (Santa-Cruz & López-Guzmán, 2017). Consequently, the tourism sector greatly benefits from people's innate desire to experience other cultures and learn about their unique identities, including World Heritage Sites. This demonstrates the critical need to highlight the motives of CHT. Historically, CHT has been analyzed from the standpoint of scientific literature in terms of its attractiveness to tourists looking to engage with local communities, further their education, or simply take in the historical structures and cultural treasures preserved for tourism. Residents of World Heritage Sites profit from an increase in visitors while also developing into an incredible resource for them. Such a cutting-edge approach to tourism that allows exposure to various cultures and familiarization with host cultures aids in developing empathy. Alongside that, it promotes economic expansion both locally and globally. Additionally, there is an opportunity for local businesses to prosper by getting involved in creative and cutting-edge entertainment activities while evaluating their tangible and intangible characteristics. As the location and brand qualities of historic properties coupled with heuristic marketing are expanding, local tourism has been upgraded. CHT also creates possibilities for intellectual and educational pursuits (Matlegiewicz, 2020).

According to Correia et al. (2013), one of the main motivators influencing travelers' decisions on where to travel is culture. As a result, CHT itself determines a traveler's decision to visit a specific area, so the market for cultural tourism may consider both the traveler's intended experience and the actual experience they have while there. With respect to CHT, experiencing and appreciating a destination's distinctive cultural features as well as learning and developing socially and personally are the driving forces behind this pattern of tourism. CHT motivation is one of the persuasive forces that make people embark on their cultural journey. Although cultural motivations play an important role in the generation of tourist flows towards CHT destinations, a tourism experience is rarely generated because of a single motivation. The act of choosing a specific destination and tourism product is influenced by multiple motivations, not only cultural heritage but also connected to relaxation or family.

Studying the motivations behind CHT can advance the field's knowledge in both theoretical and practical ways. Such a study provides academics with an understanding of the complex interplay of factors that motivate tourists to engage with cultural and heritage sites. By exposing underlying motivations, researchers can contribute to the theoretical frameworks that explain visitor behavior, expanding the body of knowledge in the area. The current chapter also increases our understanding of the multifaceted connections between human behavior and CHT exploration. This conceptual study on CHT motives offers a more nuanced perspective than existing theories, which frequently concentrate on the general characteristics of travel motivation. This chapter offers insight into several variables that shape travel choices, such as the need for personal enrichment, the need for genuine interactions, and interest in various customs. Additionally, by identifying these motives, the study opens possibilities for customizing marketing strategies and developing immersive cultural experiences that suit tourists' innate tastes.

Practically, the present work's outcomes hold significant implications for destination management organizations (DMOs), marketing managers, and policymakers. To begin with, destinations may better cater experiences to visitors' needs by comprehending the driving forces behind cultural and heritage tourists, increasing their satisfaction and encouraging return visits. Furthermore, by enabling local authorities to strike a balance between maintaining cultural authenticity and satisfying tourists' expectations, the findings lay the groundwork for the growth of sustainable tourism. With the help of these findings, targeted marketing campaigns may be created to draw in a certain demographic of passengers and increase tourism profits by showcasing a destination's distinctive cultural attractions. In addition, this research enables tourism stakeholders and governments to adjust their services more effectively by identifying the precise factors that lead people to seek out cultural experiences. Destinations can create niche marketing strategies that showcase distinctive cultural features that appeal to certain tourist demographics, drawing in a more varied and enthusiastic clientele. Insights from this study can also be used to improve tourist experiences and build a stronger connection with the local culture by directing the creation of immersive cultural activities and events. To create genuine and enriching interactions that are in line with tourists' goals, destination managers might also concentrate on developing partnerships with local populations. In the end, our research equips travel industry experts to design more fulfilling and meaningful cultural tourism experiences, resulting in higher visitor satisfaction, favorable word-of-mouth, and long-term tourism growth.

This conceptual work can also be used as a guide for future studies, allowing for more specialized investigations into specific CHT objectives. To begin with, future research is encouraged to fill the knowledge gap in the investigation of new and complicated motivating variables that influence people's interaction with CHT sites, particularly UNESCO World Heritage Sites. In this vein, future studies could concentrate on incorporating travel motivation and other pertinent variables into theoretical models like the theory of planned behavior (TPB) (see e.g., Soliman, 2021) in order to predict visitors' intention to re/visit CHT destinations. In addition, it is evident that there is a dearth of in-depth research on less-explored motives for CHT, such as emotional connections, identity formation, value co-creation, and spiritual experiences. While previous studies have looked at well-established motives or drivers like cultural exploration and education, these motives have received little attention. Consequently, further study can be done using empirical methods to investigate how spiritual experiences, identity development, and emotional relationships affect the behavior of CHT tourists. Moreover, unveiling how the adoption of some advanced technologies and electronic marketing platforms such as gamification could affect visitors' engagement and loyalty toward CHT destinations (see e.g., Abou-Shouk & Soliman, 2021) will be a wonderful piece of future academic work. A promising direction for further work is to assess the main determinant of destination evangelism (Soliman & Al Balushi, 2023) of visitors who have visited CHT destinations in different countries around the world. Future studies could also examine the role of digital content marketing (see e.g., Mathew & Soliman, 2021) in shaping consumer behavior related to CHT destinations. Moreover, future research could focus on the connection between intelligent customer experience (ICE) and customer loyalty (see e.g., Gam et al., 2023) toward CHT destinations. Another remarkable avenue for further research is to conduct a systematic literature review or comprehensive bibliometric research using multiple metrics (see e.g., de Bruyn et al., 2023; Soliman et al., 2021) to analyze and visually map the scientific production on the key motives of CHT. Furthermore, future research could thoroughly explore the unexplored region of how the motivations of CHT are influenced by changing cultural norms, technological developments, AI, and sustainability concerns.

REFERENCES

Abou-Shouk, M., & Soliman, M. (2021). The impact of gamification adoption intention on brand awareness and loyalty in tourism: The mediating effect of customer engagement. *Journal of Destination Marketing & Management, 20*, 100559. doi:10.1016/j.jdmm.2021.100559

al-Busaidi, Y. S. (2004). The protection and management of historic monuments in the Sultanate of Oman: the historic buildings of Oman. Proceedings of the Seminar for Arabian Studies, Al-Ghafri, A. S. (2008). Traditional water distribution in Aflaj irrigation systems: Case study of Oman. *What Makes Traditional Technologies Tick? A Review of Traditional Approaches for Water Management in Drylands*, 74.

Al-Marshoudi, A. S., Bin Sibly, S., & Khelghat-Doost, H. (2021). Successfully mutual cooperative collective action: Principle of institutional arrangement of the aflaj irrigation system in Sultanate of Oman. Water resources in arid lands: Management and sustainability.

Alkathiri, N. A., & Soliman, M. (2022). Examining foreign direct investment determinants of tourism industry in Oman and Egypt: The moderating role of investment environment. International Journal of Finance & Economics, 27(4), 4722–4740.

Almuhrzi, H., Hughes, K., & Ballantyne, R. (2020). Exploring Arab and Western visitors' interpretive experiences at an Omani heritage site: Does one size fit all? *Journal of Heritage Tourism, 15*(2), 180–199. doi:10.1080/1743873X.2019.1632319

Benkari, N. (2018). Archaeological site of Bat-Oman, management and public perception: Community involvement in archaeological heritage management and planning. *Journal of Cultural Heritage Management and Sustainable Development, 8*(3), 293–308. doi:10.1108/JCHMSD-10-2016-0059

Chhabra, D., Healy, R., & Sills, E. (2003). Staged authenticity and heritage tourism. *Annals of Tourism Research, 30*(3), 702–719. doi:10.1016/S0160-7383(03)00044-6

Correia, A., Kozak, M., & Ferradeira, J. (2013). From tourist motivations to tourist satisfaction. *International Journal of Culture, Tourism and Hospitality Research, 7*(4), 411–424. doi:10.1108/IJCTHR-05-2012-0022

Cravidão, F., Nossa, P., Santos, N., Vidal, M., Crous, N., Aulet, S., & Dallari, F. (2018). Tourism at world heritage sites opportunities and challenges. *Tourism Management at UNESCO World Heritage Sites*, 23-30.

Crompton, J. L. (1979). Motivations for pleasure vacation. *Annals of Tourism Research, 6*(4), 408–424. doi:10.1016/0160-7383(79)90004-5

Dann, G. M. (1977). Anomie, ego-enhancement and tourism. *Annals of Tourism Research, 4*(4), 184–194. doi:10.1016/0160-7383(77)90037-8

Dawson, E., & Jensen, E. (2011). Towards a contextual turn in visitor studies: Evaluating visitor segmentation and identity-related motivations. *Visitor Studies, 14*(2), 127–140. doi:10.1080/10645578.2011.608001

De Ascaniis, S., Gravari-Barbas, M., & Cantoni, L. (2018). *Tourism management at UNESCO world heritage sites*. Università della Svizzera italiana.

de Bruyn, C., Said, F. B., Meyer, N., & Soliman, M. (2023). Research in tourism sustainability: A comprehensive bibliometric analysis from 1990 to 2022. *Heliyon, 9*(8), e18874. doi:10.1016/j.heliyon.2023.e18874 PMID:37636413

du Cros, H. (2001). A new model to assist in planning for sustainable cultural heritage tourism. *International Journal of Tourism Research, 3*(2), 165–170. doi:10.1002/jtr.297

Elbaz, A. M., Soliman, M., Al-Alawi, A., Al-Romeedy, B. S., & Mekawy, M. (2022). Customer responses to airline companies' service failure and recovery strategies: The moderating role of service failure habit. *Tourism Review, 78*(1), 1–17. doi:10.1108/TR-03-2022-0108

Ermertz, A., Kázmér, M., Adolphs, S., Falkenroth, M., & Hoffmann, G. (2019). Geoarchaeological evidence for the decline of the medieval city of Qalhat, Oman. *Open Quaternary, 5*(8), 1–14. doi:10.5334/oq.56

Falk, J. H. (2006). An identity-centered approach to understanding museum learning. *Curator (New York, N.Y.), 49*(2), 151–166. doi:10.1111/j.2151-6952.2006.tb00209.x

Falk, J. H., & Storksdieck, M. (2010). Science learning in a leisure setting. *Journal of Research in Science Teaching: The Official Journal of the National Association for Research in Science Teaching, 47*(2), 194–212.

Fountain, J., Espiner, S., & Xie, X. (2010). A cultural framing of nature: Chinese tourists' motivations for, expectations of, and satisfaction with, their New Zealand tourist experience. *Tourism Review International, 14*(2-3), 71–83. doi:10.3727/154427211X13044361606298

Gam, A., Bahri-Ammari, N., & Soliman, M. (2023). Intelligent customer experience and behavioral brand loyalty: Empirical evidence from the commerce environment. *International Journal of Human-Computer Interaction, 39*(10), 2090–2101. doi:10.1080/10447318.2022.2074120

Gravari-Barbas, M., Cominelli, F., Condevaux, A., Jacquot, S., & Conti, A. (2018). Chapter: Concepts, Doctrines and Lists. *Tourism Management at UNESCO World Heritage Sites*, 7.

Hsu, F. C., Zhang, S., Zhang, Y., & Lee, T. J. (2022). Decision-making behavior in the sustainable development of intangible cultural heritage tourism. *International Journal of Tourism Research, 24*(6), 800–812. doi:10.1002/jtr.2546

Huang, S. (2021). Tourist motivation: a critical overview. In R. Sharpley (Ed.), *Routledge handbook of the tourist experience* (pp. 200–212). Taylor & Francis. doi:10.4324/9781003219866-18

Hughes, M., & Carlsen, J. (2010). The business of cultural heritage tourism: Critical success factors. *Journal of Heritage Tourism, 5*(1), 17–32. doi:10.1080/17438730903469805

Ismagilova, G., Safiullin, L., & Gafurov, I. (2015). Using historical heritage as a factor in tourism development. *Procedia: Social and Behavioral Sciences*, *188*, 157–162. doi:10.1016/j.sbspro.2015.03.355

Ismail, N., Masron, T., & Ahmad, A. (2014). Cultural heritage tourism in Malaysia: Issues and challenges. SHS Web of Conferences, Iso-Ahola, S. E. (1982). Toward a social psychological theory of tourism motivation: A rejoinder. *Annals of Tourism Research*, *9*(2), 256–262.

Jiang, S., Moyle, B., Yung, R., Tao, L., & Scott, N. (2023). Augmented reality and the enhancement of memorable tourism experiences at heritage sites. *Current Issues in Tourism*, *26*(2), 242–257. doi:10.1080/13683500.2022.2026303

Kondo, Y., Miki, T., Kuronuma, T., Hayakawa, Y. S., Kataoka, K., & Oguchi, T. (2016). Concurrent and sustainable development of a local-scale digital heritage inventory through action research at Bat, Oman. *Journal of Cultural Heritage Management and Sustainable Development*, *6*(2), 195–212. doi:10.1108/JCHMSD-01-2016-0005

Lima, J., Serra, J., Borges, M. D. R., & Marujo, N. (2023). Is gastronomy important for UNESCO world heritage sites' tourists? A case study from Portugal. *Anatolia*, 1-13.

Little, C., Bec, A., Moyle, B. D., & Patterson, D. (2020). Innovative methods for heritage tourism experiences: Creating windows into the past. *Journal of Heritage Tourism*, *15*(1), 1–13. doi:10.1080/1743873X.2018.1536709

López-Guzmán, T., Torres Naranjo, M., Pérez Gálvez, J. C., & Carvache Franco, W. (2019). Segmentation and motivation of foreign tourists in world heritage sites. A case study, Quito (Ecuador). *Current Issues in Tourism*, *22*(10), 1170–1189. doi:10.1080/13683500.2017.1344625

Lussetyowati, T. (2015). Preservation and conservation through cultural heritage tourism. Case study: Musi Riverside Palembang. *Procedia: Social and Behavioral Sciences*, *184*, 401–406. doi:10.1016/j.sbspro.2015.05.109

Mathew, V., & Soliman, M. (2021). Does digital content marketing affect tourism consumer behavior? An extension of t echnology acceptance model. *Journal of Consumer Behaviour*, *20*(1), 61–75. doi:10.1002/cb.1854

Matlegiewicz, M. (2020). Motivation in Cultural Tourism: The Case of the Traditional and Regional Food Products of West Pomerania. *Zarządzanie w kulturze, 21*(3), 219-237.

Matteucci, X., Koens, K., Calvi, L., & Moretti, S. (2022). Envisioning the futures of cultural tourism. *Futures*, *142*, 103013. doi:10.1016/j.futures.2022.103013 PMID:36061420

McKercher, B. (2002). Towards a classification of cultural tourists. *International Journal of Tourism Research*, *4*(1), 29–38. doi:10.1002/jtr.346

Nguyen, T. H. H., & Cheung, C. (2016). Chinese heritage tourists to heritage sites: What are the effects of heritage motivation and perceived authenticity on satisfaction? *Asia Pacific Journal of Tourism Research*, *21*(11), 1155–1168. doi:10.1080/10941665.2015.1125377

Packer, J., & Ballantyne, R. (2002). Motivational factors and the visitor experience: A comparison of three sites. *Curator (New York, N.Y.)*, *45*(3), 183–198. doi:10.1111/j.2151-6952.2002.tb00055.x

Pearce, P. L., & Lee, U.-I. (2005). Developing the travel career approach to tourist motivation. *Journal of Travel Research, 43*(3), 226–237. doi:10.1177/0047287504272020

Rasoolimanesh, S. M., Seyfi, S., Hall, C. M., & Hatamifar, P. (2021). Understanding memorable tourism experiences and behavioural intentions of heritage tourists. *Journal of Destination Marketing & Management, 21*, 100621. doi:10.1016/j.jdmm.2021.100621

Richards, G. (2018). Cultural tourism: A review of recent research and trends. *Journal of Hospitality and Tourism Management, 36*, 12–21. doi:10.1016/j.jhtm.2018.03.005

Santa-Cruz, F. G., & López-Guzmán, T. (2017). Culture, tourism and world heritage sites. *Tourism Management Perspectives, 24*, 111–116. doi:10.1016/j.tmp.2017.08.004

Secondi, L., Meseguer-Santamaría, M. L., Mondéjar-Jiménez, J., & Vargas-Vargas, M. (2011). Influence of tourist sector structure on motivations of heritage tourists. *Service Industries Journal, 31*(10), 1659–1668. doi:10.1080/02642069.2010.485641

Selmi, N., Bahri-Ammari, N., Soliman, M., & Hanafi, I. (2021). The impact of festivalscape components on festivalgoers' behavioral intentions: The case of the International Festival of Carthage. *Journal of Convention & Event Tourism.*

Seyfi, S., Michael Hall, C., & Fagnoni, E. (2019). Managing world heritage site stakeholders: A grounded theory paradigm model approach. *Journal of Heritage Tourism, 14*(4), 308–324. doi:10.1080/1743873X.2018.1527340

Smith, M. K., Pinke-Sziva, I., Berezvai, Z., & Buczkowska-Gołąbek, K. (2022). The changing nature of the cultural tourist: Motivations, profiles and experiences of cultural tourists in Budapest. *Journal of Tourism and Cultural Change, 20*(1-2), 1–19. doi:10.1080/14766825.2021.1898626

Soliman, M. (2021). Extending the theory of planned behavior to predict tourism destination revisit intention. *International Journal of Hospitality & Tourism Administration, 22*(5), 524–549. doi:10.1080/15256480.2019.1692755

Soliman, M., & Al Balushi, M. K. (2023). Unveiling destination evangelism through generative AI tools. *ROBONOMICS: The Journal of the Automated Economy, 4*, 54. https://journal.robonomics.science/index.php/rj/article/view/54.

Soliman, M., Cardoso, L., de Almeida, G. F., Araújo, A., & Vila, N. (2021). Mapping smart experiences in tourism: A bibliometric approach. *European Journal of Tourism Research, 28*, 2809–2809. doi:10.54055/ejtr.v28i.2254

Soliman, M. S. (2015). Pro-poor tourism in protected areas–opportunities and challenges:"The case of Fayoum, Egypt. *Anatolia, 26*(1), 61–72. doi:10.1080/13032917.2014.906353

Surugiu, M.-R., & Surugiu, C. (2015). Heritage tourism entrepreneurship and social media: Opportunities and challenges. *Procedia: Social and Behavioral Sciences, 188*, 74–81. doi:10.1016/j.sbspro.2015.03.340

Timothy, D. J. (2014). Contemporary cultural heritage and tourism: Development issues and emerging trends. *Public Archaeology, 13*(1-3), 30–47. doi:10.1179/1465518714Z.00000000052

Timothy, D. J., & Boyd, S. W. (2006). Heritage tourism in the 21st century: Valued traditions and new perspectives. *Journal of Heritage Tourism*, *1*(1), 1–16. doi:10.1080/17438730608668462

UNESCO. (2013). *World Heritage Resource Manual*. UNESCO. https://whc.unesco.org/en/managing-cultural-world-heritage/

UNESCO. (2023). *World Heritage List: States Parties*. UNESCO. *Oman*

UNWTO. (2018). *Tourism and Culture Synergies*. United Nations World Tourism Organization. doi:10.18111/9789284418978

Wang, M.-Y., Li, Y.-Q., Ruan, W.-Q., Zhang, S.-N., & Li, R. (2023). Cultural inheritance-based innovation at heritage tourism destinations: Conceptualization, structural dimensions and scale development. *Journal of Hospitality and Tourism Management*, *55*, 118–130. doi:10.1016/j.jhtm.2023.03.009

Wu, G.-M., Chen, S.-R., & Xu, Y.-H. (2023). Generativity and inheritance: Understanding Generation Z's intention to participate in cultural heritage tourism. *Journal of Heritage Tourism*, *18*(4), 1–18. doi: 10.1080/1743873X.2023.2178927

Zhao, Y., & Agyeiwaah, E. (2023). Exploring value-based motivations for culture and heritage tourism using the means-end chain and laddering approach. *Journal of Heritage Tourism*, *18*(5), 1–23. doi:10.1080/1743873X.2023.2215933

Chapter 6
Community's Attitude Towards the Development of Cultural and Heritage Tourism:
An Exploratory Study

Amrik Singh
https://orcid.org/0000-0003-3598-8787
Lovely Professional University, India

Supina Supina
Bunda Mulia University, Indonesia

ABSTRACT

Tourism, heritage, and culture are widely recognized as significant catalysts for economic development in various regions across the globe. Their significant contribution to fostering a thorough understanding of their rich diversity greatly enhances regional cultures. Cultural tourism is an inherent aspect of cultural heritage. Tourism possesses the capacity to generate substantial revenue for the purpose of environmental conservation while also bestowing economic significance on cultural heritage. Culture and heritage encompass various aspects such as values, beliefs, art, monuments, constructions, and other elements, which collectively serve as a manifestation of a nation's identity and integrity. The proposed chapter will discuss the emergence of cultural and heritage tourism, the cultural and heritage tourism development of six regional commissions by UNWTO, the Challenges and issues of the cultural and heritage tourism development, and the future trends of cultural and heritage tourism.

INTRODUCTION

Tourism, heritage and culture are widely recognised as significant catalysts for economic development in various regions across the globe. Their significant contribution to fostering a thorough understanding of their rich diversity greatly enhances regional cultures. Cultural tourism is an inherent aspect of

DOI: 10.4018/978-1-6684-9957-3.ch006

cultural heritage. Tourism possesses the capacity to generate substantial revenue for the purpose of environmental conservation while also bestowing economic significance on cultural heritage. Bitusikova (2021) provides an elucidation of cultural heritage as a catalyst for the development of heritage tourism, and numerous studies within the realm of social sciences have been dedicated to examining the correlation between cultural heritage and the advancement of cultural and heritage tourism in both recent years and past decades. In 2018, Richards provided an explanation regarding the growth of cultural tourism, highlighting its fragmentation into various emerging niches. One such niche is heritage tourism, which has gained momentum due to an expanding understanding of culture. This expansion, in turn, stimulated the growth of cultural tourism during the 1990s. Furthermore, the fragmentation of the cultural tourism concept itself contributed to a notable increase in the number of publications dedicated to this field.

LITERATURE REVIEW

The growth of tourism is a "two-edged sword" as it encompasses both constructive and detrimental facets for the local inhabitants. The intensity of indigenous inhabitants' inclusion in tourism is dynamically linked to tourism impacts on public perception. Tourism, without a doubt, provides the greatest social and economic benefits among the various techniques for revitalising intangible cultural heritage (ICH). According to Wei et al. (2021), residents' heritage proximity has a favourable influence on perceived positive impacts but has no effect on perceived negative impacts. Residents' ties to heritage and perceptions of consequences have an impact on their attitudes towards tourism development.Cultural heritage resources are one of the most vulnerable development sectors, notwithstanding their promise for sustainability. They are usually the first parts of a city to undergo massive planning revisions. The global World Wars, battles, and conquests have also devastated historic communities, restricting cultural heritage tourism (Mandola & Boussaa, 2023). Some severely destroyed cities were abandoned and never recovered. The UNESCO List of World Heritage in Danger includes 52 cultural, natural, mixed, and transboundary assets as of November 12, 2022. Transnational heritage sites include monuments and city centres like Vienna's historic centre, with the worst deterioration in Africa and Middle Eastern countries like Mali, Niger, Egypt, Libya, Palestine, Syria, Iraq, and Yemen.

Culture and heritage encompass various aspects such as values, beliefs, art, monuments, constructions, and other elements, which collectively serve as a manifestation of a nation's identity and integrity. The preservation of culture and heritage through generations enables nations to gain insights into their historical background and develop a sense of personal identity (7th Diversity Matters, Preserving Culture and Heritage through Generations). The significance of heritage in the context of tourism cannot be understated. The cultural heritage of a nation has the potential to draw a larger number of international tourists and serve as a driving force for the advancement of tourism within the country (Meha et al., 2020).The sector in question is centred around the objective of facilitating the gathering of individuals with the purpose of acquiring knowledge and developing an appreciation for one another, thereby cultivating a sense of mutual respect and tolerance. Well-managed tourism is recognised as a significant economic and social force globally, and it has the potential to play a crucial role in contemporary society by enhancing the prominence of heritage, both in tangible and intangible forms, where it will reinforce cultural identity and diversity, which serves as an important benchmark for development. Not to mention, the COVID-19 pandemic has significantly disrupted the domain of cultural tourism (UNESCO, 2021), the year 2022 signifies the commemoration of the 50th anniversary of the enactment of the World Heritage Convention, a framework established by UNESCO in 1972. This convention was designed to safeguard, utilise, and transmit cultural heritage through the UNESCO World Heritage Committee, with the aim of fostering constructive contributions towards the preservation and rehabilitation of the shared heritage of humanity (Zhang et al., 2022).Well-managed tourism is recognised as a significant economic and social force globally, and it has the potential to play a crucial role in contemporary society by enhancing the prominence of heritage, both in tangible and intangible forms, where it will reinforce cultural identity and diversity, which serves as an important benchmark for development. In many countries the development of cultural and heritage tourism is still needs potential improvement. The significance of cultural and heritage tourism development in numerous countries lies in its ability to address the needs of local communities, cater to the demands of tourists, and ensure the preservation of natural and cultural resources to foster sustainable tourism growth (Aisyah, 2023). One potential approach to showcasing diverse forms of regional cultural heritage on a global scale is through the integration of the cultural and heritage in tourism sectors (Yuan et al., 2022). The tourism industry has seen tremendous growth over the years in concert with the local community's perceptions and attitudes toward sustainable tourism management practices. Recognizing the consequences of tourism on local people has undergone a radical shift, and it is now a significant issue in constructing long-term and sustainable tourist initiatives. Tourism development is an essential array of financial activities for mountain economies to develop and revitalize. Mountain landscapes, on the other hand, are ecologically susceptible to alteration and tourism-induced impacts (Abdollahzadeh and Sharifzadeh, 2014). In the vast corpus of tourism research, the term 'impacts of tourism' has gained prominence. Since the 1970s, there has been a significant increase in interest in residents' perspectives on tourism, prompting academics and professionals to examine local attitudes closely (Sharpley, 2014). This phenomenon could be understood explicitly through an appraisal of residents' attitudes and perception-based assessment. Tourism is fundamentally an integrative transactional process, including a direct and indirect exchange between visitors and locals (Brida et.al, 2010; Sharpley, 2014; Abdollahzadeh and Sharifzadeh, 2014). Understanding local communities' perceptions regarding various aspects of tourism and how they vary may be critical to managing tourism activity, especially in the mountain ecosystems, which are fragile to any tourism-induced impacts. A better understanding of residents' knowledge of tourism development, their awareness of its benefits, and its perceived impact on their welfare are fundamental to developing and implementing management strategies that are both sustainable in the long term and sensitive to existing local needs. The future viability of the travel and leisure economy depends on several issues, particularly, with particular attention paid to tourism's impacts on the host community (Chandralal, 2010).

Therefore, this chapter will discuss: (1) The emergence of cultural and heritage tourism, (2) The cultural and heritage tourism development six regional commissions by UNWTO: Africa, the Americas, East Asia and the Pacific, Europe, the Middle East and South Asia as Countries are progressively pursuing partnerships with global organisations in order to promote the development of cultural and heritage tourism. National and local governments collaborate with international organisations, including UNESCO, UNWTO, and the OECD, in the realm of sustainable tourism, (3) Challenges and issues of the cultural and heritage tourism development: This section will also examine destinations that have been identified as having the lowest Human Development Index score and have been designated as cultural and heritage tourism destinations in order to leverage these areas as part of the national culture as well as enhance social welfare (Sunarminto & Mijiarto, 2022), (4) The future trends of cultural and heritage tourism: The Mount Lushan cultural landscape heritage site has implemented technology optimisation to enhance the virtual experience of cultural landscape heritage tourism and facilitate the marketing of cultural landscape tourism by integrating Internet of Things (IoT) technology and analytical models (Cai et al., 2021). The sample application of technology enhancement is to enhance the convenience of tourist movements and establish environmental subsystems where the tourist flow subsystem enables managers to effectively manage and control tourist movement based on a predetermined threshold of tourist carrying capacity.

THE EMERGENCE OF CULTURAL AND HERITAGE TOURISM

Mrca & Caric (2019) examined sustainable heritage tourism planning models that serve as a catalyst for the market positioning of destinations. From a spatial-planning perspective, the destination's competitiveness is based on its unique cultural experiences, environmental quality, and scenic landscape, which position it as a quality place for living, working, and investing as a result of (positive) tourism-heritage interaction. Models of sustainable heritage tourism planning entail harmony and equilibrium between global tourism standards and the preservation of a destination's cultural identity.Mrda & Caric (2019) examined sustainable heritage tourism planning models that serve as a catalyst for the market positioning of destinations. From a spatial-planning perspective, the destination's competitiveness is based on its unique cultural experiences, environmental quality, and scenic landscape, which position it as a quality place for living, working, and investing as a result of (positive) tourism-heritage interaction. Models of sustainable heritage tourism planning entail harmony and equilibrium between global tourism standards and the preservation of a destination's cultural identity.

In many nations, it is evident that the growth of cultural and heritage tourism is correlated with the enhancement of brand value. Heritage and cultural tourism development can help a location or attraction create a positive reputation, attract more tourists, and increase brand recognition. When contemplating cultural heritage sites as tourist destinations, a location's reputation is an important factor. It is widely acknowledged that the growth of historical and cultural tourism is an important aspect of the tourism industry. Governments and stakeholders can enhance the character of a destination and attract more tourists by investing in infrastructure development, promoting attractions, and hosting cultural events (Arumugam et al., 2023).

THE CULTURAL AND HERITAGE TOURISM DEVELOPMENT SIX REGIONAL COMMISSIONS BY UNWTO

- Africa

Madandola & Boussaa (2023) indicated that culture-driven urban regeneration is a workable method for the creation of places and for sustainable development. It has the potential to drive the necessary heritage development, for example, in Old Oyo, Nigeria, while also contributing to the city's long-term environmental, economic, and social sustainability. In addition, the potential of the project to attract archeotourists is critical to the preservation of sociocultural traditions as well as the revitalization of indigenous approaches to urban problems and practices.

- Middle East

The Middle East and North Africa (MENA) region is enormous, complicated, and diversified, and it is regarded as a major centre of cultural and travel influence. In Morocco, for example, heritage in general and the commoditization of culture for tourism in particular have evolved in a social and political atmosphere in which tradition and modernity coexist. Culture and heritage tourism account for 80% of tourism activities in Morocco, despite government officials' intention to develop seaside resorts, frequently at the expense of other forms of tourism (Seyfi & Hall, 2021).

- South Asia, East Asia and the Pacific

South Asia is a captivating and one-of-a-kind tourist destination because of its rich history, heritage, cultural variety, diverse landscapes, huge natural resources, and scenic beauty. South Asia, on the other hand, is unable to establish a foothold in the tourism business since its percentage of foreign visitor arrivals is very low in comparison to other regions (Amin, 2021).Cultural heritage is vital in the construction of creative cities, giving them new life and providing employment possibilities and local economies in the process. In the four cultural heritages of three East Asian locations (Hong Kong, Macau, and Taiwan), revitalization and development of cultural heritage and intangible cultural heritage have become key components of national development programmes globally (Ng et al., 2023).

- Europe

Cultural heritage has been one of the most important engines of tourism development in Slovakia in recent decades. It is manifested in various tangible and intangible forms (such as visits to historic cities, heritage sites, cathedrals, castles, or monuments, known as built or tangible heritage; living cultural heritage that includes people, folklore, customs, or foods, known as intangible heritage; and other forms—industrial heritage, nature-based heritage, personal heritage, or dark heritage). Observing the change from "monuments to people; from objects to functions, and therefore from preservation to sustainable use and development," Bitusikova (2021).

DISCUSSIONS AND CONLUSION

Tourism is a rapidly growing industry in the whole world. There is a great scope of tourism development and the people are also cooperative and are willing to support tourism. Various studies show that the impact of tourism has both pros and cons. On positive side, it helps to promote and preserve local culture, cuisine, crafts, arts and festivals. It also creates the opportunity to improve hospitality towards strangers as tourism provides a chance for the local residents to meet new people from different places and to learn new things and gets the opportunity to improve their social skills as well. The local community's support and involvement in the decision-making process guarantee long-term sustainability in tourism development, which is imperative in local tourist attractions. Researchers advocate that it is challenging to sustain tourism in any region that does not value and acknowledge the local inhabitants' inclusion in the active tourism function. Destination planners must recognize residents' opinions of tourism effects, as they are the primary stakeholders in perceiving the cumulative bearings throughout the process of multiplier effects of tourism development. While tourism's importance to economic and social development as well as a multiplicity of benefits to local people are well documented, even though it has culminated in many effects on which the local people are likely to suffer from waste generation, change in their traditional culture, traffic jams, crimes, and the cost of living seems to be mounting. It is also contributing towards broadening the thinking of locals and culture understanding by creating awareness regarding the diversity of culture and lifestyle by interacting and it helps in removing social barriers like caste, racism, religion etc. It is leading the community towards better living standards. Another very important finding revealed that through the literature that the local residents do not consider that tourism is increasing the crime rate, hampers the local lifestyle and is disturbing our age-old tradition which indeed is a great thing. While locals exhibit a neutral attitude towards tourism that it increases social evils such as consumption of drugs, alcohol and loss of cultural identity and adoption of western culture. It is evident that locals have a positive attitude towards the benefits of tourism but at the same time, they are also concerned about the cost of tourism. Therefore, planners and decision-makers should recognize the negative cost of it and it must be taken into consideration while decision making. Community involvement can be facilitated by involving locals in policy formation and the decision-making process. Such effective participation will help the local residents to match their expectations and a tourism-related project which will be favorably accepted by the community. Tourism has evolved as a major engine of overall development in various destinations. With a constant development of tourism, socio-cultural aspect has transformed to the great extend in tourist destinations. This chapter has proposed that tourism is like a two side of a coin with both positive and negative impact. However, there is a great scope of further tourism development and the people are also cooperative and are willing to support tourism. The results also revealed that the perceived benefits of tourism have more positive than negative impacts. The importance of support and participation from the local residents in the development of tourism has been well recognized by various practitioners. To survive tourism for the long run in any community there is a need for cooperation from local residents. Therefore, it can be possible by the development of community based tourism which will help to gain support, cooperation and participation from the local community.

PRACTICAL IMPLICATIONS

For Policy Makers

There should be proper policy planning to boost tourism and it is important to promote value tourism in such a way that village tourism needs to be encouraged which shall boost the local economy and tourists will also get an opportunity to experience the culture and tradition of village which shall promote integration among locals and tourists and it can be possible by transmitting awareness about the importance of tourism among the local residents.

For Tourists and Local Managements

Tourists must be encouraged to visit villages and ancient monuments and connect with the local community which shall create bonding between them and also travelling experience for tourists rather than just attraction. Hospitality is an important part of tourism and to bloom in this industry host community need to be well educated about the basic functioning of the hospitality industry. Therefore, locals should be trained to be professional in tourism industry.

LIMITATIONS AND RECOMMENDATIONS FOR FUTURE STUDIES

This paper serves as a comprehensive evaluation of recent advancements in a certain community's attitude towards the development of cultural heritage tourism, providing a concise overview of the current understanding and knowledge in the field. The analysis is exclusively based on existing published publications. One of the challenges encountered in this review paper is the extensive range of studies and methodologies employed to investigate community attitudes. This diversity is a significant obstacle to achieving consistency in the assessment of attitudes across various destinations. Therefore, future empirical investigations of this phenomenon will offer a clear illustration of the interconnections among established constructions.

REFERENCES

Abdollahzadeh, G., & Sharifzadeh, A. (2014). Rural residents' perceptions toward tourism development: A study from Iran. *International Journal of Tourism Research*, *16*(2), 126–136. doi:10.1002/jtr.1906

Aisyah, S. (2023). Implications of Planning and Management for the Development of Sustainable Cultural Heritage Tourism in Malaysia. *Advances in Tourism Studies, 1*(2).

Amin, S. (2021). *An Overview of South Asian Tourism Industry and Significance of the Regional Tourism Integration.* Springer Books. In *The Economy of Tourism in Bangladesh* (pp. 157–178). Springer. doi:10.1007/978-3-030-72900-4_7

Arumugam, A., Nakkeeran, S., & Subramaniam, R. (2023). Exploring the factors influencing heritage tourism development: A model development. *Sustainability (Basel)*, *15*(15), 11986. doi:10.3390u151511986

Bitušíková, A. (2021). Cultural Heritage as a means of heritage tourism development. *Muzeológia a Kultúrne Dedičstvo*, *9*(1), 81–95. doi:10.46284/mkd.2021.9.1.5

Brida, J. G., Osti, L., & Barquet, A. (2010). Segmenting resident perceptions towards tourism—A cluster analysis with a multinomial logit model of a mountain community. *International Journal of Tourism Research*, *12*(5), 591–602. doi:10.1002/jtr.778

Cai, Z., Fang, C., Zhang, Q., & Chen, F. (2021). Joint Development of Cultural Heritage Protection and Tourism: The case of mount lushan cultural landscape heritage site. *Heritage Science*, *9*(1), 86. doi:10.118640494-021-00558-5

Chandralal, K. P. L. (2010). Impacts of tourism and community attitude towards tourism: A case study in Sri Lanka. *South Asian Journal of Tourism and Heritage*, *3*(2), 41–49.

Madandola, M., & Boussaa, D. (2023). Cultural Heritage Tourism as a catalyst for sustainable development; the case of Old Oyo Town in Nigeria. *International Journal of Heritage Studies*, *29*(1–2), 21–38. doi:10.1080/13527258.2023.2169332

Meha, A., Tahiri, A., & Zhubi, M. (2020). *Importance of Cultural Heritage in tourism development: The case of Kosovo*. Research Gate. https://www.researchgate.net/publication/354223359_Importance_of_Cultural_Heritage_in_Tourism_Development_The_Case_of_Kosovo

Mıđa, A., & Carić, H. (2019). *Cultural Urban Heritage*. The Urban Book Series. doi:10.1007/978-3-030-10612-6

Ng, W.-K., Hsu, F.-T., Chao, C.-F., & Chen, C.-L. (2023). Sustainable competitive advantage of cultural heritage sites: Three destinations in East Asia. *Sustainability (Basel)*, *15*(11), 8593. doi:10.3390u15118593

Nunkoo, R., & Ramkissoon, H. (2010). Small island urban tourism: A residents' perspective. *Current Issues in Tourism*, *13*(1), 37–60. doi:10.1080/13683500802499414

Richards, G. (2018). Cultural tourism: A review of recent research and Trends. *Journal of Hospitality and Tourism Management*, *36*, 12–21. doi:10.1016/j.jhtm.2018.03.005

Seyfi, S., & Hall, C. M. (2021). Cultural heritage tourism in the MENA: introduction and background. In Hall. C.M. & Seyfi. S (Eds.). Cultural and Heritage Tourism in the Middle East and North Africa. Routledge. doi:10.4324/9780429279065-1

Sharpley, R. (2014). Host perceptions of tourism: A review of the research. *Tourism Management*, *42*, 37–49. doi:10.1016/j.tourman.2013.10.007

Sunarminto, T., & Mijiarto, J. (2022). Cultural Heritage Tourism Development Strategy in bogor regency. *Jurnal Pariwisata Pesona*, *7*(1), 63–74. doi:10.26905/jpp.v7i1.6643

UNESCO. (2021). *Cutting edge: Bringing cultural tourism back in the game*. UNESCO. https://www.unesco.org/en/articles/cutting-edge-bringing-cultural-tourism-back-game

Wei, Y., Liu, H., & Park, K.-S. (2021). Examining the structural relationships among heritage proximity, perceived impacts, attitude and residents' support in Intangible Cultural Heritage Tourism. *Sustainability (Basel)*, *13*(15), 8358. doi:10.3390u13158358

Yuan, C., Gan, L., & Zhuo, H. (2022). Coupling mechanisms and development patterns of revitalizing Intangible Cultural Heritage by integrating cultural tourism: The case of Hunan Province, China. *Sustainability (Basel)*, *14*(12), 6994. doi:10.3390u14126994

Zhang, S., Liang, J., Su, X., Chen, Y., & Wei, Q. (2023). Research on global cultural heritage tourism based on Bibliometric analysis. *Heritage Science*, *11*(1), 139. doi:10.118640494-023-00981-w

Chapter 7
Religious Tourism Areas and Cultural Conservation

Mokhlesur Rahman
https://orcid.org/0009-0006-4126-1751
Bangladesh Institute of Tourism Studies, Bangladesh

ABSTRACT

Religion plays an important role in tourism that ultimately contributes to the human development. Religious attractions belong to the cultural elements and are an important component of economic and social activities. The basic and essential elements religious tourism areas considered as more than pilgrimages. Pilgrimage has material and spiritual values which voluntarily engages with religious significance. In religious tourism, it is not necessarily involving the tourist's belief in a certain religion. The concept of culture entails the need to treat religious tourism as a manifestation of fundamental motivations. Religion, as a part of culture, can contribute to cultural conservation.

INTRODUCTION

Religious tourism is linked with theoretical approaches of pilgrimage. Earp (2015) considered religious practices as a part of cultural practices. If culture expresses how humans experience and understand the world; religion is a fundamental way in which experience of human being understand the world (Abdulla, 2018). In the contemporary world, religions have two arms, one is the practicing for the particular religion and other is the events for the society whole. So, religions have many social impacts including the touristic significance. The relationship between culture and religion is revealed in the motivation and manifestation of cultural expression. Sociology of the religions frame and shape the people, culture and livelihood. Norms, rituals and practices of religions make the society peaceful and harmonious. Discovering the fundamentals of the cultural conservation of religious areas through tourism may applly in the society. Some sociological assumptions on culture has been critiqued strongly. Many voices argue the culture, as of its dynamism and changing progressive values. However, culture and religion strengthen the peace components of the religious areas. Together with aesthetics and ethics, religion constitutes culture and society. The location of religion within culture causing religion to act as custodian of culture (Beyers, 2017).

DOI: 10.4018/978-1-6684-9957-3.ch007

OBJECTS OUTCOME

Religion is the doctrine of salvation. As well as tourism is the doctrine of safeguarding phenomena for social development. So, following objects routes the cultural conservation through religious tourism.

a. Religious attitude to the culture and tourism for safeguarding the religious areas.
b. Learning of spiritual, religious and cultural tourism as the means of peace and harmony.
c. Conservation pattern of religious occasions and pilgrimage through religious tourism.

LITERATURE REVIEW

Geertz (1973) says that religion is a system of symbols which acts to establish powerful, pervasive and long-lasting moods and motivations in men by formulating conceptions of a general order of existence and clothing the conceptions with such an aura of factuality that the moods and motivation seem uniquely realistic. That's why it is the part of cultural tourism, which triggers to change the tourist behavior. Geographically religions are located across a wide range of disciplines, but have yet to adequately contextualize within geographical debates (Slater, 2004). As Tuan (1978), certain buildings like mosques, synagogues and churches are somehow sacred and belongs to religious areas. Mulder (1985) mentioned religious implication as the expression of human culture. Rappaport (1999) observed, the true origin of religion lies in the human need to create rituals that restore the balance between people and the environment.

Geertz (1973) defines culture as an indicative and historically transmitted pattern of meanings which embodied in symbols, a system of inherited conceptions and expressed in symbolic forms by the means of communication, perpetuate, and develop their knowledge about attitudes toward life. Tourism includes visiting to a church, a shrine or a place of pilgrimage offers and all visitors clear the unambiguous attraction. Spiritual and religious tourism remains one of the most understudied areas (Ron, 2007). Management of religious and sacred places, paradigms discussed by both academics and practitioners in many articles (Olsen & Timothy, 2006; Ron, 2007; Rivera *et al.*, 2009).

Religious tourism motivates the tourists who visit the pilgrim sites and the places (Stoddard, 1997). Turner & Turner (1969) describes religious tourism runs through a ritual process. But MacCannell (1973) clearly identifies pilgrimage as a quest for the experiences. Vukonic (2002) has argued that religion and tourism have much in common even though religious institutions have traditionally attempt to minimize the theological conflict.

Studies of religious tourism can investigate the diverse destinations of devotional sites (Abbate & Di Nuovo, 2013), religious tourist attractions (Bond *et al.*, 2015), pilgrimage routes (Amaro *et al.*, 2018; Kim *et al.*, 2016), and festivals (Suntikul & Dorji, 2016). Within the broad categories, religious tourism and pilgrimage evolves into the considerations of secularity (Hung *et al.*, 2017; Nilsson & Tesfahuney, 2016). It also manifestoes the secular pilgrimage (Brown, 2016). The scholarly interest developes a better understanding of demonstrating the growing incidence of tourism. Millions of people, those who identify the sacred places to travel as religious tourists and pilgrims display remarkable loyalty. So, studies of religious tourism makes effective on the economic, social and cultural conservation (UNWTO, 2017). Furthermore, scholarly and applied perspectives on religious tourism and pilgrimage integrate the religion areas and the socio-cultural phenomena.

Ultimately, through religious tourism, religion can be a booming economic growth (Yekinni & Mercy, 2017) & Zouni & Digkas, 2019). This opportunity creates momentum, especially when a destination has various well-maintained holy sites and monuments. As the religious heritage is so great that tourism around religion is used by the regional government as a tourist attraction in different international fairs and tourism promotion campaigns.

RELIGIOUS ATTITUDE TO THE CULTURE AND TOURISM

Religion and spirituality has a positive effect on the mental and physical health of that particular religious group (Bakhshipour-Roudsari, 2008). As per Seybold & Hill (2001), religious beliefs, rituals, and obligations are related to the positive results such as improved and better quality of life, socio-psychological welfare, physical and mental health, marital satisfaction, sustainable society, and positive performances. That's why, people stay with intensely and strongly religious believers and to the religious areas. In a society, religion is not just a belief system, but also a faith realized in every aspect of life. Religion, belief and culture should be recognized as a potential source of moral and personal strength. Okulicz-Kozaryn (2009) showed the relationship between religious dimensions and life satisfactions. He also mentioned that religious people are more satisfied with theological belief.

Scholars says that religions are the central incentive for believer people to travel and visit sacred sites. Religious tourists now involve strongly the industry aimed with numerous pilgrims that attract not only believers but also general tourists (Griffiths, 2011). Religious tourism has a unique cognitive characteristics with gnoseological incentives. The spiritual process of religious tourism allow pilgrims to consume religious souvenirs. According to Nyaupane *et al.* (2015), the motivation for religious tourists is different from that of their conventional secular counterparts. Religious tourists are motivated by spiritual motivations. So, visiting sacred places may create strong feelings of religious inspiration.

Consumption of products and services during traveling perceives each religion, e.g., halal food in Islamic tourists etc. Role of religiosity influence on the attitudes of communities towards tourism. Socio-demographic variables shows the financial benefit gained from tourism. As if, religiosity exerts an important influence on community attitudes. Social Distance Theory (SDT) tells that the more extreme religious group are self-segregated from external influence and less positive towards tourism. Since tourism depends on the hospitality of residents, so perception of local populations towards tourism is a big attitudinal factor. Additional negatives increase the disruption of beliefs and faiths, and decrease in the quality of life. Sustainable policies of religious tourism, learning of different rituals, uniqueness of branding, innovative promotional tools, combining with other forms of tourism. Various infrastructure improves the connectivity and arranging events that express the attitudes of religious tourism.

SPIRITUAL, RELIGIOUS, AND CULTURAL TOURISM

Some experiences of human life belongs to profound and transformative spiritual awakening, which influences the divinity. A spiritual awakening pulls forward into a new life of conscious, growth and evolution. Religious tourism impacts specific faiths who visit the places by considering the holy thoughts. This religious journey involves all such travels and tours which are directly or indirectly motivated by religious reasons. Researchers agree with the growing of religious sites for tourism, and their importance in economic as well as socio-cultural terms as sites of current and former worship (Timothy & Olsen, 2006). Spiritual awakening represents the lifting of the veil of ignorance that is incorrect understanding. Sites of religious tourism may lead to an understanding and knowledge that reinforces spiritual prejudices and emphasizes differences in belief systems and therefore their physical manifestation in the tourism products (Andriotis, 2009). So, religious tourism can effectively help and promote the unique destinations in increasingly homogenous global tourism.

On the other hand, cultural tourism is a type of tourism in which the visitors motivates to learn, discover, experience and consume the tangible and intangible cultural attractions in a tourism destination. According to the United Nations World Tourism Organization (UNWTO), cultural tourism is the movements of persons for essentially cultural motivations such as study tours, performing arts and cultural tours, travel to festivals and other cultural events, visits to sites and monuments, travel to study nature, folklore or art, and pilgrimages. The very special point is religious tourism or pilgrimage is one kind of cultural tourism. It intermingles with some religious activities for building social phenomena and build the society a new progressive look.

In Bangladesh, there are many festivals varying from religion and culture. Eid for Muslims, Puja for Hindus, Christmas festival for Christianity and Buddha Purnima for Budhhist are the major religious as well as cultural festivals. Both religious and cultural festivals are prominent here. Literature, music, art, drama, folk dance, jatra (kind of play), traditional transport, clothing etc. are the major components of culture. Some common activities are observed countrywide by people with historical and heritage issues. All religious and cultural components intercept a point which shape the philosophy and livelihood. Cultural conservation patterns of religious occasions and pilgrimage build religious tourism. People run through social process and reach a point of excellence.

RELIGIOUS TOURISM AREAS IN BANGLADESH

About (91.04%) of Bangladeshis are Muslims, followed by Hindus, the largest-minority at (7.95%), Buddhists (0.61%) and Christians (0.30%) and others (0.12%) as per 2022 census. Hinduism, Christianity and Buddhism is other three great regions here. All the religions have their separate religious institutions, rituals and festivals. They have their own religious beliefs, but they respect each other and all rituals of other religions. It makes Bangladesh a unique place compared to others.

Islam having the most pervading and lasting impact. Bangladesh belongs to many religious areas. Most of the Islamic establishments are mosques. A local news media *bdnews24.com* published on their 1 March 2011 that 2,50,399 Mosques are in Bangladesh. Temple is the building or construction for Hindu religious practices and devotion. According to David McCutchion ((12 August 1930 – 12 January 1972), there are few types of temples in Bengal, traditional style and this style is divided into two types: Rekha and Pidha deul. Wikipedia says that, there are 144 prominent Hindu temples and 6 Shakti Piths in Bangladesh. Buddhism is the third-largest religious affiliation in Bangladesh. In 11th century, Buddhism started here from Noagaon district, the northern part of Bangladesh. There are few Buddhist temple, vihara or monastery, stupa, wat, gompa, kyaung and pagoda. Few Catholic and Protestine both live in Bangladesh having some Churches and 16 Christian theological institutions.

DEVELOPMENT PATTERN OF RELIGIOUS TOURISM

Generally, concept of religious tourism was introduced with the essence of the complex relationship between pilgrimage and tourism. Religious tourism includes a range of spiritual sites and associated services, which are visited for both secular and religious reasons (Raj *et. al.,* 2015). Religious tourism is being analyzed as an economic leverage and providing a major contributor to the society. Because, religion and tourism are closely related to the social process and the relationship among religion, tourism, and economics. So, religion has to develop a marketable commodity keeping its all gravities and meanings. In the present consumer society, pilgrimage occupies the space with different spatial behaviors.

Development of religious tourism is considered as the development of alternative or specific form of tourism, due to its socio-cultural impacts and contribution to the local community. According to Smith (1992), the pilgrim-tourist path should ultimately be redefined as two parallel, interchangeable lanes, one of which is the secular knowledge-based route of western science; and other is the sacred road of faith and belief. Then, every religious tourist travels worldwide for personal needs or motivation, and as appropriate to time, place, and cultural circumstances. Smith also concludes that tourists and pilgrims seek the same leisure, income, and sanction, as well as, they share the same infrastructure and other resources. Iliev (2020) shows that the previous development of the religious tourism has been characterized by a partial overlapping between elements of mass and alternative tourism. But the rapid growth of religious tourism on the international market, changes the requirements of future tourists.

The multi-religious and multi-ethnic diversity may differentiate the community interests and significant benefit of locals the business. This situation amplifies the awareness and tolerance of eliminating radicalization. The growth of the process of religious tourism is to know and respect the host communities of tourist area for the growth of pluralism. This chapter indicates the ethnic and religious tourism that needs to develop the community intolerance. Human experience is different because it is social diversity that makes life harmonious.

In the recent years, the scholars have begun to think about other forms of religious tourists, such as spiritual tourists (Stausberg, 2014), 'New Age' spiritual travelers (Redden, 2005; Timothy & Conover, 2006), 'cyberpilgrims' (Digance, 2006). Digance noted that 'cyberpilgrims may routinely undertake for certain rituals (e.g. making a cup of coffee, lighting a candle or incense, and logging on only at a certain time and issues.

CULTURAL CONSERVATION THROUGH RELIGIOUS TOURISM

It is known to all that the religion is the set of faiths, attitudes, beliefs, and practices. Worship of the supernatural power, commitment and devotion to religious faith is the observance of religion. But the Sociological approach of religion belongs to the beliefs, values, and practices related to sacred or spiritual concerns. Tourism can unify the belief system and practices to the sacred things. Resources in the globe is limited. So, tourism activities conserve the resources by the all religious faithful people. People will practice the spiritual-religious tourism which acts as the ambassador of social peace and harmony. This approach may save the mankind and civilization. Conceptual proposition of religious tourism along with in-depth discussions will make the cultural conserved the society.

Religious faith and beliefs play a great role in religious tourism. On the one hand, tourists visit religious places with holy perception. Santos (2000) says, this type of tourism is a special display of cultural tourism followed with a promotion of different cultural products. It also leads to a richness in thought and spirit. This is also an important part of cultural tourism lies around holy places.

Robertson and Giddens often emphasize the cultural dimension of globalization. According to Moaven (2017), religious tourism is the part of process of globalization requires a deeper understanding of the concept of tourists. Furthermore, he says that religious tourism acts as a unique type of cognitive approach that satisfies passengers and gives them opportunity to go through the process of religious cults, rituals and rites, observe and live with them to obtain religious property or purchase souvenirs.

CONCLUSION

Religions have a distinct and receptive role to tourism with multiple motivations for traditions and secularity. The consumer shows the need for culture and accumulating new experiences. Furthermore, the cultural needs are interweaved with the needs for entertainment, and visiting religious sites is the source of knowledge and active relaxation. Religions promote peace by using cultural factors in the society. It may supposed to unite the society in terms of cultural parts of religion as said to be secularism. Religious tolerance among the people are accepting the faith and belief of others. It makes the belief that all religions are equally valid within their own culture. Making religious areas into tourism destination may be an ultimate approach for the upcoming world. Riya Mishra (2019) wrote her blog that it will reduce the religious intolerance, establish human rights and extend religious freedom to people of all faith and traditions without any discrimination. Religions are the big mix which can perform universally for economics, social harmony and a quality globe through tourism. It is the power house of socio-biological ecology that ultimately conserve the future mankind. Cultural conservation is a factor that intervene and effect the attitude towards religious tourism areas especially in the sacred places.

REFERENCES

Abbate, C. S., & Di Nuovo, S. (2013). Motivation and personality traits for choosing religious tourism. A research on the case of Medjugorje. *Current Issues in Tourism, 16*(5), 501–506. doi:10.1080/13683500.2012.749844

Abdullah, M. R. (2018).. . *Culture, Religion, and Freedom of Religion or Belief., 16*(4), 102–115. doi:10.1080/15570274.2018.1535033

Amaro, S., Antunes, A., & Henriques, C. (2018). A closer look at Santiago de Compostela's pilgrims through the lens of motivations. *Tourism Management, 64*, 271–280. doi:10.1016/j.tourman.2017.09.007

Andriotis, K. (2009). Sacred Site Experience: A Phenomenological Study. *Annals of Tourism Research, 36*(1), 64–84. doi:10.1016/j.annals.2008.10.003

Bakhshipour-Roudasri, A. (2008). Relationship between the life satisfaction, problem solving style and religious attitude with mental health. Journal of Psychology. *University of Tabriz, 9*, 8–26.

Beyers, J. (2017). Religion and culture: Revisiting a close relative'. *Hervormde Teologiese Studies, 73*(1), a3864. doi:10.4102/hts.v73i1.3864

Bond, N., Packer, J., & Ballantyne, R. (2015). Exploring visitor experiences, activities and benefits at three religious tourism sites. *International Journal of Tourism Research*, *17*(5), 471–481. doi:10.1002/jtr.2014

Brown, L. (2016). Tourism and pilgrimage: Paying homage to literary heroes. *International Journal of Tourism Research*, *18*(2), 167–175. doi:10.1002/jtr.2043

Digance, J. (2006). Religious and secular pilgrimages: Journeys redolent with meaning. In D. Timothy & D. H. Olsen (Eds.), *Tourism, religion and spiritual journeys* (pp. 36–48). Routledge.

Earp, B. D. (2015). On the Supposed Distinction between Culture and Religion: A Brief Comment on Sir James Munby's Decision in the Matter of B and G (Children). *University of Oxford Practical Ethics blog*. University of Oxford.

Geertz, C. (1973). *The interpretation of cultures: Selected essays*. Basic Books.

Griffiths, M. (2011). Those who come to pray and those who come to look: Interactions between visitors and congregations. *Journal of Heritage Tourism*, *6*(1), 63–72. doi:10.1080/1743873X.2010.536234

Hung, K., Yang, X., Wassler, P., Wang, D., Lin, P., & Liu, Z. (2017). Contesting the Commercialization and Sanctity of Religious Tourism in the Shaolin Monastery, China. *International Journal of Tourism Research*, *19*(2), 145–159. doi:10.1002/jtr.2093

Iliev, D. (2020). The evolution of religious tourism: Concept, segmentation and development of new identities. *Journal of Hospitality and Tourism Management*, *45*, 131–140. doi:10.1016/j.jhtm.2020.07.012

Kim, B., Kim, S. S., & King, B. (2016). The sacred and the profane: Identifying pilgrim traveler value orientations using means-end theory. *Tourism Management*, *56*, 142–155. doi:10.1016/j.tourman.2016.04.003

MacCannell, D. (1973). Staged authenticity: Arrangements of social space in tourist settings. *American Journal of Sociology*, *793*(3), 589–603. doi:10.1086/225585

Moaven, Z. (2017). Globalization of Culture and Religious Tourism. *European Online Journal of Natural and Social Sciences*, *6*(1). www.european-science.com

Mulder, D. C. (1985). Het vak godsdienstwetenskap. In D. J. Hoens, J. H. Kamstra, & D. C. Mulder (Eds.), *Inleiding tot de studie van godsdiensten* (pp. 35–40). J.H. Kok, Kampen.

Nilsson, M., & Tesfahuney, M. (2016). Performing the "post-secular" in Santiago de Compostela. *Annals of Tourism Research*, *57*, 18–30. doi:10.1016/j.annals.2015.11.001

Nyaupane, G. P., Timothy, D. J., & Poudel, S. (2015). Understanding tourists in religious destinations: A social distance perspective. *Tourism Management*, *48*, 343–353. doi:10.1016/j.tourman.2014.12.009

Okulicz-Kozaryn, A. (2009). *Religiosity and life satisfaction (A multilevel investigation across nations)*. Institute for Quantitative Social Science, Harvard University.

Olsen, D. H., & Timothy, D. J. (2006). *Tourism and religious journeys in Tourism, Religion & Spiritual Journeys*. Routledge.

Raj, R., Griffin, K., & Blackwell, R. (2015). Motivations for religious tourism, pilgrimage, festivals and events. In R. Raj & K. Griffin (Eds.), *Religious tourism and pilgrimage D. Iliev Journal of Hospitality and Tourism Management 45 (2020) 131–140 management: An international perspective* (pp. 103–117). CAB International. doi:10.1079/9781780645230.0103

Rappaport, R. A. (1999). *Ritual and religion in the making of humanity*. Cambridge University Press. doi:10.1017/CBO9780511814686

Redden, G. (2005). The New Age: Towards a market model. *Journal of Contemporary Religion, 20*(2), 231–246. doi:10.1080/13537900500067851

Rivera, M. A., Shani, A., & Severt, D. (2009). Perceptions of service attributes in a religious theme site: An importance satisfaction analysis. *Journal of Heritage Tourism, 4*(3), 227–243. doi:10.1080/17438730902822939

Ron, A. (2007). Book Review: Tourism, Religion & Spiritual Journeys. *Annals of Tourism Research, 34*(2), 547–548. doi:10.1016/j.annals.2006.10.007

Santos, M. (2000). Fátima: religious tourism in a sanctuarycity. *World Leisure Congress*. Bilbau.

Seybold, K. S., & Hill, P. C. (2001). The role of religion and spirituality in mental and physical health. *Journal of Current Directions in Psychological Science, 10*(1), 21–24. doi:10.1111/1467-8721.00106

Slater, T. R. (2004). Encountering God: Personal reflections on 'geographer as pilgrim'. *Area, 36*(3), 245–253. doi:10.1111/j.0004-0894.2004.00221.x

Smith, V. L. (1992). Introduction: The quest in guest. *Annals of Tourism Research, 19*(1), 1–17. doi:10.1016/0160-7383(92)90103-V

Stausberg, M. (2014). Religion and spirituality in tourism. In A. A. Lew, C. M. Hall, & A. M. Williams (Eds.), *The Wiley Blackwell companion to tourism* (pp. 349–360). John Wiley & Sons, Ltd. doi:10.1002/9781118474648.ch28

Stoddard, R. H. (1997). Defining and classifying pilgrimages. In R. H. Stoddard & A. Morinis (Eds.), *Sacred places, sacred spaces: The geography of pilgrimages* (pp. 41–60). Louisiana State University: Geoscience Publications.

Suntikul, W., & Dorji, U. (2016). Local Perspectives on the Impact of Tourism on Religious Festivals in Bhutan. *Asia Pacific Journal of Tourism Research, 21*(7), 741–762. doi:10.1080/10941665.2015.1069365

Timothy, D. J., & Conover, P. J. (2006). Nature religion, self-spirituality and New Age tourism. In D. J. Timothy & D. H. Olsen (Eds.), *Tourism, religion and spiritual journeys* (pp. 139–155). Routledge. doi:10.4324/9780203001073-21

Timothy, D. J., & Olsen, D. H. (2006). *Tourism, Religion & Spiritual Journeys*. Routledge. doi:10.4324/9780203001073

Tuan, Y. F. (1978). Sacred Space: explorations of an idea. In K. W. Butzer (Ed.), *Dimensions of Human Geography: Essays on some familiar and neglected themes* (pp. 84–99). University of Chicago Press.

Turner, V., & Turner, E. (1969). *The ritual process*. Routledge.

UNWTO. (2017). *International Congress on Religious Tourism and Pilgrimage*. UNESCO. http://europe.unwto.org/event/international-congress-religious-tourism-and-pilgrimage

Vukonic, B. (2002). Religion, tourism and economics: A convenient symbiosis. *Tourism Recreation Research*, *27*(2), 59–64. doi:10.1080/02508281.2002.11081221

Yekinni, O. B., & Mercy, B. B. (2017). Exploring the Relationship between Religion Tourism and Economic Development of a Host Community. *Int. J. Bus Manag Invent*, *6*(9), 41–51.

Zouni, G., & Digkas, D. (2019). Marketing suggestions for multi-religious tourism development: The case of Thessaloniki. *J Tour Herit Serv Mark.*, *5*(2), 36–42.

Chapter 8
Harmony and Holiness:
Navigating the Challenges of Religious Tourism

Bhavna Taneja
https://orcid.org/0000-0002-5447-7758
Amity University, Ranchi, India

ABSTRACT

Religious tourism is a rapidly growing global phenomenon that involves the travel of individuals or groups to sacred destinations for spiritual or religious purposes. This research paper aims to explore the complex interplay between harmony and holiness in religious tourism contexts. It explores the various issues and challenges associated with religious tourism. It examines the socio-economic impacts, environmental concerns, cultural clashes, and management challenges that arise from the influx of pilgrims and tourists visiting sacred sites. By analyzing case studies and examining best practices, this paper seeks to propose sustainable strategies to navigate the challenges of religious tourism while preserving the sanctity and cultural integrity of sacred sites. Promoting harmony and holiness in religious tourism requires a collective effort from all stakeholders. Responsible tourism practices that prioritize sustainability, cultural preservation, and interfaith dialogue are essential for ensuring a positive impact on both visitors and local communities.

INTRODUCTION

Pilgrimages to holy sites are considered the oldest form of tourism (Saidovna, 2022). Religious tourism is a dynamic and burgeoning segment of the global tourism industry (Swarbrooke, 2020; Olsen, 2019), serving as a bridge between diverse cultures (Timothy, 2019), fostering economic development, and providing spiritual enrichment (Shinde, & Olsen, 2020) to travellers. It encompasses pilgrimages to sacred sites, visits to places of worship, and participation in religious festivals, becoming integral aspects of many people's travel experiences.

Religious heritage sites, often characterized by their historical and spiritual significance, hold a unique place in the cultural and social structure of communities worldwide. Bhardwaj (2021) says that these

DOI: 10.4018/978-1-6684-9957-3.ch008

sites encompass a wide range of religious structures, artifacts, and landscapes, from ancient temples, churches, and mosques to sacred groves, pilgrimage routes, and natural formations. The importance of religious heritage extends beyond religious communities; it influences tourism, conservation, and community dynamics.

A growing market in the tourism industry that offers major economic benefits is religious tourism according to Kim et al., (2020). It involves traveling to spiritual locations that meet both the tourists' spiritual expectations and the business objectives of those locations. Tsironis (2022) research shows that pilgrimage and religious tourism maintain a resilient dynamic and inspire people from diverse backgrounds to visit religious destinations year-round not just during holidays. The globalization of religious events, from Varanasi to Mecca and from Jerusalem to Rome, highlights the public dimension of religious practices. Religious tourism is among the fastest-growing forms of global tourism, influenced by various factors such as shrine promotion, self-discovery quests, revitalization of pilgrimage routes, and infrastructure development in pilgrimage centres.

This section provides an overview of the key dimensions of religious heritage.

IMPACT OF RELIGIOUS HERITAGE

Religious heritage sites have a profound impact on both the spiritual and socio-economic dimensions of communities. They serve as centres of worship, pilgrimage, and cultural expression. Furthermore, they often attract tourists and pilgrims, contributing significantly to the local and regional economy (Ghimire et al. 2022). However, this dual role can sometimes lead to challenges, as the influx of visitors can strain local resources, disrupt daily life, and even dilute the spiritual essence of these sites (Cole et al., 2020). Understanding and managing these impacts is crucial for preserving the sanctity of religious heritage.

COMMUNITY PERCEPTIONS

Local communities play a vital role in preserving and promoting religious heritage. Moshin et al., (2020) say that the Local community's perceptions and attitudes towards these sites are integral to their long-term sustainability. Communities often have a deep emotional and cultural connection to these sites, and their perceptions can range from viewing them as sources of pride and identity to potential sources of conflict due to the presence of tourists or differing interpretations of heritage. Managing community perceptions and involving them in decision-making processes are key factors in ensuring the harmonious coexistence of religious heritage and tourism according to Rastegar et al., (2021).

ENVIRONMENTAL CONCERNS

Many religious heritage sites are in ecologically sensitive areas, such as forests, mountains, or coastal regions. The environmental impact of religious tourism can be significant. Increased foot traffic, construction, and waste generation can harm fragile ecosystems, lead to deforestation, and cause pollution according to Gawlik et al., (2022). Shinde, (2021) says that preserving the natural environment surrounding these sites is essential not only for ecological reasons but also to maintain the spiritual and aesthetic qualities that often draw visitors in the first place. Balancing conservation and tourism are a critical challenge according to Mandić, (2020).

Incorporating these aspects into discussions and management strategies for religious heritage is crucial for ensuring their preservation and sustainable development. This comprehensive understanding acknowledges that religious heritage is not just a matter of history and culture but an intricate interplay of spiritual significance, community dynamics, and environmental responsibility.

Pilgrimage extends beyond the realm of religious experiences and carries social, political, and economic implications. The rapid expansion of religious tourism has brought to the forefront a host of issues and challenges that demand careful consideration and responsible management. Various forms of pilgrimage, from spiritual tourism to religious protest, shape the public dimension of pilgrimage and influence societies in diverse ways.

OBJECTIVE AND METHODOLOGY

This research paper aims to delve into the multifaceted realm of religious tourism, shedding light on the challenges faced by sanctified destinations while seeking solutions for sustainable and harmonious religious tourism practices. It emphasizes the increasing significance of religious tourism, its associated complexities, and the importance of responsible management practices. As the industry evolves, stakeholders must proactively address socio-economic, environmental, cultural concerns and management challenges.

The research methodology for this study involves case studies from different religious tourism destinations. The case study method is justified when research questions demand a detailed, context-specific, and in-depth analysis. It is a flexible and versatile approach that can be adapted to various research contexts and offers valuable insights into complex and real-world phenomena. The Researcher has chosen this method because it aligns with her research objectives and the nature of the subject being studied.

These case studies will be analyzed to understand the challenges faced and successful practices employed to maintain harmony and holiness. A diverse range of religious tourism destinations, including sacred cities, temples, churches, mosques, and other places of religious significance, are selected based on their significance, popularity, and geographical spread.

LITERATURE REVIEW

Religious tourism has garnered substantial interest from scholars, researchers, and practitioners owing to its expanding significance in contemporary society. This literature review seeks to explore the extensive body of research concerning religious tourism, encompassing diverse facets such as its economic implications, typologies, and the profound experiences it offers to travellers.

Economic Significance of Religious Tourism

The sheer magnitude of religious tourism is highlighted by staggering statistics. Over 300 million individuals embark on journeys to religious tourist destinations annually, underlining the scale of this phenomenon (Gabor, 2016). Moreover, this sector is estimated to contribute a substantial $20 billion annually to the global economy (Askin, 2017). These figures underscore the economic relevance of religious tourism, emphasizing its potential for local economic development and sustainability.

Interdisciplinary Research Interest

Scholars from diverse academic domains have shown a intense interest in exploring the unique characteristics of religious travel. Research emanates from fields such as religious studies, psychology, leisure studies, sociology, and tourism (Argyle, 2002; Harris, 2004). This interdisciplinary engagement underscores the multifaceted nature of religious tourism and its relevance in the contemporary context.

Classification and Typologies

Researchers have grappled with the challenge of distinguishing between pilgrimages and tourist-related activities within the realm of religious travel. Classification and typology concerns have received considerable attention (Oktadiana, Pearce, & Chon, 2016; Liutikas, 2017); Wong, Ryan, & McIntosh, 2013). These initiatives seek to offer a precise framework for comprehending the various motives and experiences of tourists who partake in religious tourism.

The Transformative Nature of Pilgrimage

Collins-Kreiner (2019) describes modern pilgrimage as a form of mobility driven by the quest for meaning, often characterized by deep and enduring transformation. This perspective highlights the profound impact that religious travel can have on individuals, extending beyond mere tourism to become a life-altering journey of self-discovery.

Spiritual Experiences and Travel

The intertwining of travel and spirituality has historical roots (Wang et al., 2021; Zargham Boroujeny, 2017; Zahra, 2016). This dimension of religious tourism transcends traditional paradigms, encompassing various forms of spiritual experiences that individuals seek through travel (Cheer et al., 2017). A common thread among pilgrims, regardless of their specific classification, is the pursuit of mystical or miraculous religious experiences. These experiences often transcend reason and can be transformative, enlightening, and life-changing, leading to altered states of consciousness (Kim and Kim, 2018; Liutikas, 2015).

Blurred Identities and Converging Phenomena

With terms like secular pilgrimage, spiritual tourism, religious tourism, church tourism, dark tourism, and transformational tourism emerging, the modern landscape of religious tourism presents muddled identities (Collins-Kreiner, 2016). Distinguishing between these categories has become increasingly challenging as they all contribute to the broader phenomenon of tourists seeking meaningful experiences (Amaro et al., 2018; Terzidou et al., 2017; Collins-Kreiner, 2016).

This literature review provides an overview of the extensive research landscape surrounding religious tourism, offering insights into its economic significance, interdisciplinary nature, classification challenges, and the profound experiences it offers. It lays the foundation for further exploration of themes related to religious tourism, local communities, cultural heritage, interfaith relations, harmony, and holiness, underscoring the complexities and opportunities within this growing field of study.

Impact on Local Communities

The impact of religious tourism on local communities is a complex issue with both positive and negative aspects. Some studies (e.g., Nunkoo & Ramkissoon, 2012) emphasize the economic benefits that religious tourism can bring to local communities, including job creation and increased revenue for local businesses. Additionally, religious tourism may lead to the development of infrastructure and amenities, improving the overall quality of life for residents. Negative impacts can range from rising property prices and displacement of residents, altering the social fabric of the community.

Impact on Cultural Heritage

Religious tourism can have both positive and negative effects on cultural heritage. On one hand, it can contribute to the preservation and promotion of cultural heritage (Hede & Kellett, 2017). Tourists interest in sacred sites can lead to increased funding for conservation efforts and raise awareness of the historical and cultural significance of the sites. One fundamental aspect of religious tourism is the encounter between the sacred and the secular (Cohen, 1982). Pilgrimage, a central aspect of religious tourism, involves the journey to a sacred destination to seek spiritual transformation. Pilgrims often perceive these sites as places of holiness and power, where they can connect with the divine. Tourists often seek spiritual experiences in sacred sites, while the host communities must strike a balance between religious practices and accommodating visitors' needs.

However, other researcher like Hall, (2008) highlights potential negative impacts, such as the commodification and restoration of sacred spaces. Concerns about cultural appropriation and commercialization have been raised (Sofield & Sivan, 2003). The interaction between pilgrims and tourists can lead to a tension between them for holiness and the commercialization of pilgrimage sites (Cole, 2012). Mass tourism, driven by commercial interests, can lead to over-tourism and the degradation of sacred spaces (Farsari, 2023). The influx of tourists can disrupt the sacredness of sites and undermine the spiritual experiences of pilgrims. Additionally, the commodification of religious artifacts and practices can raise ethical concerns and jeopardize the authenticity of sacred rituals (Bowman, 2010; Oslen 2014).

Impact on Interfaith Relations

Interfaith harmony is an essential goal of religious tourism, as tourists of different faiths come together to explore sacred sites (Crick, 2004). Religious tourism has the potential to foster interfaith understanding and harmony (Timothy & Boyd, 2003). Visiting sacred sites of different religions can promote dialogue and mutual respect among pilgrims of diverse faiths. Many studies have highlighted the role of religious tourism in breaking down stereotypes and promoting tolerance and cooperation. The encounter between religious traditions can lead to mutual understanding, empathy, and respect. Studies have shown that religious tourism experiences can play a significant role in promoting interfaith dialogue and harmony (McCool & Martin, 1994).

However, tensions can arise when religious tourism encounters sensitive political and religious contexts (Richards, 2011). In some cases, religious tourism may exacerbate existing interfaith conflicts, particularly in regions with historical religious rivalries or contested religious sites.

Role of Religious Authorities

Religious authorities play a critical role in maintaining harmony and holiness in religious tourism destinations (Johnston & Weller, 2022). They may set rules and guidelines to protect sacred spaces and regulate visitor behaviour. Striking a balance between accommodating tourists and preserving the purity of the sites is often a delicate task for religious leaders.

Sustainable Tourism Practices

The promotion of sustainable tourism practices is vital to ensuring harmony and holiness in religious tourism (Aref & Redzuan, 2009). Integrating the principles of responsible tourism can help minimize the negative impacts on sacred sites and local communities while enhancing the overall visitor experience.

The application of artificial intelligence in the tourist and hospitality business is emphasized by authors like Rawal et al. (2022) in order to solve obstacles, provide quick service delivery, improve the guest experience, and maintain competition.

The existing literature on religious tourism reveals a diverse range of perspectives regarding its impact on local communities, cultural heritage, and interfaith relations. While there are potential benefits in terms of economic development, cultural preservation, and interfaith understanding, there are also concerns about over-tourism, cultural appropriation, and potential conflicts. Responsible and sustainable tourism management and the use of AI is essential to maximize the positive impacts of religious tourism while minimizing the negative effects on the destinations and communities involved.

CASE STUDIES

Several case studies from different religious tourism destinations around the world will be presented to illustrate the challenges and solutions discussed in earlier sections. Selecting cities like Varanasi, Jerusalem, Mecca, and Amritsar as case studies for this paper can be justified for several reasons, each tied to the unique characteristics and significance of these cities.

Case Study One: Varanasi, Uttar Pradesh—The City of Ghats

Varanasi, known as the spiritual capital of India, is famous for its ghats on the Ganga River, where pilgrims perform religious rituals. It is one of world's oldest inhabited cities and a prominent Hindu and Buddhist pilgrimage destination. Varanasi is a multifaceted city with a deep-rooted religious and cultural heritage. Its shrines and religious destinations offer a diverse and enriching experience for pilgrims and visitors alike, making it a fascinating case study in the realm of spirituality and religious tourism. The city attracts millions of domestic and international tourists seeking spiritual experiences and cultural immersion.

Here are some key points on the importance of Varanasi.

Kashi Vishwanath Temple: This is one of the most famous and revered temples in Varanasi. It is dedicated to Lord Shiva and is believed to be one of the twelve Jyotirlingas (sacred representations of Lord Shiva) in India. The temple is an important pilgrimage site for Hindus, and thousands of devotees visit it daily.

River Ganges and the Ghats: The River Ganges is considered sacred in Hinduism, and its waters are believed to have purifying properties. Varanasi is known for its numerous ghats along the banks of the river Ganges. These ghats are used for various religious activities, including bathing, cremation, and rituals. Dashashwamedh Ghat and Manikarnika Ghat are among the most prominent and frequented ones.

Sarnath: Located just a few kilometres from Varanasi, Sarnath is an important Buddhist pilgrimage site. It is where Lord Buddha is said to have delivered his first sermon after attaining enlightenment. The Dhamek Stupa and the Sarnath Museum are notable attractions here.

Tulsi Manas Temple: This temple is dedicated to Lord Rama and is known for its exquisite white marble architecture. It is not only a place of worship but also a cultural and historical attraction.

Annapurna Temple: This temple is dedicated to Goddess Annapurna, the deity of food. It is believed that those who worship here will never go hungry. The temple's architecture and religious significance make it a popular stop for pilgrims.

Kaal Bhairav Temple: This temple is dedicated to Lord Bhairav, a fierce manifestation of Lord Shiva. It is known for its unique rituals, and devotees offer alcohol to the deity as part of their worship.

Tulsi Ghat: Named after the famous poet-saint Tulsidas, this ghat has a special significance for followers of the Ramcharitmanas, an epic poem narrating the life of Lord Rama.

Jain Temples: Varanasi is also home to several Jain temples, including the Parshvanath Temple and the Suparshvanath Temple. These temples are significant pilgrimage sites for Jains.

Music and Culture: Varanasi is a hub for classical music, particularly Hindustani classical music. It has produced many renowned musicians and continues to be a center for musical performances, especially during festivals and religious events.

Spiritual Tourism: Beyond its religious significance, Varanasi is a hub for spiritual tourism. Many seekers and practitioners of yoga, meditation, and Ayurveda visit the city to deepen their spiritual knowledge and practice.

Challenges

Pollution and Sanitation: The city faces challenges of waste management and pollution due to the large number of visitors and religious activities. The influx of tourists can strain the natural environment, particularly the Ganga River, leading to pollution and degradation of the sacred river.

Cultural Preservation: Balancing the preservation of ancient cultural practices with the demands of modern tourism can be complex.

Over tourism: Varanasi faces challenges of over tourism, especially during major religious festivals, which strains the cities infrastructure and threatens the sanctity of its sacred sites.

Cultural Appropriation: The influx of tourists has led to the commercialization of religious practices, raising concerns about cultural appropriation and commodification of sacred rituals.

Solutions

Clean Ganga Initiative: The Indian government has launched the Clean Ganga Initiative, focusing on cleaning, and conserving the Ganges River, which is vital to Varanasi's religious significance. Local authorities and NGOs have initiated campaigns to clean the river Ganga and promote eco-friendly tourism practices, including waste management and plastic reduction.

Heritage Conservation: Local organizations work to preserve the city's cultural heritage and promote sustainable tourism practices that respect the sanctity of the ghats.

Sustainable Tourism Initiatives: Local authorities have implemented sustainable tourism practices, such as regulating the number of boats on the Ganges and promoting eco-friendly alternatives for pilgrims and tourists.

Interfaith Dialogue: Varanasi has become a melting pot of different religious traditions. Interfaith dialogues and cultural exchange programs have been organized to foster mutual understanding and harmony.

Case Study Two: Jerusalem, Israel/Palestine

Jerusalem holds immense religious significance for three major Abrahamic religions: Judaism, Christianity, and Islam. It is often referred to as the "Holy City" due to its association with important religious events and figures. The city is deeply intertwined with the histories and beliefs of these faiths, making it a complex and multifaceted religious destination. Here are some key points to consider on the importance of Jerusalem-

Religious Diversity: There are many holy landmarks in Jerusalem, including the Western Wall. For Jews, the Church of the Holy Sepulchre, for Christians, the Wailing Wall; for Muslims, the Dome of the Rock and the Al-Aqsa Mosque. Visitors and pilgrims from all over the world flock to these locations.

For Jews, the Church of the Holy Sepulchre, for Christians, the Wailing Wall; for Muslims, the Dome of the Rock and the Al-Aqsa Mosque. Visitors and pilgrims from all over the world flock to these locations.Historical Significance: The city's history spans thousands of years, with mentions in ancient texts and scriptures. It is believed to be the place where King Solomon built the First Temple and where Jesus was crucified and resurrected.

Religious Tensions: Jerusalem has been a focal point of religious and political tensions between Israelis and Palestinians due to conflicting claims and aspirations. The city's status is a contentious issue in the Israeli-Palestinian conflict.

Cultural Heritage: Beyond its religious significance, Jerusalem is a treasure trove of cultural heritage. Its Old City is a UNESCO World Heritage Site, featuring narrow winding streets, historic architecture, and vibrant marketplaces.

Challenges:

Conflicting Claims: The competing religious and political claims over Jerusalem have resulted in tensions and conflicts between religious groups, impacting religious tourism experiences.

Cultural Heritage Conservation: The preservation of cultural heritage sites faces challenges due to political and religious sensitivities.

Cultural Sensitivity: Maintaining cultural authenticity and respect for local traditions can be challenging amidst commercialization.

Solutions:

Interfaith Tourism Initiatives: Several organizations promote interfaith tourism programs, encouraging visitors to engage in respectful dialogue and understand the various perspectives on the city's significance.

Heritage Preservation Projects: Collaborative efforts between religious authorities, heritage organizations, and governments work to preserve and protect the city's religious and cultural heritage.

Responsible Tourism Guidelines: The local government and community organizations collaborate to develop responsible tourism guidelines, emphasizing respect for local customs and traditions.

Case Study Three: Mecca, Saudi Arabia

Mecca, the holiest city in Islam, is significant to Muslims all around the world. It is the location of the Prophet Muhammad's birthplace and the Masjid al-Haram, or "Grand Mosque," which houses the Kaaba, Islam's holiest shrine. Millions of Muslims from all over the world go to Mecca for the annual Hajj pilgrimage, which is one of Islam's Five Pillars. Here are some key points to consider on the importance of Mecca:

The Kaaba: The Kaaba is a cube-shaped structure located within the Masjid al-Haram. Muslims believe it was built by the Prophet Ibrahim (Abraham) and is the House of Allah. Pilgrims circumambulate the Kaaba during the Hajj and Umrah pilgrimages.

Hajj and Umrah: Mecca is the focal point for the annual Hajj pilgrimage, which is obligatory for every physically and financially capable Muslim to undertake at least once in their lifetime. Additionally, Muslims perform Umrah, a lesser pilgrimage, throughout the year.

Religious Significance: Mecca is not only the location of the Kaaba but also the place where the Quran, the holy book of Islam, was first revealed to the Prophet Muhammad. This city is deeply significant in Islamic history and theology.

Economic and Cultural Impact: The influx of pilgrims to Mecca has a profound impact on the city's economy and culture. The hospitality industry, transportation, and infrastructure development are crucial for accommodating millions of visitors annually.

Challenges:

Mass Gathering Management: Ensuring the safety and well-being of the large number of pilgrims during the Hajj poses significant logistical challenges.

Commercialization: The rapid expansion of hotels and commercial activities around Mecca has raised concerns about the commercialization of the pilgrimage experience.

Solutions:

Improved Infrastructure: Saudi Arabian authorities have invested in improving transportation, accommodation, and medical facilities to better manage the annual influx of pilgrims.

Sustainable Development: The Saudi government has initiated projects to develop Mecca sustainably, focusing on preserving the city's heritage and environmental resources.

Case Study Four: Amritsar, Punjab—The Golden Temple

The Golden Temple, known as Harmandir Sahib in Punjabi, is one of the most revered and iconic shrines in Sikhism and a major religious destination in India. Located in Amritsar, Punjab, this shrine holds immense spiritual and cultural significance for Sikhs and is also a symbol of India's rich religious diversity. Here are some key points to consider on the importance of Amritsar:

Religious Significance: The Golden Temple is the holiest shrine in Sikhism, housing the Guru Granth Sahib, the central religious scripture of Sikhism. It is a place of worship and reflection for Sikhs around the world.

Historical Importance: The foundation of the Golden Temple was laid by Guru Ram Das, the fourth Guru of Sikhism, in the 16th century. Over time, it has evolved into a symbol of Sikh resilience and spirituality.

Architectural Marvel: The Golden Temple is renowned for its stunning architecture. Its distinctive gold-plated dome and marble structure, surrounded by a sacred pool known as the Amrit Sarovar, make it a visually striking religious site.

Langar Tradition: The Golden Temple is famous for its langar, a community kitchen that serves free meals to thousands of visitors daily, regardless of their caste, creed, or social status. This practice embodies the Sikh principles of equality and selfless service.

Tourist Attraction: Beyond its religious significance, the Golden Temple attracts tourists from all over the world who are captivated by its beauty, spirituality, and the unique experience it offers.

Challenges

Managing crowds: The Golden Temple sees a massive influx of devotees and tourists, leading to overcrowding especially during religious festivals.

Preservation of Sanctity: the balance between accommodating tourists and preserving the sacredness of the site can be delicate.

Solutions

Crowd Management: The management committee of the Golden Temple has implemented measures like timed entry slots, separate queues for devotees and tourists, and crowd control personnel to ensure a smooth experience for all visitors.

Community Participation: Local Sikh community members actively engage in the management and maintenance of the temple, reinforcing a sense of ownership and responsibility for the site's sanctity.

These case studies illustrate the complexities and challenges of religious tourism in diverse destinations. While there are common issues such as over-tourism, cultural appropriation, and conflicts, each location requires context-specific solutions that consider the religious, cultural, and environmental aspects of each destination. By learning from these experiences, religious tourism stakeholders can create sustainable and enriching experiences for pilgrims and tourists, while safeguarding the sanctity and cultural heritage of these significant religious destinations and the well-being of local communities.

KEY FINDINGS

Balancing the Sacred and the Secular: Religious tourism involves a delicate balance between the sacred and the secular, where travellers seek spiritual experiences while local communities manage the influx of tourists. Maintaining the sanctity of sacred sites while accommodating visitors' needs is a central challenge.

Importance of Interfaith Dialogue: Religious tourism offers a unique opportunity for interfaith dialogue and understanding as visitors from diverse religious backgrounds come together to explore sacred sites. Interfaith initiatives can foster harmony and respect among different religious communities.

Challenges of Over-Tourism: Popular religious tourism destinations often face issues of over-tourism, leading to overcrowding, environmental degradation, and strain on local resources. This threatens the holiness of the sites and the well-being of the local communities.

Cultural Appropriation and Commercialization: The commodification of religious practices and artifacts can lead to cultural appropriation and compromise the authenticity of sacred rituals, undermining the sense of holiness for both pilgrims and tourists.

RECOMMENDATIONS FOR PROMOTING HARMONY AND HOLINESS IN RELIGIOUS TOURISM

This section emphasizes the significance of responsible tourism practices to ensure a positive impact on both visitors and local communities. Effective crowd management, community involvement, eco-friendly initiatives, Sustainable tourism practices, interfaith dialogue, and collaboration between stakeholders emerge as key strategies to maintain harmony and holiness in these sacred sites.

Implement Responsible Tourism Practices: Stakeholders involved in religious tourism, including governments, religious authorities, tour operators, and visitors, should prioritize responsible tourism practices. This includes promoting sustainable tourism, managing visitor numbers, and minimizing the environmental impact on sacred sites.

Preserve Cultural Heritage: Emphasize the preservation of the cultural heritage of religious destinations. Encourage local communities to actively participate in the management and preservation of sacred sites, fostering a sense of ownership and responsibility.

Foster Interfaith Understanding: Encourage interfaith dialogue and cultural exchange programs among tourists and local communities. This can lead to a deeper understanding of diverse religious traditions and promote mutual respect.

Develop Community-Based Tourism Initiatives: Support community-based tourism initiatives that empower residents economically and socially. This can help create a positive relationship between tourists and local communities, leading to more harmonious interactions.

Educate Tourists: Raise awareness among tourists about the significance of sacred sites, local customs, and traditions. Providing information on responsible behaviour and cultural sensitivity can enhance visitors' experiences while respecting the holiness of the destinations.

Engage Religious Authorities: Involve religious leaders in promoting responsible and sustainable tourism practices. Their influence can encourage visitors to be respectful and mindful of the sanctity of religious sites.

Collaborate with Governments and NGOs: Encourage collaboration between governments, non-governmental organizations, and religious institutions to develop and implement policies and initiatives that support harmony and holiness in religious tourism.

Promoting harmony and holiness in religious tourism requires a collective effort from all stakeholders. Responsible tourism practices that prioritize sustainability, cultural preservation, and interfaith dialogue are essential for ensuring a positive impact on both visitors and local communities. By embracing these recommendations, religious tourism destinations can maintain their sacredness, preserve their cultural heritage, and create enriching experiences for travellers while contributing positively to the well-being of host communities.

FUTURE DIRECTIONS FOR RESEARCH IN RELIGIOUS TOURISM

Technology and Religious Tourism: Advancements in technology have significantly impacted the travel industry, and religious tourism is no exception. Future research can explore the integration of technology in religious tourism experiences, such as virtual reality tours of sacred sites, and mobile apps providing real-time information to visitors. Additionally, the study can investigate the potential benefits and challenges of using technology to manage visitor flows, enhance accessibility for people with disabilities, and preserve the sanctity of religious sites in the digital age.

Limitations: The findings of specific case studies, have been generalised to other religious tourism destinations.

CONCLUSION

Religious tourism has experienced significant growth contributing to the economic development and cultural exchange. However, this growth has brought forth numerous challenges that necessitate attention and careful management. This research paper has explored the impact on local communities, environmental concerns, cultural clashes, and the management challenges associated with religious tourism.

To address these challenges, the implementation of sustainable and responsible tourism practices is crucial. Only then religious tourism can flourish, while respecting the sanctity of sacred sites and nurturing harmonious relationships between visitors and local communities.

Recommendations for promoting harmony and holiness in religious tourism include implementing responsible tourism practices, preserving cultural heritage, fostering interfaith understanding, developing community-based tourism initiatives, educating tourists, engaging religious authorities, and fostering collaboration between governments, NGOs and religious institutions.

REFERENCES

Amaro, S., Antunes, A., & Henriques, C. (2018). A closer look at Santiago de Compostela's pilgrims through the lens of motivations. *Tourism Management, 64*, 271-280..

Aref, F., & Redzuan, M. (2009). Religious Tourism and Sustainable Tourism Principles: A Case Study of Iran. Tourism Management, 30(3), 465–472.

AŞKIN, R. A. D. (2017). New Consumption Trends Among Muslims In Turkey: Understanding Muslims In Terms Of Modern Consumptions. *PROCEEDINGS BOOKS*, 299.

Bharadwaj, R. M. (2021). *Trees, Flowers and Waters: Nature and Environment in the Hindu Vrata Religious Tradition-AD. 400-1200*. Indraprastha Shodha Sandarsh.

Cheer, J. M., Belhassen, Y., & Kujawa, J. (2017). The search for spirituality in tourism: Toward a conceptual framework for spiritual tourism. *Tourism Management Perspectives*, *24*, 252–256. doi:10.1016/j.tmp.2017.07.018

Cole, S. (2012). A political ecology of water equity and tourism: A case study from Bali. *Annals of Tourism Research*, *39*(2), 1221–1241. doi:10.1016/j.annals.2012.01.003

Cole, S. K., Mullor, E. C., Ma, Y., & Sandang, Y. (2020). *Tourism, water, and gender—An international review of an unexplored nexus*. Wiley.

Collins-Kreiner, N. (2016a). Dark tourism as/is pilgrimage. *Current Issues in Tourism*, *19*(12), 1185–1189. doi:10.1080/13683500.2015.1078299

Collins-Kreiner, N. (2016b). The lifecycle of concepts: The case of 'Pilgrimage Tourism'. *Tourism Geographies*, *18*(3), 322–334. doi:10.1080/14616688.2016.1155077

Crick, H. Q. (2004). The impact of climate change on birds. *The Ibis*, *146*(s1), 48–56. doi:10.1111/j.1474-919X.2004.00327.x

Farsari, I. (2023). Exploring the nexus between sustainable tourism governance, resilience and complexity research. *Tourism Recreation Research*, *48*(3), 352-367.

Fedele, A. (2014). Energy and transformation in alternative pilgrimages to Catholic shrines: Deconstructing the tourist/pilgrim divide. *Journal of Tourism and Cultural Change*, *12*(2), 150–165. doi:10.1080/14766825.2014.915091

Gabor, D., Colombo, U., & King, A. S. (2016). *Beyond the age of waste: a report to the Club of Rome*. Elsevier.

Gawlik, A., Apollo, M., Andreychouk, V., & Wengel, Y. (2022). 11 Pilgrimage Tourism and Sacred Places in the Himalaya. *Tourism and Development in the Himalaya: Social, Environmental, and Economic Forces*, 53.

Ghimire, D. K., Gautam, P., Karki, S. K., Ghimire, J., & Takagi, I. (2022). Small business and livelihood: A study of Pashupatinath UNESCO heritage site of Nepal. *Sustainability (Basel)*, *15*(1), 612. doi:10.3390u15010612

Hall, C. M. (2008). *Tourism Planning: Policies, Processes, and Relationships*. Pearson Education.

Hede, A. M., & Kellett, P. (2017). Religious tourism and cultural sustainability. *Journal of Sustainable Tourism*, *25*(1), 1–16.

Johnston, J., Weiler, A., & Baumann, S. (2022). The cultural imaginary of ethical meat: A study of producer perceptions. *Journal of Rural Studies, 89*, 186–198. doi:10.1016/j.jrurstud.2021.11.021

Jørgensen, N. N. K. D., Eade, J., Ekeland, T. J., & Lorentzen, C. A. N. (2020). *The processes, effects and therapeutics of pilgrimage walking the St. Olav Way.*

Kim, B., & Kim, S. (2018). Hierarchical value map of religious tourists visiting the vatican city/rome. *Tourism Geographies.*

Kim, B., Kim, S., & King, B. (2020). Religious tourism studies: Evolution, progress, and future prospects. *Tourism Recreation Research, 45*(2), 185–203. doi:10.1080/02508281.2019.1664084

Liutikas, D. (2017). The manifestation of values and identity in travelling: The social engagement of pilgrimage. *Tourism Management Perspectives, 24*, 217–224. doi:10.1016/j.tmp.2017.07.014

Mandić, A. (2020). Structuring challenges of sustainable tourism development in protected natural areas with driving force–pressure–state–impact–response (DPSIR) framework. *Environment Systems & Decisions, 40*(4), 560–576. doi:10.100710669-020-09759-y

McCool, S. F., & Martin, S. R. (1994). Community attachment and attitudes toward tourism development. *Journal of Travel Research, 32*(3), 29–34. doi:10.1177/004728759403200305

Moshin, A., Brochado, A., & Rodrigues, H. (2020). Halal tourism is traveling fast: Community perceptions and implications. *Journal of Destination Marketing & Management, 18*, 100503. doi:10.1016/j.jdmm.2020.100503

Nilsson, M., & Tesfahuney, M. (2018). The post-secular tourist: Re-thinking pilgrimage tourism. *Tourist Studies, 18*(2), 159–176. doi:10.1177/1468797617723467

Nunkoo, R., & Ramkissoon, H. (2012). Power, trust, social exchange and community support. *Annals of Tourism Research, 39*(2), 997–1023. doi:10.1016/j.annals.2011.11.017

Oktadiana, H., Pearce, P. L., & Chon, K. (2016). Muslim travellers' needs: What don't we know? *Tourism Management Perspectives, 20*, 124–130. doi:10.1016/j.tmp.2016.08.004

Olsen, D. H. (2014). A scalar comparison of motivations and expectations of experience within the religious tourism market. *International Journal of Religious Tourism and Pilgrimage.*

Olsen, D. H. (2019). 23. Religion, spirituality, and pilgrimage in a globalising world. Handbook of globalisation and tourism.

Plate, S. B. (2009). Introduction: The varieties of contemporary pilgrimage. *Cross Currents, 59*(3), 260–267. doi:10.1353/cro.2009.a782452

Rastegar, R., Zarezadeh, Z. Z., & Gretzel, U. (2021). World heritage and social justice: Insights from the inscription of Yazd, Iran. In *Justice and Tourism.* Routledge. doi:10.4324/9781003143055-22

Rawal, Y. S., Soni, H., Dani, R., & Bagchi, P. (2022).A Review on Service Delivery in Tourism and Hospitality Industry Through Artificial Intelligence. Role of Religious Authorities in Sustainable Tourism Practices. *In Proceedings of Third International Conference on Computing, Communications, and Cyber-Security.* Research Gate.

Richards, G. (2011). Religious tourism and social change: Expressions of faith in contemporary tourism. *Tourism Recreation Research.*

Saidovna, V. F. (2022). Pilgrimage Tour in Tourism. *Miasto Przyszłości, 28*, 342–345.

Sharpley, R. (2016). Tourism and Spirituality: An Evolving Relationship. *International journal of Tourism & Spirituality.*

Shinde, K. A. (2021). The environmental impacts of religious and spiritual tourism. The Routledge handbook of religious and spiritual tourism.

Shinde, K. A., & Olsen, D. H. (2020). The environmental impacts of religious tourism. *Religious tourism and the environment.*

Sofield, T., & Sivan, A. (2003). From cultural tourism to creative tourism: A case study of the Jewish community of Rosh Pina, Israel. *Tourism Recreation Research.*

Swarbrooke, J. (2020). *Postscript-The COVID-19 Pandemic.*

Terzidou, M., Scarles, C., & Saunders, M. N. (2017). Religiousness as tourist performances: A case study of Greek orthodox pilgrimage. *Annals of Tourism Research, 66*, 116–129. doi:10.1016/j.annals.2017.06.009

Thoresen, C. E., & Harris, A. H. (2004). *Spirituality, religion, and health: A scientific perspective.* Research Gate.

Timothy, D. J., & Boyd, S. W. (2003). *Heritage tourism.* Pearson Education.

Tsironis, C. N. (2022). *Pilgrimage and religious tourism in society, in the wake of the COVID-19 pandemic: A paradigmatic focus on 'St. Paul's Route'in the Central Macedonia Region.* Religions.

Vistad, O. I., Øian, H., Williams, D. R., & Stokowski, P. (2020). Long-distance hikers and their inner journeys: On motives and pilgrimage to Nidaros, Norway. *Journal of Outdoor Recreation and Tourism, 31*, 100326. doi:10.1016/j.jort.2020.100326

Wang, Y. C., Chen, P. J., Shi, H., & Shi, W. (2021). Travel for mindfulness through Zen retreat experience: A case study at Donghua Zen Temple. *Tourism Management, 83*, 104211. doi:10.1016/j.tourman.2020.104211 PMID:32895588

Willson, G. B., McIntosh, A. J., & Zahra, A. L. (2013). Tourism and spirituality: A phenomenological analysis. *Annals of Tourism Research, 42*, 150–168. doi:10.1016/j.annals.2013.01.016

Wong, C. U. I., McIntosh, A., & Ryan, C. (2016). Visitor management at a Buddhist sacred site. *Journal of Travel Research, 55*(5), 675–687. doi:10.1177/0047287514563164

Zahra, A., & McIntosh, A. J. (2016). Volunteer Tourism: Evidence of Cathartic Tourist Experiences. *Tourism Recreation Research.*

Zargham Boroujeny, H. (2017). Tourism: A tool for spirituality development. *International Journal of Tourism & Spirituality.*

Chapter 9
Institutionalizing Cultural Tourism for Heritage Preservation:
Comprehending Shri Kashi Vishwanath Temple Corridor (SKVTC)

Atul Tripathi
Banaras Hindu University, India

Sujay Vikram Singh
https://orcid.org/0000-0002-7113-2698
Banaras Hindu University, India

Rashmi Jha
Banaras Hindu University, India

ABSTRACT

The study attempts to analyze the temple Corridor Project in detail and understand its impact on cultural tourism. It focuses on the scope and challenges of heritage preservation and tourism infrastructure development, had been conducted through field visits and observations. During the investigation, different sources have been taken into account, including document review, personal interviews with the experts, local community stakeholders, and those who are/were directly or indirectly involved or affected by the SKVTC project. The study proposed unique way of using different but related set of factors of different natures in cultuaral heritage tourism, and hence offers an important empirical contribution to the literature of heritage management.

INTRODUCTION

The postulation of immersive heritage experiences at the 'sacred' city of Varanasi has been supported by the development of Shri Kashi Vishwanath Temple Corridor in rhythm with 21st century tourism

DOI: 10.4018/978-1-6684-9957-3.ch009

infrastructure that allows room for cultural reflection (Bec *et al.*, 2019). Cultural inheritance in the contemporary urban landscape of Varanasi entails determining maximum efficiency in visitor-experience that enables them to perceive heritage as a proponent of collective identity and source of understanding the present (UNESCO 2014). The ambitious Shri Kashi Vishwanath Temple Corridor (SKVTC) Project, inaugurated on March 08, 2019, by the Prime Minister of India, Shri Narendra Modi, completed its Phase I on December 13, 2021. This Project gains considerable momentum as a driving force for emboldening aspects of heritage preservation, its connection with the geographical landscape of the city, the related historical mythologies, and the impact of a sustainable approach to religious infrastructure development in a generally perceived 'sacred' city like Varanasi (Divya 2021; Verma, 2021).

The attraction of cultural tourism has become a common strategy for countries and cities seeking to enhance the value of their tourism. This is mostly because cultural tourists tend to interact with tourist sites more comprehensively through participation in cultural activities, visits to museum, and an overall tendency to be 'culturally motivated' and therefore more active through their engagement with sites (Richards 2010). Needless to say, cultural tourism enhances the economics related with tourism, however, the present study focuses on discerning the impact of cultural tourism over aspects of preservation of heritage. The mainspring of this study is the Shri Kashi Vishwanath Temple Corridor (SKVTC) Project of Varanasi, for a few fundamental reasons: this case is an ongoing Project which is witnessing first-hand implications of cultural tourism in the post-pandemic era; and because it is characteristically a highly-rich cultural and historical icon in a city famous for its heritage: Varanasi.

An interesting outcome of the Project has been the 'rediscovery' of multiple historical temples in the vicinity of the Corridor, which promulgates the dense heritage of the area. Shri Kashi Vishwanath Special Area Development Board (SKVSADB) has been conserving the rediscovered historical temples, which represent the artistic and cultural legacy of the city in the last three hundred years.

The paper purports to present how the SKVTC Project has propositioned its roadmap of 'Extension and Beautification' of a densely-populated cultural hotspot (*Kashi Vishwanath Mandir Visistikaran-Saundrikaran Yojana*) deciphering its cultural heritage glory. The scope of this study includes the rediscovered historical temples of the region that were saved from encroachment as a result of this Corridor. These rediscovered temples are placed in the milieu of Varanasi's growing religious and cultural heritage. The Corridor's planning has been corroborated with a Detailed Project Report, and current status of the Project's Phase. A systematic understanding of new developments at the site, along with its efficiency and impact is required, and the paper attempts at filling this research gap. The heritage conservation process of the Corridor has been evaluated in the contexts of transforming congested lanes, modern-amenities based architecture and overall tourism infrastructure (Rani, *et al.* 2018). Tourism infrastructure, amenities, etc. are criterion for cultural attraction and contribute towards propensity for developing positive destination image, which explains the constant rise in tourism at the Kashi Vishwanath Corridor (Bertacchini *et al.* 2021).

Heritage and Its Types

UNESCO (United Nations Educational, Scientific and Cultural Organization), defines heritage as our legacy from the past, what we live with today, and what we pass on to future generations (2003). It encompasses cultural, natural, and mixed properties of outstanding value, which are considered irreplaceable sources of life and inspiration. Heritage reflects the achievements, struggles, traditions, and aspirations of societies and holds significant meaning for the continuity of human civilizations. UNESCO's recognition and protection of heritage aim to preserve its diversity and promote mutual understanding and dialogue among different cultures. According to UNESCO, tangible and intangible heritage are two distinct categories used to classify and safeguard different aspects of cultural and natural heritage. These categories are outlined in the UNESCO World Heritage Convention and the UNESCO Intangible Cultural Heritage Convention.

Tangible Heritage

Tangible heritage refers to physical objects, sites, structures, and landscapes that hold historical, cultural, artistic, scientific, or architectural significance. It includes the physical manifestations of a society's cultural identity and history. Examples of tangible heritage include historical buildings, archaeological sites, monuments, museums, artworks, artifacts, and natural sites with cultural value. The preservation and conservation of tangible heritage are crucial to maintaining a connection with the past and passing on cultural knowledge to future generations.

Intangible Heritage

Intangible heritage, on the other hand, encompasses cultural practices, traditions, knowledge, skills, and expressions that are transmitted orally, socially, or through performance. It includes elements such as rituals, languages, music, dance, traditional craftsmanship, festivals, oral traditions, and knowledge systems. Intangible heritage is often deeply rooted in communities and plays a vital role in shaping their identities and fostering a sense of belonging. UNESCO's focus on intangible heritage aims to ensure the continuity of these living traditions and practices, acknowledging their significance in maintaining cultural diversity and promoting cultural sustainability.

Both tangible and intangible heritage are essential components of the world's cultural diversity and contribute to the understanding and appreciation of human creativity and history. UNESCO's efforts to protect and promote these forms of heritage help raise awareness about their value and encourage their sustainable preservation for future generations.

BACKGROUND

The 'Heritage' That Needs Preservation

The interlinkages between heritage and tourism have been fairly established as a system that works in mutual advancement. At the 22nd session of the UNWTO General Assembly, the definition of cultural tourism and its link with heritage preservation has been identified as, "(relating to) a set of distinctive material, intellectual, spiritual and emotional features of a society that encompasses arts and architecture, historical and cultural heritage, and the living cultures with their lifestyles, value systems, beliefs and traditions" (2017). The advocacy of SKVTC Project as a heritage icon has its basis in the forms and identities represented across this Corridor. These identity forms narrate the historicity of Varanasi and the various cultic renditions regarding the heritage of the city. In order to ascertain its legitimacy, it is important to discuss them in detail.

The notion of antiquity ascribed to the city is one of the primary facets for its heritage. Repeatedly, in classical and modern literature, Varanasi has been labelled as the "oldest living city" (Twain, 1897). Academic history has been working towards producing the dynamic history of Varanasi, however, in popular parlance, the city is often synonymous with being ahistorical. Multiple luminaries from India's ancient past have been known to praise the city, its environment, making the city an assembling place for priests and philosophers of the highest order- Adi Shankaracharya, Ramananda, Kabirdas, Tulsidas, Chaitanya Mahaprabhu, etc (Motichandra 1962). According to the Kashi Khand of Skanda Purana, Varanasi is among the *Saptpuris,* i.e., 'seven sacred cities' of ancient India (Bhatt 2004). Pilgrimage to the place accorded unique reverence, and various kinds of pilgrimage, including the multiple temples where these could be performed have been listed in the Kashi Khanda. Archaeological excavations at Rajghat and Sarnath have revealed the settlement history of the region from at least 8th century BCE (Narain and Roy 1977). Temple formations and monastic complexes existed at Sarnath since Mauryan period (3rd century BCE) (Sahni 1914). Foreign travellers to the city have recorded the cultural heritage in alluring details- in his writings, Faxian (4th century CE) described Varanasi as a significant religious and cultural center (Chung 1970). He noted its importance as a prominent location for Buddhist monasteries, as well as a place of great religious activity for Hindus. He detailed the presence of stupas, monasteries, and temples in the city. Hiuen Tsang (7th century CE) recorded the existence of several Buddhist monasteries and stupas in and around Varanasi (1970). He documented his visits to these places and the interactions he had with the local Buddhist community. Hiuen Tsang also recorded Shaiv worship in Varanasi:

'Moreover, there exist about a hundred Deva temples, hosting around 10,000 followers of this sect (Sammatiya). These temples are predominantly dedicated to the worship of Maheshwara, also known as *Ta-tseu-tsai.* The religious practices vary, with some adherents shaving their heads, others tying their hair in knots, and some even choosing to go naked (Nirgranthas). These ascetics, known as Pashupatas, cover their bodies with ashes as part of their austerities, seeking liberation from the cycle of birth and death. In the capital city, there are twenty Deva temples, distinguished by their sculpted stone towers and wooden halls. These sites are adorned with foliage that provides shade, while pristine streams encircle them. Notably, a statue of the Deva Maheshwara, constructed from native copper, stands at nearly 100 feet in height. The statue emanates an aura of gravity and majesty, giving the impression of a living deity' (1970). This account by Hiuen Tsang amplifies that Varanasi had an active culture of Shiva worship in around 7th century.

With the crystallization of Varanasi as a religious centre, every ruler established their own set of temples which became major shrines during their respective rule. The Gahadvalas for instance patronized the Adi Keshav temple when they shifted their capital from Kannauj to Kashi in 12th century (Kumar 2015). The Sharqi and Mughal dynasty built multiple Indo-Islamic architecture in the 14th – 17th centuries, including the Alamgir Mosque at Panchganga Ghat (2015). The decline of Mughal dynasty led to the rebuilding of Hindu temples and pilgrimages at Varanasi by various provincial rulers, zamindars, nawabs, etc. (Desai 2017). This was also a tumultuous time politically for Varanasi as the East India Company had been sieging control from multiple provinces across erstwhile India, and in order to establish legitimacy, the diverse set of patrons reinvigorated Hindu heritage at Varanasi till independent India's government took over in 20th century (2017). During these years, the principle shrine of the city, the seat of the jyotirlinga, Kashi Vishwanath was built in 1777-1780 CE under the patronage by Maharani of Indore, Ahilyabai Holkar (Eck 1982). A rather curious donation of one tonne of pure gold was given by Maharaj Ranjit Singh for the dome (Freitag, 1800). It is imperative to discuss rich patronage emerged from all centres of the country and Varanasi rose as the ultimate Hindu pilgrimage once again.

The region is renowned for its rich cultural heritage, with the Ganga River and the Kashi Vishwanath Temple playing integral roles. The city of Varanasi attracts a substantial number of tourists, both from within the country and internationally, who visit for purposes of pilgrimage, leisure, or a combination of both. These visitors engage with both the tangible and intangible heritage of Varanasi, seeking to immerse themselves in the historical essence of this ancient city. Interestingly, much of the existing tangible heritage of the city took shape during the early 18th century CE, following the decline of the Mughal Empire and the rise of the East India Company as a governing power. During this period, various artistic communities, including architects, musicians, and dancers, migrated in search of patronage. Varanasi, being a significant kingdom and later a princely state under British rule, emerged as a focal point for what many scholars consider a cultural renaissance (Sandria, 1989).

Particularly notable are the narrow lanes (locally known as "Gali") within the pakka mahal area, which draw numerous tourists. However, on festive occasions, accessibility to these lanes remains a notable challenge. The central focus of the corridor project, titled "Beautification and Extension of Temple Premises," is to honour the tangible and intangible cultural heritage associated with the Kashi Vishwanath Temple, the Ganga River, and the sacred city itself. The Shri Kashi Vishwanath Temple Corridor (SKVTC) initiative has upheld the religious significance outlined in Hindu scriptures. The 6.4-kilometer-long riverfront along the Ganga, featuring 84 ghats (steps leading to the river), aligns with the historically evolved socio-religious concepts, values, and the deep connection pilgrims hold with their faith. Together, these elements contribute to the formation of a distinct faith-centered landscape.

The Centrality of Kashi Vishwanath Temple

As has been mentioned in various ancient literary records, Varanasi has been a temple town, especially of Shaiva affiliation. In the Taittiriya Upanishad, a jyotirlinga is defined as an all-pervading Brahmatmalinga or all-pervading light (Sastri 1901). The Jyotirlinga is a profound symbol of the axis mundi, an ancient concept that represents the formless and supreme reality (nirguna) that resides at the core of creation (1901). From this reality emerges the manifested form (saguna) of Shiva. The Jyotirlinga shrines serve as locations where Shiva manifested as a luminous column of light. It's important to note that the Jyotirlinga concept is distinct from the 64 forms of Shiva (1901). There are twelve Jyotirlinga sites, and each site is associated with a specific presiding deity. Each of these sites is regarded as a distinct manifestation of Shiva. The term "Jyotirlinga" derives from "jyoti," meaning light, and "linga," symbolizing the divine presence (1901). These shrines hold deep significance within the context of Shiva worship and offer devotees a means to connect with the divine in its transcendent and manifest aspects. It is integral to the religiosity and practice of Shaiva community adherents.

For Varanasi, it has been a source of cultural identity with the earliest shrine for a jyotirlinga being assumed to be from the Gupta period ($4^{th} – 6^{th}$ century CE) (Motichandra 1962). The conquests between political rulers have resulted in shifts of the location of the jyotirlinga, but culminating at the Dashashwamedh Ghat in the late 18th century (Eck 1982). The present jyotirlinga temple is therefore seen as an overarching emblem of Varanasi's religious heritage. It symbolizes the city's evolution with relation to time, politics, social conditions, and developments (Eck 1982). In modern literature, closely all travel records have included, at least mentioned the Kashi Vishwanath temple. In present times, the temple has become one of the key attractions for cultural tourism in eastern Uttar Pradesh (Divya 2021). The temple complex was comprised of a sequence of smaller shrines, situated along a narrow lane known as Vishwanatha Galli, close to the river. The primary deity's linga within the shrine stands at 60 centimeters (24 inches) in height and 90 centimeters (35 inches) in circumference. The central temple takes on a quadrangular shape and is encircled by shrines dedicated to other deities. Among these, there are small temples dedicated to Kala Bhairava, Kartikeya, Avimukteshwara, Vishnu, Ganesha, Shani, Shiva, and Parvati within the complex. The temple encompasses a Sabha Griha, or Congregation Hall, which leads to the inner Garbha Griha, or Sanctum Sanctorum. The temple's architecture is structured into three parts: a spire atop the temple, a golden dome, and a golden spire crowning the sanctum. The latter holds a flag and a trident.

The significance and prominence of the temple are continually growing among both pilgrims and tourists, leading to a requirement for more expansive areas that can provide essential facilities and appropriate spaces for various spiritual rituals and customary practices. The Corridor project addresses this need by extending the temple premises and ensuring convenient accessibility through the Ganga River. The primary goal of the Corridor initiative is to enhance the temple's infrastructure to meet current necessities and effectively respond to the rising demand for religious and cultural tourism. The responsibility for planning the temple corridor, which encompasses renovation, expansion, and enhancement of existing structures, has been entrusted to the Shri Kashi Vishwanath Special Area Development Board (SKVSADB). This effort aims to accommodate the evolving requirements of the devotees and visitors while preserving the sanctity and cultural heritage associated with the temple.

Role of Cultural Tourism in Heritage Preservation

Cultural tourism is a fundamental element in the preservation of history, serving as a vital mechanism for protecting the historical, cultural, and architectural assets of a given community. Cultural tourism is commonly regarded as a constituent aspect of the broader tourism industry; yet, its profound influence on the conservation of heritage should not be underestimated. The fundamental nature of cultural tourism is in the investigation and admiration of the cultural and historical legacy of a certain history. When peoples engage in travel to visit historical monuments, museums, cultural events, and other significant cultural points of interest, they assume an active role in the conservation and safeguarding of these irreplaceable resources.

The contribution of cultural tourism to economic sustainability is readily apparent and instantaneous. Tourists are attracted to locations that provide genuine cultural encounters. The generation of income from tourists is facilitated through multiple avenues, encompassing entry fees, lodging, sustenance, and the purchase of indigenous handicrafts. The generated revenues have the potential to be allocated towards the preservation and refurbishment of cultural heritage sites, so guaranteeing their enduring viability. As a result, cultural tourism plays a significant role in promoting economic incentives for the preservation of heritage.

Additionally, cultural tourism functions as a powerful mechanism for enhancing knowledge and promoting the significance of safeguarding heritage. Tourists that partake in cultural tourism frequently endeavor to fully immerse themselves in the indigenous culture and historical context. Through the process of visiting heritage places, engaging with local communities, and actively participating in cultural activities, individuals acquire significant insights and develop a profound respect for the heritage they encounter. The direct encounter with heritage sites has the potential to convert tourists into advocates for the conservation and safeguarding of cultural and historical assets. Upon returning to their respective home countries, individuals bring with them their recently acquired knowledge and enhanced appreciation, thereby facilitating the global dissemination of these valuable insights. The heightened level of knowledge can potentially result in a greater level of support for conservation initiatives, originating from both governmental bodies and non-governmental organizations. Cultural tourism additionally serves as a catalyst for cultural interchange and promotes the cultivation of mutual comprehension among individuals hailing from diverse socio-cultural contexts. When tourists engage in visits to heritage sites, they are afforded the opportunity to gain exposure to the practices, traditions, and lifestyles that are inherent to the local population. The process of intercultural communication has the potential to dismantle prejudices and cultivate a collective consciousness of global citizenship. The sharing of ideas and viewpoints has the potential to foster peaceful coexistence and promote international cooperation, thereby positioning cultural tourism as a catalyst for the preservation of heritage and the cultivation of global harmony. Furthermore, cultural tourism frequently contributes to the preservation of cultural heritage by ensuring its commercial sustainability. A significant number of heritage assets face the risk of deterioration as a result of insufficient resources allocated towards its upkeep and conservation. Cultural tourism has the potential to enhance the economic viability of these locations. When heritage monuments are transformed into tourist attractions, they frequently attract expenditures aimed at enhancing infrastructure and ensuring proper care. The continuous arrival of visitors guarantees the ongoing economic significance of these places, so averting any potential neglect or deterioration. In addition, cultural tourism serves to strengthen local populations through the provision of job and entrepreneurship prospects. The transformation of heritage monuments into tourist destinations engenders employment opportunities across multiple sectors, encompassing hospitality, transportation, and handicraft industries. This not only enhances the well-being of individuals within the community but also fosters their active engagement in the conservation of their cultural heritage. The local community frequently exhibits a sense of pride in their cultural resources, diligently striving to preserve and safeguard them for the benefit of future cohorts. Nevertheless, it is imperative to recognize the need of striking a delicate equilibrium between accommodating tourists and safeguarding the authentic essence of heritage places. The erosion of authenticity and the destruction of heritage sites can occur as a result of over-commercialization and the proliferation of tourism activities. In order to ensure that cultural tourism effectively contributes to heritage preservation, it is imperative to use sustainable tourism practices. These practices encompass several measures, including but not limited to restricting tourist numbers, enforcing conservation principles, and actively engaging with local populations. Hence, cultural tourism plays a crucial role in the preservation of heritage. The activity in question offers economic support, enhances knowledge and understanding, promotes the sharing of cultural values, guarantees financial feasibility, empowers communities at the local level, and safeguards the integrity and genuineness of the subject matter. The interdependent association between cultural tourism and heritage preservation is of utmost importance in safeguarding and commemorating the global cultural and historical assets. The implementation of responsible and sustainable tourist practices is crucial for fully leveraging the potential of cultural tourism in the continued preservation of our collective heritage.

RESEARCH OBJECTIVES

The primary objective of this paper is to understand the relevance of SKVTC in heritage preservation, particularly in the context of the discovered ancient temples, with a focus on infrastructure development to support the revival of the tangible and intangible heritage of Varanasi.

The research study was conceptual and exploratory in which both primary and secondary sources were used; however, the data is primarily qualitative with the inductive research approach. The primary sources used are personal interviews undertaken with the residents, temple priests, pilgrims, and authorities directly or indirectly affected by the temple corridor project, the artists of the Banaras Gharana, and those having lifelong experience with the cultural traditions of Varanasi. The secondary data mainly consists of literary source materials collected from books, chapters, newspapers, journals, magazines, and internet sources.

THE QUEST FOR HERITAGE PRESERVATION

According to the World Heritage Convention Article 1, "Monuments: architectural works, works of monumental sculpture and painting, elements or structures of an archaeological nature (inscriptions, cave dwellings and combinations of features) which are of outstanding universal value from the point of view of history, art or science" are a critical dimension of cultural heritage (1972). As these monuments, and in this case the Kashi Vishwanath Temple, reflects the living expressions of traditions and knowledge inherited from ancestors, they validate the notion for preserving and protecting the heritage. The consumption of this heritage is a unique dimension which renders intangibility to the Kashi Vishwanath temple: the purposes of value and custom, adhered to its historical nature, are significant vehicles for the cultural discourse. This discourse has wide-ranged uses: commercial, ideological, and educational (Schouten 1995). In order to develop and manage such heritage institutions, certain sectors need to be evaluated: access to the resources essential for the preservation, enhancement, and upkeep of a historical site or object. A general analytical tool for managing and preserving heritage sites develops a holistic framework to evaluate the site's future potential in relation to tourism development.

The 'Cult Tour Analysis Tool' utilises the empirical results which suggests that heritage sites should go beyond their distinctive characteristics and focus on improving their visitor services (Martin *et al.*, 2021). This includes enhancing restroom facilities, providing access to interpretation materials, and the capability to host events, all aimed at creating a fulfilling and enjoyable visitor experience. Taking from Mason in assessing greater cultural preservation, participation in the planning process is a significant aspect in heritage studies, emphasizing the importance of involving local communities in the development of rural or other areas. The approach of involving locals in planning and executing heritage projects is valuable. However, historically, the planning process has been primarily controlled by entities within the heritage sector itself (Mason, 2002). Adopting bottom-up approaches is rooted in understanding the particular community and the environment in which values are embedded. Consequently, these approaches place a stronger emphasis on comprehending the unique community and context. By incorporating bottom-up approaches, it's possible to create long-term initiatives that are specifically tailored to local circumstances and requirements.

Sponsorship schemes represent another avenue through which external entities can engage in the cultural realm. These arrangements have grown in significance, particularly for the operational sustainability of cultural institutions such as museums, libraries, and archives, as well as for the preservation of heritage sites (Martin 2021). Prominent companies like State Bank of India and Times of India, have extended their support to the arts over many years. Consequently, the stigma once associated with sponsorship has diminished. Receiving institutions no longer perceive such partnerships as encroachments on artistic integrity; instead, they view them as opportunities to foster economic growth and development (2021).

Religious sites have increasingly adopted comprehensive management plans that encompass not only the religious aspect but also tourist related facilities and activities designed to inform, educate, and engage secular tourists (Alexis 2022). Incorporating concepts of public engagement and audience development, these efforts align with a modern conservation approach that recognizes interpretation as a fundamental component of the preservation process (ICOMOS 2008). In the present context, interpretation is regarded as a crucial element in the overall management of religious sites. In recent times, there has been significant advancement in how religious sites communicate their narratives. This involves the utilization of various mediums such as printed materials, informative placards, and interactive technologies (Alexis 2022). The objective is to offer visitors personalized information that allows them to better understand and connect with the site's historical, cultural, and religious significance. This approach helps to create a more meaningful and enriching experience for tourists while maintaining the sanctity of the religious site (2022).

ROLE OF CULTURAL TOURISM

In the field of cultural tourism research, the concept of "cultural contact" is frequently employed to gauge tourists' inclination toward the culture of a specific cultural destination (Gnoth 2013). Cultural contact refers to instances where groups of individuals interact with or are exposed to a different culture, spanning various durations, from short-term to long-term interactions. It involves a sustained and direct exchange between individuals from societies with distinct identities. Within tourism literature, the notion of cultural contact assesses the purpose and depth of tourists' experiences when they travel to immerse themselves in a diverse cultural environment (Cusick 2015). It also encompasses tourists' readiness to engage with external cultures and their aspiration to manage such interactions (2015). The original cultural contact scale, considering its dimensions (cultural tendency, depth of contact, involvement, and overall level of contact), was closely examined for this study. In the context of this research, cultural tendency, which assesses the extent to which tourists positively engage with the destination culture, is the focal point. In prior cultural tourism research, cultural tendency was found to impact specific behaviours at cultural tourism sites (Shinde, 2023). For instance, it influences tourists' yearning for authentic experiences and their level of engagement in the tourism process. Moreover, it plays a role in shaping the tourism experience, aiding tourists in forming cultural memories of the destination, and cultivating a sense of attachment to the place (Chen, 2018). Delving deeper into the culture of the destination and actively participating in the lives of local communities can render the experience more authentic and lasting for tourists. In the realm of heritage tourism destinations, tourists are exposed to the tangible historical culture that has been preserved and constructed.

ROLE OF KASHI VISHWANATH CORRIDOR IN CULTURAL TOURISM

The Kashi Vishwanath Corridor, a monumental urban development project situated in Varanasi, India, has emerged as a pivotal force in reinvigorating heritage and cultural tourism in the region. The Kashi Vishwanath Corridor, a visionary project initiated by the government, has sought to revitalize and rejuvenate the city's heritage while enhancing its appeal to tourists. This article aims to dissect the manifold ways in which the corridor has contributed to heritage preservation and cultural tourism promotion.

The Kashi Vishwanath Corridor has played a pivotal role in the restoration and preservation of these architectural marvels. By widening lanes, cleaning up the Ganga ghats, and renovating adjacent temples, the corridor has restored the city's historical charm. Tourists and pilgrims alike can now experience the authentic, centuries-old Varanasi.

Varanasi's cultural richness extends beyond its historical sites to encompass classical music, dance, and crafts. The Kashi Vishwanath Corridor has provided a platform for local artisans and craftsmen to showcase their talents. Visitors can now witness the creation of intricate stone sculptures, vibrant paintings, and exquisite textiles, all of which reflect India's diverse cultural heritage. The city's musical traditions have also flourished. With an influx of tourists, musicians and dancers now have a broader audience. The melodious tunes of sitar and tabla, along with the graceful movements of Kathak dancers, reverberate through the corridor, creating an immersive cultural experience. Both pilgrims and tourists are attracted to the spiritual appeal of this destination, which offers a profound immersion into Hindu rites and devotion. The temple's exuberant festivals and the widely acclaimed Ganga Aarti provide tourists with an insight into the dynamic cultural traditions of India. The preservation of traditional arts by surrounding artisans and craftsmen, together with the tantalizing local food that showcases the culinary culture of the city, collectively contribute to the cultural richness and diversity of the region (Batabyal *et al.*, 2023). The temple serves as a significant cultural hub, contributing to the enhancement of visitor encounters, the stimulation of the local economy, and the promotion of the conservation of India's multifaceted traditions.

CULTURAL CORRIDOR AND TOURISM ENHANCEMENT

The Kashi Vishwanath Corridor exerts a substantial impact on the tourism sector. The current state of temple visitation offers improved accessibility to visitors through the provision of well-maintained pathways, sanitary amenities, and efficiently arranged queues. The reconstruction has significantly enhanced the overall quality of the tourist experience, rendering Varanasi more accessible and attractive to visitors from various parts of the globe. The Kashi Vishwanath Corridor serves as a compelling illustration of the harmonious coexistence and prosperous synergy that may be achieved via the integration of tradition and modernity. In addition to safeguarding the heritage of Varanasi, it has effectively disseminated knowledge about the city's vibrant culture, thereby transforming it into a prominent hub for cultural tourism (Shukla, 2022). The present study sheds light on the transformative function fulfilled by the corridor in revitalizing historical and cultural tourism in Varanasi. This measure aids in the preservation of the historical city's prosperity while effectively addressing the demands of contemporary society.

SOCIAL IMPLICATIONS

The dynamic nature of cultural landscapes across the globe necessitates the implementation of sustainable strategies and mechanisms to ensure their continued relevance. The ongoing institutionalisation of cultural tourism through the Shri Kashi Vishwanath Temple Corridor Project is expected to contribute significantly to the preservation and conservation of Varanasi's rich history. The financial advantages obtained will contribute to the promotion of artisanal activity and the maintenance of local crafts. Moreover, cultural tourism is expected to have a favourable influence on several social aspects, ultimately contributing to the formation of a globalised society.

MANAGERIAL IMPLICATIONS

The service sector responsible for maintaining cultural tourism along the Shri Kashi Vishwanath Temple Corridor will have the chance to enhance their skills and, consequently, enhance the quality of services offered within the city due to increased interaction with a larger audience. The implementation of infrastructure improvements would enhance the operational efficiency of hosting events of this scale, hence generating advantages for the local stakeholders and enhancing operational performance. The Kashi Vishwanath Corridor, an infrastructure development project of great significance in Varanasi, has several important management consequences for cultural heritage tourism. First and foremost, it is crucial to prioritise the implementation of efficient crowd management strategies and visitor flow systems in order to guarantee a smooth and respectful experience for both pilgrims and tourists, all the while safeguarding the sacredness of the temple. Furthermore, it is imperative to prioritise the preservation and repair of the corridor and its surrounding heritage sites in order to uphold its historical and architectural authenticity (Batabyal *et al.*, 2022). This necessitates the implementation of rigorous methods to ensure their integrity is maintained. Furthermore, the implementation of skill development and training initiatives targeted towards local guides, artists, and vendors has the potential to significantly improve the overall quality of visitor experiences and contribute to the promotion of sustainable tourism. In addition, it is imperative to implement comprehensive waste management strategies and environmental sustainability measures in order to protect the cultural and ecological elements of the region. In conclusion, it is imperative to develop marketing and promotional tactics that effectively appeal to a wide array of tourists, encompassing both local and foreign visitors, with the aim of positioning Varanasi as a prominent cultural tourism hub. Efficiently overseeing these elements can guarantee that the Kashi Vishwanath Corridor optimises its capacity as a cultural heritage tourism destination while safeguarding its historical and spiritual importance.

CONCLUSION

The case study of the Kashi Vishwanath Corridor in Varanasi exemplifies the significant contribution of cultural tourism towards the conservation and rejuvenation of historical regions. The manifestation of this phenomenon is evident in the context of the Kashi Vishwanath Corridor. The primary objective of this extensive study is twofold: to enhance the overall visitor experience and to safeguard and rejuvenate the abundant cultural and historical legacy inherent in the region. The corridor serves as a catalyst for economic incentives, so facilitating the conservation of the unique cultural treasures of the sacred city through its ability to attract tourists from all parts of the globe. The Kashi Vishwanath Corridor exemplifies the interdependence and mutual advantages of cultural tourism and the preservation of historical landmarks. This statement demonstrates how responsible tourist development has the capacity to help the preservation and upkeep of historically significant sites, while also contributing to the promotion of local culture and customs (Batabyal *et al.*, 2023). This project exemplifies a noteworthy model for cultural sites worldwide due to its meticulous strategic planning, use of sustainable methodologies, and active engagement of the local community. The recognition of cultural tourism's potential is growing in significance for numerous sites. Consequently, it is imperative for these areas to draw inspiration from successful existing efforts, such as the Kashi Vishwanath Corridor. The utilization of tourism as a strategic tool for environmental preservation has promise in safeguarding cultural heritage for future generations, so ensuring their perpetual enjoyment. The utilization of this strategy has the potential to foster a deep-seated sense of self-worth, a collective sense of belonging, and an awareness of the interdependence that exists within common human heritage.

FUTURE RESEARCH DIRECTIONS

Within the context of cultural tourism and with the goal of safeguarding a historical heritage, future research avenues for the Shri Kashi Vishwanath Corridor should encompass a wide range of significant locations. There exists a pressing demand for extensive scholarly investigation into the methodologies employed in the management of visitors at heritage sites, as well as the resultant impact of these techniques on the sites' preservation capacity. This study aims to explore environmentally sustainable approaches that can effectively manage the increasing influx of tourists, while simultaneously safeguarding the architectural heritage and spiritual sanctity of the temple. Furthermore, it is imperative for scholarly inquiry to explore state-of-the-art technologies such as augmented reality and virtual tours, with the aim of enhancing the tourist experience while concurrently mitigating potential physical degradation of historical landmarks. The integration of the two research objectives enables the achievement of this outcome. Furthermore, conducting thorough research on the subject matter of community engagement, as well as the involvement of local stakeholders in safeguarding historical heritage and fostering tourism, is of paramount significance. The findings of this study will make a valuable contribution to the advancement of a sense of ownership and the adoption of sustainable practises in safeguarding cultural assets. To summarise, the integration of cultural studies, architecture, environmental science, and tourism management in interdisciplinary research has the potential to provide holistic insights into the conservation of the cultural and spiritual significance of the Kashi Vishwanath Corridor. Moreover, this approach can establish it as a noteworthy exemplar for cultural heritage tourism worldwide. The preservation of the Kashi Vishwanath Corridor as a UNESCO World Heritage Site can facilitate the achievement of this objective.

REFERENCES

Alexis, T. (2022). Heritagization of religious sites: In search of visitor agency and the dialectics underlying heritage planning assemblages. *International Journal of Heritage Studies*, *28*(9), 1036–1065. doi:10.1080/13527258.2022.2122535

Ashworth, G., & Tunbridge, J. (2000). The tourist city. *The Tourist-Historic City*, 53–81. doi:10.1016/B978-0-08-043675-3.50007-7

Batabyal, D., Das, D. K., Sen, S. K., Porya, A., Halam, H., & Chatterjee, P. (2022). Evolution of Shakti Peetham for religious, mythological and dark tourism in Indian sub-continent. *Revista Turismo & Desenvolvimento*, *39*, 291–300. doi:10.34624/rtd.v39i0.30366

Batabyal, D., Rawal, Y. S., Chatterjee, P., & Goswami, S. (2023). Temple service quality model for future pilgrimages in Indian sub-continent through user-generated contents (UGC) in TripAdvisor. *International Journal of Tourism Anthropology*, *9*(2), 113–135. doi:10.1504/IJTA.2023.130849

Batabyal, D., Ray, N., Bag, S., & Nag, K. (2023). Tourists' Harassment During Pilgrimage: A Case Study of TripAdvisor's Review for the Hindu Pilgrim Centers in India. In J. Kumar, G. E. Bayram, & A. Sharma (Eds.), *Resilient and Sustainable Destinations After Disaster (Tourism Security-Safety and Post Conflict Destinations)* (pp. 139–156). Emerald Publishing Limited. doi:10.1108/978-1-80382-021-720231011

Bec, A., Moyle, B., Timms, K., Schaffer, V., Skavronskaya, L., & Little, Ch. (2019). Management of immersive heritage tourism experiences: A conceptual model. *Tourism Management*, *72*, 117–120. doi:10.1016/j.tourman.2018.10.033

Bhatt, G. P. (2004). The Skanda Purana. Part X. AITM, 58. Motilal Banarasidass Publishing House.

Braslavsky, C. Fernǹdez Ana La̓, & Cox Cristiǹ. (2002). A Educaô Secundr̀ia: mudanȧ ouimutabilidade? Brasil: UNESCO Brasil.

Byrne, D. (2022). *Heritage Corridors: transnational flows and the built environment of migration. Journal of ethnic and migration studies, 42 (14)* (1st ed.). Routledge.

Chen, H., & Rahman, I. (2018). Cultural Tourism: An Analysis of Engagement, Cultural Contact, Memorable Tourism Experience and Destination Loyalty. *Tour. Management Perspective*, *26*, 153–163. doi:10.1016/j.tmp.2017.10.006

Chung, T. (1970). ANCIENT INDIAN LIFE THROUGH CHINESE EYES. *Proceedings of the Indian History Congress*, *32*, 137–149. https://www.jstor.org/stable/44141059

Conzen, M. P., & Wulfestieg, B. M. (2001). Metropolitan Chicago's Regional Cultural Park: Assessing the Development of the Illinois & Michigan Canal National Heritage Corridor. *The Journal of Geography*, *100*(3), 111–117. doi:10.1080/00221340108978426

Conzen, M. P., & Wulfestieg, B. M. (2001). Metropolitan Chicago's Regional Cultural Park: Assessing the Development of the Illinois & Michigan Canal National Heritage Corridor. *The Journal of Geography*, *100*(3), 111–117. doi:10.1080/00221340108978426

Corridor, K. (2018). India and Pakistan sign a deal on the Sikh temple project. *BBC news*. https://www.bbc.com/news/world-asia-india-50138275.

Cusick, J. G. (2015). *Studies in Culture Contact: Interaction, Culture Change, and Archaeology*. Southern Illinois University Press.

Desai, M. (2017). *Banaras Reconstructed. Architecture and Sacred Space in a Hindu Holy City*. University of Washington Press.

Dikshit, R. (2018). 7.35 crore visit Kashi Vishwanath Dham in opening year. *The Times of India [TOI]*. https://timesofindia.indiatimes.com/city/varanasi/7-35-crore-visit-kashi-vishwanath-dham-in-opening-year/articleshow/96211539.cms

Divya, A. (2021, December 14). Explained: What is changing at the ancient Kashi Vishwanath temple complex? *Indian Express*. https://indianexpress.com/article/explained/kashi-vishwanath-temple-narendramodi-varanasi-7669709

Eck, D. L. (1982). *Banares, City of Light*. Columbia University Press.

Enrico, B., Nuccio, M., & Durio, A. (2021). Proximity tourism and cultural amenities: Evidence from a regional museum card. *Tourism Economics*, *27*(1), 187–204. doi:10.1177/1354816619890230

Freitag, S. B. (1800). Introduction: The History and Political Economy of Bañaras. Sandria B. Freitag (ed.) Cultural and Power in Banaras. University of California Press. doi:10.1525/9780520313392-005

Gnoth, J., & Zins, A. H. (2013). Developing Tourism Cultural Contact Scale. *Journal of Business Research*, *66*(1), 738–744. doi:10.1016/j.jbusres.2011.09.012

Graham, B. (2000). The past in place: historical geographies of identity. In B. Graham & C. Nash (Eds.), *Modern Historical Geographies* (pp. 70–79). Pearson.

Graham, B. (2002). Heritage as Knowledge: Capital of Culture. *Urban Studies (Edinburgh, Scotland)*, *39*(56), 1003–1017. doi:10.1080/00420980220128426

Harkness, T. & Sinha, A. (2004). Taj Heritage Corridor: Intersection between History and culture on the Yamuna riverfront. *Places*, *16*(2).

Henderson, J. (2000). Attracting tourists to Singapore's Chinatown: A case study in conservation and promotion. *Tourism Management*, *21*(5), 525–534. doi:10.1016/S0261-5177(99)00108-9

Kermani, S. (2018). Kartarpur corridor: A road to peace between India and Pakistan. *BBC News*. https://www.bbc.com/news/world-asia-46382657

Kongjian, Y., Li, W., Li, D., Li, C.B., Huang, G. & Liu, H.L. (2005). Suitability analysis of heritage corridor in the rapidly urbanising region: A case study of Taizhou City. *Geographical Research*. Pecking University, Beijing.

Kumar, S. (2015). Rural Society and Rural Economy in the Ganga Valley during the Gahadavalas. *Social Scientist*, *43*(5/6), 29–45.

Martin, J., Dominica, W., Klara, Ł., & John, A. (2021). Development of the My Cult-Rural Toolkit. *Sustainability*, *13*(1), 7128. doi:10.3390u13137128

Mason, R. (2002). Assessing Values in Conservation Planning: Methodological Issues and Choices. In Assessing the Values of Cultural Heritage Los Angeles: Research Report. The Getty Conservation Institute.

Meyer, D. (2004). *Tourism routes and getaways: key issues for the development of tourism routes and gateways and their potential for pro-poor tourism*.

Motichandra. (1962). *Kashi Ka Itihas*. Mumbai: Hindi Granth Ratnakar Pvt. Ltd.

Narain, A. K., & Roy, T. N. (1977). *Excavations at Rājghāt, 1957-1958, 1960-1965: The pottery (Vol. 2)*. Department of Ancient Indian History, Culture and Archaeology, Banaras Hindu University.

Pandey, V., Agarwal, M. K., & Singh, S. V. (2021). A tourism prospect of religious destination of Uttar Pradesh with special reference to Naimisaranya (Neemsaar). *ZENITH International Journal of Multidisciplinary Research*, *11*(7), 42–54.

Pearce, D. G. (1997). *The roles of the Public sector in conservation and tourism planning. Tourism and Heritage Management*. Gadjahmada University Press.

Pearce, J. (1997). Commentary by J. Pearce: Reality check—but how real should we get? *Bulletin de l'Institut Pasteur*, *5*(95), 182. doi:10.1016/S0020-2452(97)81374-7

Pearson, M., & Sullivan, S. (2013). *Looking after heritage places: The basics of heritage planning for managers, landowners and administrators*. Melbourne Univ. Publishing.

Prasad, N. (2019). Kashi Vishwanath Temple Corridor's Final DPR Ready! PM Narendra Modi's Dream Project is ready Pilgrims. *Financial Express*.

Rangaswami, C. M. (Ed.). (2017). Ramanathaswamy temple. In Ramanathaswamy temple, 201–207. New Delhi, Delhi: Bharti publications.

Rani, W. M., Wan, T., Shakir, M., & Wahab, M. H. (2018). Governance of Heritage Conservation: Overview on Malaysian Practice. *The Journal of Social Sciences Research*, *847-851*. doi:10.32861/jssr.spi6.847.851

Report, D. V. (2020), Government of Uttar Pradesh. *Varanasi tourism statistics*. https://varanasi.nic.in/

NDTV. (2019). PM Modi Lays the Foundation stone of kasha Vishwanath Temple Corridor. *NDTV Corridor*. www.ndtv.com

Richards, G., & Munsters, W. (2010). Cultural tourism research methods. CAB International. Research Gate. doi:10.1079/9781845935184.0000

Richards (1996). *Journal of Environment Quality*, *25*(3). doi:10.2134/jeq1996.253npx

Roche, E. (2019). India and Pakistan signed a historic agreement to open the Kartarpur corridor. *Live mint*.

Sahni, D. R. (1914). *Catalogue of the Museum of Archaeology at Sārnāth*. Superintendent Government printing press.

Sastri, S. A. (1901). *The Taittiriya Upanishad*. GTA Printing Works.

Schouten, F. F. (1995). Heritage as historical reality. *Heritage, tourism and society*, 21-31.

Shinde, K., & Singh, R. P. (2023). Still on UNESCO's "Tentative List of World Heritage"? Heritage, Tourism, and Stunted Growth in Sarnath (Varanasi), India. *Heritage*, *6*(7), 5051–5065. doi:10.3390/heritage6070267

Shri Kashi Viswanath Temple Trust. (n.d.). *About*. Shri Kashi Viswanath Temple Trust. https://www.shrikashivishwanath.org/

Shukla, S. (2022). *Reimagining socio-spatial Arrangements of a cultural neighbourhood by tourism: a case of silk weaving community of Varanasi* [Doctoral dissertation, SPA Bhopal].

Sigala, M., & Leslie, D. (2016). *International cultural tourism: management, implications, and cases*. Routledge.

Singh, R. P. B. (2011). *Varanasi, India's cultural heritage city: Contestation, Conservation, and Planning*. Surbhi Publications.

Singh, S. V., & Ranjan, R. (2019). Online travel portal and their effect on travel agency: A study on outbound visitors of Varanasi. [IJRAR]. *International Journal of Research and Analytical Reviews*, *6*(2), 387–393.

Sinha, A., Harkness, T., & Harkness, T. (2009). Views of the Taj—Figure in the Landscape. *Landscape Journal*, *28*(2), 198-217. https://www.jstor.org/stable/43323846

Swain, M. (1989). Developing ethnic tourism in Yunnan, China. *Tourism Recreation Research*, *14*(1), 33–39. doi:10.1080/02508281.1989.11014541

Teo, P., & Huang, S. (2000). Singapore: Tourism development in a planned context. In M. Hall & S. Page (Eds.), *Tourism in South and Southeast Asia: Issues and Cases*. Butterworth Heinemann.

Timothy, D. J. (2000). Sacred Journeys: Religious heritage and tourism. *Tourism Recreation Research.*

Twain, M. (1897). *Following the Equator: A journey around the world.* Archive. https://archive.org/details/followingequator00twaiuoft

UNESCO. (2003). *Convention for the safeguarding of the intangible Cultural Heritage*. UNESCO.

UNESCO. (2014). Heritage Sustainability. *CDIS Methodology Manual V7*. UNESCO. https://en.unesco.org/creativity/sites/creativity/files/cdis/heritage_dimension.pdf

UNESCO (n.d.) *Guidelines for the Establishment of National "Living Human Treasures" Systems.* Paris: UNESCO.

Varanasi tourism report. (2020) UP Tourism. http://www.varanasi.gov.in/en/destination

Verma, L. (2021). *PM Modi to inaugurate Kashi Vishwanath corridor in Varanasi today. The New Indian Express*. Express Publications. https://indianexpress.com/article/india/pm-modi-to-inaugurate-kashi-vishwanath-corridor-in-varanasi-7669259/#:~:text=%E2%80%9CTomorrow%2C%2013th%20December%20is%20a,temple%20town%2C%20tweeted%20on%20Sunday

West, A. (1985). *Heritage. Heritage*. Coronet. www.collinsdictionary.com

Yang, W., & Kang, J. (2005). Soundscape and sound preferences in urban squares: A case study in Sheffield. *Journal of Urban Design*, *10*(1), 61–80. doi:10.1080/13574800500062395

KEY TERMS AND DEFINITIONS

Bottom-Up Approaches: This refer to strategies, initiatives, and development processes that prioritize the active participation and engagement of local communities, individuals, and grassroots organizations in shaping and benefiting from tourism activities

Cultural Tourism: This involves travellers seeking to engage with and understand the unique cultural aspects of a place, including its customs, language, cuisine, architecture, art, festivals, and historical sites.

Heritage Preservation: Also known as cultural heritage preservation or conservation, refers to the proactive efforts and practices aimed at safeguarding and protecting the cultural, historical, natural, and architectural heritage of a particular region, community, or place for the benefit of present and future generations.

Institutionalizing: This refers to the process of establishing or formalizing something as a fundamental and enduring part of an organization, society, or system

Tourism Infrastructure: It encompasses a wide range of components and services designed to enhance the travel experience for tourists and ensure the smooth operation of tourism-related activities in a destination.

Chapter 10
Animal Cruelty and Sacrifice in Religious and Mythological Tourism:
Perspective and Challenges in South Asia

Pratim Chatterjee
https://orcid.org/0000-0002-4136-8157
Amity University, Kolkata, India

Yashwant Singh Rawal
https://orcid.org/0000-0001-5367-7443
Parul University Vadodara, India

Debasish Batabyal
https://orcid.org/0000-0003-0191-9804
Amity University, Kolkata, India

Sankar Kumar Mukherjee
https://orcid.org/0000-0002-3667-001X
Amity University, Kolkata, India

Namrata Chakrabortty
Guru Nanak Institute of Technology, India

Soumyadeep Roy
Amity School of Hospitality, Amity University, Kolkata, India

ABSTRACT

Given the growing importance of the conservation and preservation of animals and other species, the purpose of the chapter is to review the ritualistic killing and cruelty of wild and/or domestic animals in religious tourism in South Asia. Purposively, the search engine Google Scholar has been considered for searching the most cited paid and unpaid articles based on the keywords. Key themes from the literature focus on varying region-specific cruelty-conservation paradoxes for future policy issues, business scopes, and animal utilitarianism in tourism, along with new responsible and sustainable orientation have been highlighted. Also, a total of twenty-three statements derived from such reviews are verified through an opinion survey of 200 pilgrims in Hindu temple sites, keeping in view the concentration of temple sites where animal sacrifice is still in vogue. Through the factor-wise mean scores reform-oriented contemporary outlook of temple authority and priest association and socio-religious background are of immediate recommendations for the policy issues.

DOI: 10.4018/978-1-6684-9957-3.ch010

INTRODUCTION

It has been common in almost all communities and cultures for companion animals, or pets, to share living quarters with people around the world (Savishinsky, 1983; Serpell, 1987, 1996; Gray &Young, 2011). While attachment theory exhibits emotional relationships between the infant and caregiver (Bowlby, 1958; Ainsworth &Bowlby, 1991), social support theory exhibits evidence that humans with good social connections are healthier and that there are significant medical outcomes of loneliness (Lynch, 1977; 2000). The concept of animal sacrifice in any religion is believed to have been inhibited bythe pain of sacrificing the nearest and dear ones (Batabyal et al., 2022; Batabyal et al., 2023). On the other hand, social support theory suggests and strengthens a human-animal bonding of coexistence since the primitive age.

In Islam, the sacrifice of animals is accepted as complete obedience and demonstration of submission from the morning of the 10th to the sunset of the 13th *Dhual-Hijjah*, the 12th lunar month of the Islamic calendar. In Hinduism, animal killing on specific occasions in most of the Shakti temples is based on faith and belief that it stimulates her violent vengeance against the buffalo demon. Unlike in Islam, different types of animals or birds are sacrificed or, as inBaliYatra,the slaying of goats, chickens, pigeons, ducks, fishes, male Water buffalos, and such other animals is performed(Fuller, C.J,2004). In Bhutan and remote Sikkim Himalaya as a whole, Bonismand shamanism, pre-Buddhist religions of Tibet and its adjacent regions, animals are sacrificed to local deities (YulLhas). As winter spreads across the country, people move from their residences to mountains, caves, ruined houses, streams, and giant trees to worship their local deities, mostly associated with animal sacrifice. With the acceptance of Buddhism, the practice of killing animals is now replaced by offerings of joints and cuts of meat (ShaNga-not a fresh or sacrificed meat but naturally dead animal's meat).

The23rd Tirthankara Parswanath accelerated *Jainism* and *Ahimsha*in in the 9th century BC which raised voices in favor of animal rights in religious practices on theIndian subcontinent. Also, the *Dravidian* philosopher *Valluvar* in his *Tirukkural* emphasized and invoked*Ahimsha*(tolerance) and moral vegetarianism at the individual level. Interestingly, the Parsees or *Parsis*, one of the oldest extant ethnic religious groups of the Asian sub-continent practicing Zoroastrianism, perform their funerals dedicated mostly to birds and animals. A dog is essential in the funeral process because it can see death and the body is taken to the tower of death where the vultures feed on it.

Residents generally show interestto portray local culture, traditions, and heritage to tourists (Rakitovac et al., 2021). Religious and pilgrimage treks are one of the most popular tourist activities on the Indian subcontinent but killing animals for rituals, catching birds, fishing, and cutting trees is a matter of concern (Gupta, 1999). In ritual slaughter, animal suffering is very likely so this kind of act can never be morally justified on the grounds of tradition, involvement of culture, religiously required practice, or a source of income from tourism. Domestic animals which make the most noise while dying are mostly offered as a part of tradition(Behrens,2009).

Some experts such as Kuznets (1955)and Morris (2013) worked on an animal protection index, reducing cruelty, and encouraging vegetarianism from a social-economic perspective, not necessarily focusing absolutely on community cultural issues incorporating animal stress, cruelty, entertainment and sacrifice directly. However, Holst &Martens (2016) found a strong voice against animal consumption and killings in greater democracy and Kuznets et. al (1955) evidenced that vegetarianism prevailed over the very poor and very rich. Fennell &Sheppard (2020) focused directly on utilitarianism in tourism and developed a scales of justice frameworkon two different normative ethics models, varying from deep, intermediate,andshallowto no justice. Their study dealt with more contemporary issues than the

Figure 1. Places of mythological and religious interest in South Asia associated with animal sacrifice in Hinduism
(Purana, Purana, & Tantra, n.d.)

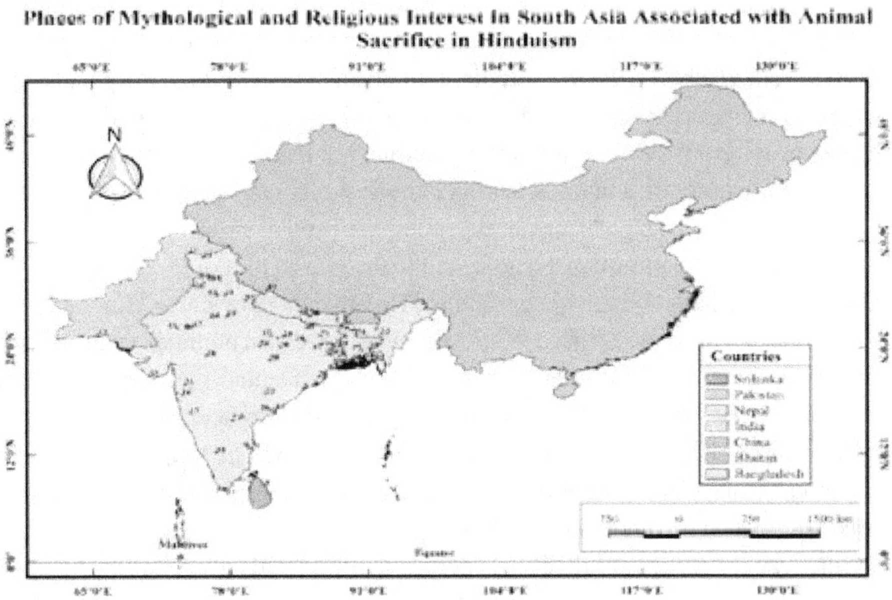

core area of grantingritualisticcruelty inhibited in culture and heritage. Some are animal-specific discussing elephants, pandas, etc.,and identifying and explaining animal consumption, killings,and stress as subsequent effects of economic and socio-political issues (Fennell &Sheppard, 2020; Cohen, 2010; Bansiddhi, 2020; Duffyet.al., 2011). This school of scholars found how the common tourists are critically varying and compromising in nature. Therefore, a systematic understanding and body of knowledge of the human-animal interrelation and interdependence in future tourism should be studied for ultimate responsible and sustainable tourism practices. This study is an extensive review of using and abusing animals in tourism along the continuum of primitive cruelty to modern conservation with a broader timeframe of changing values, cultural rigidities, and modern perception, acceptance, and awareness towards responsibility and sustainability (Duffy& Moore, 2011; Shaheeret.al., 2021).

Many studies identified, reviews and analysis animal cruelty-conservation paradox through the millennial perspective for responsible and sustainable tourism in the backdrop of socio-religious and cultural rigidity and social acceptability for sustainable religious tourism in the study area.

REVIEW OF LITERATURE

Yajur Veda mentions that no person should kill animals helpful to all (Dwivedi & Tiwari, 1987). Again, animal sacrifice was allowed in the Vedic rituals, mostly for a special occasions. Later, frequent animal sacrifice superseded the Vedic philosophy of love and care for them. The same is evident in Islam and other religions. It may bethat other important geophysical and socio-political, natural reasons are involved. However, animal sacrifice, killing, etc. have become a vehicle of exchange where it is offered to God in return for something. In South and SoutheastAsian regions, the gender, age, number of animals to be sacrificed, and even how to kill that animal are prescribed and practiced in culture and religion. community and religious affiliations have a direct association with the ritual slaughter of animals in many countries and, as such, the concerned communities have been successful in getting their views enacted into law (Lerner&Rabello, 2007). InIndia, tourists also visit and pay respect to or view shrines and temples where animal sacrifice is offered to gods or goddesses, mainly the goddess Kali (Tul, 2015). This is how places or religious centers of animal sacrifice are sources of attraction. The cults of Saint Anthony of Padua at Puliyampatti, a famous Tamil Catholic pilgrimage area, follow the Hindu culture of animal sacrifice as offerings to the gods. Generally, goats are used as offerings and are beheaded outside the church, after which the pilgrims offer the goat skin to the god as part of their Tamil Catholic tradition (Sébastia, 2004). On the South Asian subcontinent, Muslims and low-caste Hindus are mostly found performing rituals through animal sacrifice. Kali worship is based on a tantric tradition where sacrifice is necessary to satisfy the god. Sometimes animals, their presence,and their sacrifice are an extensive part of black magic and dark tourism. Even if the courts stop the sacrifice, itsstill secretly takes place in many states in India (Moodie, 2019). Animal sacrifice is mentioned as the barbaric and tribal form of sacred power. Domestic animals like buffalo, sheep, or goats are used mainly as offerings and as part of the main rituals to gods in different parts of South and Southeast Asia. Moreover, additional animals are also often sacrificed by others to fulfill their personal vows (Hitlerite, 1980). The modern transformation of tantric tradition as part of a tourist destination is observed on the Indian subcontinent, also incorporating Bhutan and Nepal. Countless goats, pigeons, fish, buffalo, and other animals are brought for slaughter by pilgrims and tourists. Different non-Vedic offerings are made as a part of rituals on different pujas like Durga puja and Kali puja (Urban, 2019). The goddess Kali or Durga and their different consorts (*Bhairava*)are worshipped throughout Asia, including some places in Tibet in China as seen in *figure 1*above. According to Hindu mythology, these are the main places that have had specific rituals invoking Shakti through animal sacrifice for several thousand years. The rituals of animal sacrifice are even popularized in many other places on this sub-continentderiving mostly from these places.

Animal offerings to gods as a means of sacrifice also induce tourism which is observed in different religions. At the Kamakhya temple in Assam, India, and Gadhimai temple in Nepal, buffalos are beheaded and offered to the goddess. The ritual attracts many locals in addition to national and international tourists, who come to behold the tradition (Urban, 2016). Animal sacrifice is part of the old shamanic tradition in Sikkim. As part of the ritual, these animal sacrifices are made to worship or show loyalty to the gods and to calm or tame the supernatural. Though at present animal sacrifices can create some problems, these rituals are still practiced as a part of the heritage from forefathers, and it also depicts one's own tradition and identity (Marlene, 2019). Snake charming is a traditional matrix of the culture, practiced throughout the entire Indian subcontinent where two hundred different species of snakes are capturedin baskets and used for entertaining local people and tourists. Though this type of entertainment is justified by the snake charmers as a part of their livelihood but catching snakes from the jungle, removing their poisonous teeth in unhealthy ways which can cause scarring or even death to snakes, and finally keepingthem in small baskets for sports and entertainment as a part of local tradition or tourism is not to be entertained (Mohanty et al., 2004).Animal slaughter or injuring animals for sports, entertainment, or for the encouragement of tourism can also be observed in Tamil Nadu with bull racing where bulls are injured or killed while taking part in the race. Injuries to their knees, abdomen, horns, and fractured legs are common after each race (Ashokan et al., 2000).Animal fairsalso cause animal suffering in many ways. Foot and mouth disease and other animal illnessesare only treated locally in those fairs and lack of proper treatment claims the lives of hundreds of cattle in traditional cattle fairs in India (Galavet al., 2010). A number of festivals are also considered a tourist commodity (Virloget&Kavrečič, 2020) and attract a lot of tourists and locals at the expense of these animals. Camel wrestling in Turkey is one such source of incomefor the locals through tourism though the camels are very likely to be injured or killed in this violent sport (Donlon et al., 2010).

To develop pilgrimages and sacred tourism in the Indian Himalayas, a strict code of conduct isrecommended for rituals, rites, and sacrifices. Himalayan environments and pilgrimage centers are under siege from strong commercial and cultural interests like poaching, mining, and animal sacrificing as rituals (Singh, 2006). Harsh training is also very common for animals that are extensively used in different tourism activitiessuch as trekking, animal shows, and painting. Camel festivals, desert tours, camel fairs, and camel tourism attracts millions of tourists each year in Jaisalmer, Rajasthan. Camels which aregenerally owned by the locals are extensively used in tourism throughout the year, especially between November to March. The rest of the time they are used forwork or as carriers (Shackley, 1996). Animal slaughters are performed in different cultures for different purposes. Like in Islam and Judaism, ritual slaughters are performed per guidelines and to make the meat fit for human consumption. In Hinduism, the blood of the animal is offered to the deities per the rituals which are performed in temple and festivals. Individual devotes also offer animal sacrificesto show gratitude for fulfilling certain prayers. In Brazil and Sri Lanka, bans of animal sacrifice are observed not only due to animal welfare concerns. In these countries the strong advocacy against animal cruelty is mainly demonstrated by those people who are also responsible for carrying out human cruelty against the minority devotes of that country (Boaz, 2019). One of the most powerful and vulnerable wild creatures is the brown or black bear and exploiting them in circuses for entertainment and attracting tourist endanger thespecies. Killing the mother bears to capture the cubs andtrain them for the circus is proving to be very harmful for the long-term bear population (Bunnelet al., 1993). From economic, social, and ecological perspectives, the long-term effect of poaching on tourism is devastating as the number of tourists will tend to decline when there will be hardly any wildlife to attract them. The experience of tourists is affected by poaching and anti-poaching activitiesthat have an impact on their choice of destination or for future visits. Improper safeguarding of animals will cause a decline in tourist visits which will, in turn, decrease the revenue from tourism that funds the conservation and protection of endangered animals (Lubbe et al., 2017).In the domain of cultural tourism, abuse of animals is observed on almost all continents of the world. Animals in a contest against each other, animals in a contest against men, animals used as religious sacrifices, animals being hunted, animals used as means of transportation, animals used as circus or street attractions, animals used as tourist picture props, animals as tourist food, etc. are surely a threat to the safety and security of these affected animals. However, mainly in developing countries, animal rights and their well-being arenot the primary concern of the local residents as they are financially dependent on the animals for their livelihood (Malchrowicz et al., 2020).

METHODOLOGY

This study analyses and reviews a range of literature that providesinsight into the unethical use and/or abuse of animals in tourism from a wider perspective. Till April 2021 reviews are carried out and all earlier cited papers meeting the criteria were also included. This was the last time the search results were accessed (sorted by citation) to decide whether they should be part of the review. Purposively, the search engine Google Scholar has been considered for searching the most cited paid and unpaid articles on the basis of the selected keywordssuch as 'animal sacrifice (and) tourism', 'tourism animal relation', 'animals (in) religious tourism' and 'abuse (and) utilitarianisms (of) animal (in) tourism' respectively. In each case, a maximum of five pages of resultswere considered, and only cited articles have been examined. A total of 250 documents represent the population size in which only 80 are found cited as well as relevant for the study. Due to financial constraints and partial availability of some articles, a final total of sixty-six articleswere usedin this study. More than fifty percent of relevant and cited articles are found to be on or from 2010, representing contemporary literature for more than a decade.

Therefore, the area of the study is extensive and broad and unfolds many areas of new research.As a concerted and simplistic approach, the discussion in this study is rather confined to the largest religion Hinduism in South Asia with a typical concentration of Hindu religious sites where animal sacrifice is still in vogue. In this study, primary data are also collectedfrom such areas in West Bengal as are already found the most popular and with maximum Hindu temple sites now a days. Birbhum district in West Bengal with maximum number of Shakti Peeths or Peethams is found the best place for data collections for ritualistic animal sacrifice (Batabyal, et al.,2022). Muchevidenceindicates that average daily footfalls of the most popular such site Tarapeeth is 12000 (Office of the District Magistrate and Collector, Birbhum; 2016). Considering the domination of the local pilgrims who are not within the purview of tourists technically, the sample size is restricted to two hundred and one hundred fifty-two are found usable. At the time of the survey, pilgrims are told that this study is absolutely for academic study and policy implications only.

For identifying and analyzing the most sensitive opinions of the pilgrims the principal component analysis (PCA) has been chosen. This techniqueusually discusses in terms of scores of factor and loadings i.e. the weight by which eachstandardized original variable needs to be multiplied to get the factor score. If these scores arestandardized to the square of standard deviation, loadings must incorporate the data variance inthem (Panneerselvam, R., 2014).

If component scores are not standardized, then loadings must have to be unit-scaled or normalized and these weights are called eigenvectors. If a multivariate dataset is envisioned as a set of coordinates in a high-dimensional data space (1 axis per variable), PCA can provide the user with a lower-dimensional picture, a protrusion of this object when considered from its most informative viewpoint. PCA also upgrades the data to a new coordinate system in such a way that the greatest variance by some scalar projection of the data comes to lie on the first coordinate. This is called the first principal component and it goes on in this way from first to second and so on.

MAP LAYOUT DESIGN

After successful verification and marking the Shakti Peeths or Peetham in Google Earth Pro the place marks have been converted and saved in KML file. These Shakti Peeths or Peethams are usually the places of ritualistic animal sacrifice in Hinduism. Points marked are based on ancient Scanda Purana and Kalika Purana (Bhattacharyya, N. N.,1974 ; Banerjee,S.C.,1992).

Then using ArcMap10.5 this KML file is converted into Layer file format. Then Data frame is created, and shape file of the South Asian country is added and those points are marked for the final Map.

DATA ANALYSIS AND INTERPRETATION

The primary data collected in this study shows that 59.60% of respondents were male whereas 40.40% were female, and most of the respondents were from the age group between 18-25 (91.39%) years. Therefore, this opinion survey represents the millennial perspective of the study area that is most important for the future policy of India where the majority of the population belongs to the young generation and more specifically unmarried, young, students and professionals. This survey also represents the opinion of such young professionals who are college graduates or above and their average daily expenses is less than Rs. 2000(89.40%).The respondents were mostly either in a small group of fewer than seven persons (50.33%) or solo (34.44%). Interestingly millennial perspective represents a wide gap between rituals in vogue and what they actually think as the majority of respondents (70.86%) agree that ritualistic animal killing to be banned immediately. Only 0.66% of respondents support ritualistic animal killing.

For analyzing twenty-three statements directly associated with ritualistic animal cruelty and killings factor analysis was used. These twenty-three statements were given to respondents to rate on a five-point Likert scale, varying from strongly agree to strongly disagree. The KMO test of sampling adequacy was applied to check whether our data set is appropriate or not for applying factor analysis and Bartlett's test was applied to check whether variables in the population are uncorrelated or not.

The KMO test value was 0.834 which is greater than 0.5 showing presence of strong partial correlation. Bartletts test of sphericity was significant ($p < 0.000$) indicating that correlation matrix is not identity matrix and variables are sufficiently correlated. Hence, both KMO and Bartlett's test indicate that our data set was appropriate to apply factor analysis.

After checking the validity of the data. Factor analysis was applied using the Principal component method. Varimax rotation was used to rotate factors. Total of six factors were extracted. The criterion for determining factors was eigne values. Only those factors are retained whose eigne values were one or more. The total variance explained by these six factors was 66.38%. Eigne values and other statistics are given below Table 1.

In this table 1 above, the six components are incorporated on the basis of characteristic they represent and F1 is thesocio-religious background (Lanternari, V., 1974; Abbasi et al., 2020) of all stakeholders, F2 is reform oriented contemporary outlook of the temple authority and priest association, F3 is the true explanation of recognized religious scholars associated, F4 is the awareness among pilgrims, F5 is

Table 1. Percentage variance explained

Component	Initial Eigenvalues	Rotation Sums of Squared Loadings		
	Total	Total	% Of Variance	Cumulative %
Socio-religious background (F1)	7.41	3.94	17.11	17.11
Reform oriented contemporary outlook of temple authority and priest association (F2)	2.50	3.10	13.47	30.58
True explanation of recognized religious scholars associated (F3)	1.67	2.51	10.90	41.48
Awareness among pilgrims (F4)	1.36	2.05	8.92	50.40
Promotion of alternative Hindu religious philosophy viz. Vaishnavism etc. (F5)	1.24	1.89	8.20	58.60
Ritualistic rigidity (F6)	1.10	1.79	7.79	66.38

Table 2. Factor-wise mean scores

Factors	Mean	SD	Rank
Socio-religious background (F1)	3.51	0.84	2
Reform oriented contemporary outlook of temple authority and priest association (F2)	3.62	0.95	1
True explanation of recognized religious scholars associated (F3)	3.17	0.84	5
Awareness among pilgrims (F4)	3.18	0.79	4
Promotion of alternative Hindu religious philosophy viz. Vaishnavism etc. (F5)	3.27	0.80	3
Ritualistic rigidity (F6)	1.63	0.85	6
Overall	**3.22**	**0.59**	

the promotion of alternative Hindu religious philosophy viz. Vaishnavism etc. and F6 is the ritualistic rigidity respectively.

On the basis of the mean score, the ranking has been given to all six factors. The majority of the respondents showed their agreement with regard to F1 and F2 whereas the majority of respondents showed disagreement with regard to F 6. Respondents were found on a moderate level with regard to F3, F4, and F5 respectively.

FINDINGS AND DISCUSSIONS

There are many instances of religious and cultural amusement through animal killings, trafficking, and fighting as a means of promoting a temple, festival, or place of attraction. This concept is further strengthened in table 3 showing factor-wise mean scores on the basis of which reform-oriented contemporary outlook of temple authority and priest association and socio-religious background are ranked first and second respectively. Therefore, policymakers may initiate a mandatory accreditation of responsible religious tourism destinations keeping in view this paradox. As a remedial measure, the socio-religious background of the pilgrims is to be improved, and a reform-oriented contemporary outlook of temple authority and priest association is suggested. Both these issues are, so far, not found in any religious tourism policy in the study area. Promotion and adoption of correct rituals, religious scripture-oriented or alternative schools of scholarly thoughts are imperative as well. These will not draw any conflict of interest or misinterpretation and thereby conglomerate modern sustainable principles with the values of the predecessors. The world we live in today should not only require but rather demand that we implement policies that bring about a peaceful coexistence for all.

CONCLUSION AND LIMITATION

The purpose of this paper was to provide an overview of the animal cruelty conservation paradox as an upcoming piece of literature contributing to a widely discussed responsible approach to all next-generation tourism development. The extensive review justifies a two-fold aspect of policy issues from culture and community practices to conservation efforts of reserve areas. This is a growing field incorporating a wide variety of tourism practices. It is crucial that academics and practitioners realize the dire need to expand the type of research being conducted in this area. It is surprising that, despite a huge socio-political relevance, cultural rigidity, religious belief, and, above all, a paradoxical behavior is ignored while studying animals in tourism and hospitality literature in South and Southeast Asia.

Business houses and entrepreneurs can identify and utilize many profitable opportunities in different forms of tourism with an understanding of this body of knowledge, e.g. non-superstitious tourism, promotion of animal welfare with hymns, slokas, and sacred words, concerted and coordinated planning for rehabilitation of animals and pets in destinations, etc.

Animal cruelty, sacrifice, trafficking, consumption, etc. may be one of the parameters while promoting and accrediting any destination and thereby reducing the possibility of boycotting destinations (Shaheeret al., 2019). Moreover, this article advocates responsible religious practices and application of ethical standards in tourism (Winter, 2020). Scholars may conduct further research on the reasons forthe concentration of the Hindu pilgrim centers sacrificing animals in East India as mentioned in Figure 1. Academics can contribute by proposing and studying both the theoretical and practical impacts of this relatively hidden and much-avoided area of study for an immediate sustainable future.

The limitations of this study are varying levels of acceptance, tolerance, awareness, and implications of utilitarianism from society to society, community to community, and country to country in the region. Also, this area lacks even the preliminary records of animal killing, sacrifice, diseases of animals due to abuse, etc. in the study region.

REFERENCES

Ainsworth, M. D. S., & Bowlby, J. (1991). An ethological approach to personality development. *The American Psychologist*, *46*(4), 331–341. doi:10.1037/0003-066X.46.4.333

Asokan, G. V., Srihari, K. R., & Kumar, S. S. (2000). Bull racing injuries in southern India. *Veterinary*, *147*(19), 556. PMID:11095398

Banerjee, S. C. (1992). *Tantra in Bengal: A study of its origin, development, and influence*. South Asia Books.

Bansiddhi, P., Brown, J. L., Thitaram, C., Punyapornwithaya, V., & Nganvongpanit, K. (2020). Elephant Tourism in Thailand: A Review of Animal Welfare Practices and Needs. *Journal of Applied Animal Welfare Science*, *23*(2), 164–177. doi:10.1080/10888705.2019.1569522 PMID:30689409

Batabyal, D., Das, D. K., Sen, S. K., Porya, A., Halam, H., & Chatterjee, P. (2022). Evolution of Shakti Peetham for religious, mythological and dark tourism in Indian sub-continent. *Revista Turismo & Desenvolvimento (Aveiro)*, *39*, 291–300.

Batabyal, D., Rawal, Y. S., Chatterjee, P., & Goswami, S. (2023). Temple service quality model for future pilgrimages in Indian Sub-continent through user generated contents (UGC) in Trip Advisor. *International Journal of Tourism Anthropology*, *10*(1), 113–135. doi:10.1504/IJTA.2023.130849

Behrens, K. (2009). Tony Yengeni's Ritual Slaughter: Animal Anti-Cruelty vs. Culture. *South African Journal of Philosophy*, *28*(3), 271–289. doi:10.4314ajpem.v28i3.47859

Bertella, G. (2014). The Co-creation of Animal-based Tourism Experience. *Tourism Recreation Research, 39*(1), 115–125. https://www.trrworld.org/. doi:10.1080/02508281.2014.11081330

Bhattacharyya, N. N. (1974). *History of the tantric religion*. MunshiramManoharlal Publisher.

Boaz, D. N. (2019). The "Abhorrent" Practice of Animal Sacrifice and Religious Discrimination in the Global South. *Religions, 10*(3), 160. doi:10.3390/rel10030160

Bowlby, J. (1958). The nature of the child's tie to his mother. *The International Journal of Psycho-Analysis, 39*, 350–373. PMID:13610508

Bunnell, F. L., & McCann, R. (1993). The Brown or Grizzly Bear. In Bears: Majestic Creatures of the Wild. Rodale Press.

Cape Town Declaration on Responsible Tourism. (2002). *International Conference on Responsible Tourism in Destinations,* Cape Town, South Africa.

Dalal, R. (2014). *Hinduism: An Alphabetical Guide*. Penguin.

Donlon, J. G., Donlon, J. H., & Agrusa, J. (2010). Cultural tourism, camel wrestling, and the tourism 'bubble' in Turkey. *Anatolia. International Journal of Tourism and Hospitality Research, 21*(1), 29–39.

Duffy, R., & Moore, L. (2011). Global regulations and local practices: The politics and governance of animal welfare in elephant tourism. *Journal of Sustainable Tourism, 19*(4), 589–604. doi:10.1080/09669582.2011.566927

Dwivedi, O. P., & Tiwari, B. H. (1987). *Environmental Crisis and Hindu Religion*. Gitanjali Publishing House.

Dycka, M. G., & Baydack, R. K. (2004). Vigilance behavior of polar bears (Ursus maritimus) in the context of wildlife-viewing activities at Churchill, Manitoba, Canada. *Biological Conservation, 116*(3), 343–350. doi:10.1016/S0006-3207(03)00204-0

Erschbamer, M. (2019). Taming of supernatural entities and animal sacrifice, The synthesis of Tibetan Buddhism and local shamanistic traditions in Northern Sikkim (India). *Études mongoles et sibériennes, centrasiatiques et tibétaines, 50*. . doi:10.4000/emscat.3915

Fennel, D. A. (2014). Exploring the boundaries of a new moral order for tourism's global code of ethics: An opinion piece on the position of animals in the tourism industry. *Journal of Sustainable Tourism, 22*(7), 983–996. doi:10.1080/09669582.2014.918137

Fennell, D. A. (2012). Tourism, animals, and utilitarianism. *Tourism Recreation Research, 37*(3), 239–249. doi:10.1080/02508281.2012.11081712

Fennell, D. A., & Sheppard, V. (2020). Tourism, animals and the scales of justice. *Journal of Sustainable Tourism, 29*(3), 314–335. doi:10.1080/09669582.2020.1768263

Fuller, C. J. (2004). The camphor flame: popular Hinduism and society in India (Revised and Expanded ed.). Princeton University Press.

Galav, P., Jain, A., Katewa, S. S., & Nag, A. (2010). Animal healthcare practices by livestock owners at Puskar animal fairs, Rajasthan. *Indian Journal of Traditional Knowledge*, *9*(3), 581–584.

Goodwin, H. (2016). *Responsible tourism: using tourism for sustainable development.* Goodfellow. doi:10.23912/978-1-910158-84-5-3101

Goodwin, H. (2021). *Responsible Tourism in India. Government of India Institute of Secretariat Training & Management.* ISTM.

Gray, P. B., & Young, S. M. (2011). Human–pet dynamics in cross-cultural perspective. *Anthropos*, *24*(1), 17–30.

Gupta, V. (1999). Sustainable tourism: learning from Indian religious traditions. *International Journal of Contemporary Hospitality Management*, *11*, 91–95. //. doi:10.1108/09596119910250751

Hiltebeitel, A. (1980). The Sacrifices of the Water Buffalo and the Bull of Heaven. *History of Religions*, *19*(3), 187–223. doi:10.1086/462845

Holst, A., &Martens, P. (2016). Determinants of animal protection policy: a cross country empirical study. *Politics and Animals*, 1-14. Olju.

Jasrotia, S. S., & Koul, S. (2020). Do young consumers buy country made products? Evidence from India. *International Journal of Management and Enterprise Development*, *19*(4), 378–397. doi:10.1504/IJMED.2020.110830

Kavrečič, P., & Virloget, K. H. (2020). Traditional Festivities in North Istria in Terms of Authenticity and (dis) continuity. *Academica Turistica-Tourism and Innovation Journal*, *12*(2), 109–120.

Kerala. (2008). *Kerala declaration on responsible tourism.* 2nd International Conference on responsible tourism in destinations, Kerala, India.

Kuznets, S. (1955). Economic growth and income inequality. *The American Economic Review*, *45*, 1–28.

Lerner, P., & Rabello, A. M. (2007). The Prohibition of Ritual Slaughtering (Kosher Shechita and Halal) and Freedom of Religion of Minorities. *The Journal of Law and Religion*, *22*(1), 1–6. doi:10.1017/S0748081400003210

Lubbe, B. A., Preez, E. A., Douglas, A., & Wessels, F. F. (2017). The impact of rhino poaching on tourist experiences and future visitation to National Parks in South Africa. *Current Issues in Tourism*, *22*(1), 8–15. doi:10.1080/13683500.2017.1343807

Lynch, J. J. (1977). *The broken heart: The medical consequences of loneliness.* Basic Books.

Lynch, J. J. (2000). *A cry unheard: New insights into the medical consequences of loneliness.* Bancroft Press.

Malchrowicz-M, Ewa., Munsters, W., Korzeniewska-Nowakowska, P., & Gravelle, F. (2020). Controversial Animal Tourism Considered from a Cultural Perspective. *Tourism / Turyzm*, *30*(1,5), 21-30. . doi:10.18778/0867-5856.30.1.14

Mohanty, G. N., & Mohanty, S. N. (2004). Living with Snakes: The Lifestyle of the Snake Charmers (SapuaKela). *Adivasi*, *44*(12), 24–40.

Moodie, D. (2019). On blood, power, and public interest: The concealment of Hindu sacrificial rites under Indian law. *The Journal of Law and Religion*, *34*(2), 1–18. doi:10.1017/jlr.2019.24

Moorhouse, T. P., D'Cruze, N. C., & Macdonald, D. W. (2016). Unethical use of wildlife in tourism: What is the problem, who is responsible, and what can be done? *Journal of Sustainable Tourism*, 1–12. doi:10.1080/09669582.2016.1223087

Morris, M. C. (2013). Improved animal welfare is more related to income equality than it is to income. *Journal of Applied Animal Science*, *16*, 272–293. doi:10.1080/10888705.2013.768921 PMID:23795689

Negi, P. C., Sondhi, S., Asotra, S., Mahajan, K., & Mehta, A. (2019). Current status of rheumatic heart disease in India. *Indian Heart Journal*, *71*(1), 85–90. doi:10.1016/j.ihj.2018.12.007 PMID:31000189

Office of the District Magistrate and Collector, Birbhum (2016). *Mega Tourism Project at Tarapeeth, Birbhum*. Office of the District Magistrate and Collector. file:///C:/Users/it/Downloads/121114121012 17EWVJGJustificationfortheProjectinforestland.pdf

Panneerselvam, R. (2014). *Research methodology*. PHI Learning Pvt. Ltd.

Rakitovac, K. A., Urošević, N., & Vojnović, N. (2021). Community Participation in Sustainable Valorisation of Cultural Heritage: The Case of the Municipality of Vrsar. *Academica Turistica*, *14*(1), 7–22. doi:10.26493/2335-4194.14.7-22

Savishinsky, J. S. (1983). Pet ideas: The domestication of animals, human behavior, and human emotions. In A. H. Katcher & A. M. Beck (Eds.), *New perspectives on our lives with companion animals* (pp. 112–131). University of Pennsylvania Press.

Sébastia, B. (2004). *Like an Indian God, Hinduisation of the cult of Saint Anthony of Padua in Tamil Nadu*, Auxlieuxd'Etre.

Serpell, J. (1987). Pet keeping in non-western societies: Some popular misconceptions. *Anthrozoos*, *1*(3), 166–174. doi:10.2752/089279388787058443

Shackley, M. (1996). Community impact of the camel safari industry in Jaisalmar, Rajasthan. *Tourism Management*, *17*(3), 213–218. doi:10.1016/0261-5177(96)00008-8

Shaheer, I., Carr, N., & Insch, A. (2021). Spatial distribution of participation in boycott calls: A study of tourism destination boycotts associated with animal abuse. *Anatolia*. *Anatolia*, *33*(3), 323–334. doi: 10.1080/13032917.2021.1931896

Shastri, B. (1994). *KalikapuraneMurtivinirdesah*. Motilal Banarsidass Publications.

Shastri, J. L. (1970). *The Siva Puraṇa*. Motilal Banarasidass.

Singh, S. (2006). Secular pilgrimages and sacred tourism in the Indian Himalayas. *GeoJournal*, *64*(3), 215–223. doi:10.100710708-005-5649-8

Tul, H. (2015). Kali's Tongue: Shame, Disgust, and the Rejection of Blood and Violence in Vedic and Hindu Thought. *International Journal of Hindu Studies*, *19*(3), 301–332. https://doi.org/. doi:10.1007/s11407-015-9181-2

Urban, H. B. (2016). Desire, Blood and Power– Georges Bataille and the Study of Hindu Tantra in Northeast India. *RevistaCientíficaArbitrada de la Fundación MenteClara*, *1*(3), 26–48. doi:10.32351/rca.v1.3.22

Urban, H. B. (2019). 'The Cradle of Tantra': Modern transformations of a Tantric Centre in Northeast India from Nationalist Symbol to Tourist Destination. *South Asia*, *42*(2), 256–277. doi:10.1080/00856401.2019.1570609

Winter, C. (2020). A review of research into animal ethics in tourism: Launching the annals of tourism research curated collection on animal ethics in tourism. *Annals of Tourism Research*, *84*(10), 29–89. doi:10.1016/j.annals.2020.102989

APPENDIX ONE

Table 3. Factor and factor loadings

Statement	Component					
	1	2	3	4	5	6
Educational level of the pilgrims has direct relationship with the perception of animal cruelty, killing, torture etc	0.797					
Educational level of priests has direct relationship with the perception of animal cruelty, killing, torture etc.	0.756					
More are the followers more will be the ritualistic animal sacrifice, torture etc.	0.630					
It causes disrespect for a religion	0.627					
Ritualistic animal killing, cruelty etc. exuberate a negative image for a pilgrim center	0.614					
Promoting debate, talk show etc. of true religious scholars on animal sacrifice, cruelty, trafficking, harsh training etc.	0.609					
Promoting animal welfare through religious scriptures can be a good option to reduce animal cruelty, killings, torture etc.	0.471					
Temple authority has direct control over increase, decrease or fully stop of this ritual		0.832				
As pilgrims follow temple priests, these priests can check or control animal sacrifice		0.828				
Ritualistic animal sacrifice can be significantly checked if priests or alike are willing to do so		0.786				
Religion does not always promote animal killing, rather held responsible for worship, love and care			0.823			
Religious scriptures are not directly promoting more animal sacrifice, rather very less in number			0.703			
Religion does not always promote animal cruelty, killing etc., rather it is symbolic, rarely occurred event with almost no significant impact on overall daily consumption of meat, etc.			0.678			
Conservation awareness has been increasing, though not able to influence ritualistic rigidity of animal sacrifice, cruelty etc.			0.582			
Promoting Vaishnab philosophy, Jainism etc. near around Shakti Peeths is another way of checking animal sacrifice.				0.661		
Priests cannot control as they are orthodox and follow rituals strictly				0.623		
Ritualistic killing has been decreasing day by day and will stop in near future				0.574		
Conservation awareness has been increasing, and able to influence ritualistic rigidity of animal sacrifice, cruelty etc., though more such awareness required				0.542		

Chapter 11
Community Participation and Government's Role in Religious Tourism Development in Kurukshetra

Lakhvinder Singh
https://orcid.org/0000-0002-0778-1132
Government College, Kaithal, India

ABSTRACT

The present study confined to Kurukshetra and empirically investigate participation of local community and role of government towards development of religious tourism in area. The study used 234 responses conveniently collected by using self-administered questionnaire. On the analysis of data using different statistical tools, the study reports sufficient potential of religious tourism in the area along with availability of basic and limited facilities. From an empirical perspective, the study reports moderate participation level of local community in tourism development and government as major decision-making body for tourism development in the area. The study concludes to implement a participatory approach in tourism planning as a mechanism to accelerate the overall development of the area.

INTRODUCTION

The religious tourism is an integral part of human travel and among the most ancient of travel destination such as holy shrines, temples, rivers, festivals and other sacred sites. The visit to these places (pilgrimage) has been major motive of movement and migration of people along with seeking outdoor recreation (Shinde,2008) and thus progressively attaining resurgence throughout the world and has become an essential component of tourism industry. Thus, every region of the country is identified with its unique culture and religious heritage and hence, is a paradise for those seeking religious and pilgrimage places along with unique features of modernity and traditional hospitality (Dasgupta, 2011). India is dotted with thousands of prominent religious tourism places such as sacred four dham including Badrinath,

DOI: 10.4018/978-1-6684-9957-3.ch011

Puri, Rameshwaram and Dwarka, besides Ajmer, Goa, Bodhgaya, Amritsar and Mount Abu are among some famous pilgrimage places (Singh,2002). The incredible tourism potential of country, succeed to receive more than 1.5 million foreign tourist arrivals and above 6.7 million domestic tourists visits in the country during 2021 (Ministry of Tourism, Govt. of India, 2022).

Among tourism potential in the country, the Kurukshetra has been emerging as a popular destination for religious tourism in the region (Singhal, 2009). Kurukshetra is located in northern state of Haryana is considered as important among tourists. The destination is famous as the battle of '*Mahabharata*' was fought and lord '*Krishna*' had preached his philosophy of '*Karma*' at this place as well as it is richly dotted with many important religious tourist attractions (Kumar & Singh, 2014). Religious tourism development in Kurukshetra is going well and numerous religious places are scattered in the city. Every year millions of tourists visited different attractions Kurukshetra to places such as Sacred water tank of Brahm Sarovar, Bhadra Kali temple, Bhishma Kund, Sthaneshwara Mahadev Temple and very famous International Gita festival and Sun eclipse celebrated at sacred water tank 'Brahma Sarovar' received huge influx of visitors to the area.

The number of tourist arrivals to Kurukshetra was notices as 11.71 Lakhs during the year 2021-22 from 11.50 lakhs during 2011-12 (Statistical Abstract of Haryana, 2021-22) and thus, considering the touristic importance of Kurukshetra, number of efforts has been taken by the Indian Government and concerned authorities for integrated development of mega destinations for suitability addressing the existing infrastructure gaps and to speed up the pilgrimage tourism in the area. For instance, the region is considered under integrated circuit scheme to develop the position of tourism in the area. Pilgrimage Rejuvenation and Spiritual Augmentation Drive (PRASAD) of the country under the aegis of Ministry of Tourism has selected Kurukshetra under Krishna circuit and its surrounding tourist spots for development (FICCI,2015). Resultantly, with the improvement of basic facilities and amenities at the destination, the tourist footfalls increased in the area. For instance, nearly 1.2 million tourists have been visited Kurukshetra during the year 2021-22 as per report of the Statistical Abstract of Haryana (2021-22).

In respect to development of tourism at the destination, World Tourism Organization (UNWTO) reports an improvement in economy such as generation of employment, entrepreneurial and other earning activities following by socio-cultural development as well as protection of environment (UNTWO, 2015). Consequently, it has given rise to better understanding and management of host community and stakeholders' perspective to achieve overall development (George & Swain, 2005). But from another side, it has been considered that, although, the development of religious tourism led to development of destination through revenue generation, employment & business growth beside promoting culture-heritage (Vijayanand, 2013; Obasola,2014), but, not infrequently also bring certain challenges such as economic leakage due to involvement of outsiders, deterioration of culture and damage to environment as consequence of unplanned tourism development (Pendit, 2002; Ketmuang,2005). Therefore, a question whether tourism is beneficial or harmful for a society is a much debated one because the host communities are directly affected by tourism at destination; therefore there is a need to include opinion and views of local community and other stakeholders regarding impact of tourism to their life. This present research examines local community and government engagement in development of tourism in Kurukshetra.

OBJECTIVES AND HYPOTHESIS OF THE STUDY

The aims of the present research are to:

Identify the community perception of potential of religious tourism in study area.
Assess the community participation in religious tourism development in the area.
To discover the role of government in religious tourism development in the area.

The study also proposes to test the following proposed:

H_1: There is no significant difference in community perception about potential of religious tourism in Kurukshetra as per occupational status of respondents.
H_2: There is no significant difference in community perception about participation in religious tourism development in Kurukshetra as per occupational status of respondents
H_3: There is no significant difference in community perception towards role of government in religious tourism development in Kurukshetra as per occupational status of respondents.

THEORETICAL BACKGROUND

Religious Tourism

There are many kinds of tourist attractions available including travelling to religious places. The religious tourism can be interpreted as visits to sacred places, to participate in religious ceremonies and pilgrimages in the form of visits or activities to perform worship in evaluating tourism understanding purposes such as to get a blessing, inner strength, firmness of faith, and even abundant wealth (Jalilvand & Heidari, 2017). Thus, visit to religious tourism destination is among emerging segment of tourism sector. The growing improvement in infrastructural and basic amenities along with advancement in technology triggered the visit of tourists to religious tourism destinations For instance, World Tourism Organization in a report ascertained that nearly sixty percent people of world practices travelling to religious and pilgrimage tourism places. (UNWTO, 2011).

Travelling to religious places is increasing and resulted in to several benefits to local community and other stakeholders at the destination (Wall & Mathieson, 2006). Religious tourism is found to generate economic development opportunities such as revenue through tourist spending (Holloway & Valins, 2002), employment generation (Vijayanand, 2012), entrepreneurial opportunities (Wall & Mathieson,2006), facilitated investment and infrastructure development (Akal, 2010). However, to some extent due to excess and unplanned development, few of challenges also associated with religious tourism such as, lack of economic diversification, economic strain due to inflation of prices of goods & services (Hall,2000), undesirable loss of social values and cultural ethnicity, pollution in physical and ecological resources (Samira, Mondal & Basu,2006). Thus, it can be ascertained that tourism has become a tool for change, development and reconstruction of the destination (Claiborne, 2010) and hence, it become crucial to document the tourism from viewpoint of its different stakeholders to induce development through adequate tourism planning and management.

Community Participation and Government Role

Tourism development involves participation of different stakeholders in planning and formulation of certain plans regarding growth of tourism at the destination. However, the most significant role played by government, regional authorities, and local community and event tourists themselves in working out concrete measures for development of a particular destination (Dredge & Jenkins, 2007). This has considerably documented researchers (such as, Mohammadi, Khalifah, & Hosseini, 2010) that ascertained that the perception of local community may have both positive and negative agreement with the development of tourism at religious tourism site. Resultantly, it has been suggested by existing literature (Mishra,2000; James & Robinson,2009) to adopt measures like restructuring tourism approach and supportive role of its different stakeholder including local community, tourism organizations and other important stakeholders such as government being directly associated to the formulation of policies, plans, strategies and laws about tourism. Similar to this, some of the authors (such as, Sharma, 2000; Wilson & Venes, 2001; Hall & Richards, 2002; Kandari & Chandra, 2004) identified that participation of local community and regional government particularly responsible to work out a balanced and overall development of destination. More explicitly, from the perspective of religious tourism, some research (Shinde, 2003; Saravanan & Rao, 2012) also suggested to understand the opinions of local community including other stakeholders (including government) to help drive sustainable and community-led tourism growth.

Therefore, in view of the above discussion, it has been observed by author that, participation of local community and government involvement is crucial for development of tourist at a particular destination, thus has been noticed as important aspect to consider. But, despite in tourism development, but despite this, very little or no attention has been given to this research domain particularly when it pertains to religious tourism in Indian context, which has been considered as a research gap. More, to develop a better approach to management and comprehensive development of religious tourism, it is justified to conduct a meaningful study to understand local community participation and role of government at religious destination of Kurukshetra in order to increase the effectiveness of development of tourism in the area.

RESEARCH APPROACH

The present chapter covers an overview of religious tourism potential in Kurukshetra as well as empirically investigates the participation of local community and government role in tourism development in the area. For providing an overview of religious tourism potential of the destination, the author used secondary data sources such as books, magazines, government reports and internet sources to draw a list of important religious tourism sites and present in the chapter.

The study used primary data sources and consider two groups of respondents, the first covering who are working in tourism industry and later includes those who are not working in this industry. For collection of data from groups of respondents, the study applied field based survey among local residents and government stakeholders of tourism including government tourism resorts, accommodation, information centres, museums staff and other relevant boards. Field bases survey was conducted by using convenient sampling method to contact respondents including local residents near tourist sites, working professional in the tourism sector, public and private tourism organizations, shopkeepers & retailers in the area. A self-structured questionnaire was used in the survey to collect data and distributed to 400 respondents and only, 234 completely filled and usable responses were returned with response rate of 58.5 percent.

The survey was conducted from October 2022 March 2023. The questionnaire was designed in three parts. The first part contained information related demographic background of the respondent. The second part covers questions related to perception of local community about tourism potential, level of their participation in tourism development and government-sponsored tourism development plans. The last part of the questionnaire contains questions related to role played by government in development of tourism in the area. The second and third part of the questionnaire was asked to rate their responses to specified statements based on a 5-point Likert-type scale from "strongly agree (1)" to "strongly disagree (5)". Data analysis was conducted using suitable statistical tools such as mean, frequencies, standard deviation, and Chi- square test. The result of data analysis presented through tables in order to make findings more meaningful and presentable.

MAIN FOCUS OF THE CHAPTER

This section of the chapter is presented in two sections. The first part of the section covers explanation of major tourist places and tourist facilities in Kurukshetra. The later portray analysis of empirical result regarding community perception and role of government in religious tourism development at the destination.

Religious Tourism in Kurukshetra

Kurukshetra is situated in the state of Haryana and well connected with all the major destinations of country through rail and road network. The nearby airports are located at Chandigarh (100 Km) and New Delhi (160 Km). It is among the major tourist destination of Haryana. The tourists during their visit to Kurukshetra do not find any difficulty regarding arrangement of accommodation. Besides government run tourist complexes at Pipli, Thanesar, and Pehowa, numbers of budget hotels, guest houses, lodges, inns and charitable accommodation are existed in the city for staying there. However, there is lack of big hotels or star category hotels in the area. A wide variety of foods and drinks are easily available in the city for tourists and general population. The basic amenities such as bathing places (ghats) on sacred water tanks, toilets and washrooms, drinking water, medical facilities, banking & ATM facilities, arrangement of tourist police and shopping & souvenir opportunities are easily spotted in the area. For understanding the real picture of religious tourism, a brief overview of different tourist attractions of the area presented in following paragraphs.

The holy *Brahmasarovar*, the sacred water tank is located on south-western corner of the town and claimed to be the largest man made tank in the country. It is among the most sacred site in Kurukshetra and consecutively the most visited one, as well. The sanctity of the sarovar is lies in the belief that anybody taking bath in the tank especially on the occasion of Solar and Lunar eclipse gets blessings of Gods and Goddesses and bring salvation from all sins. Thus lakhs of pilgrims from all over the country assemble here to take holy bath in its holy water and to visit its nearby pilgrimage sites.

The religious fragrance of Kurukshetra is enriched by the magnificent Birla temple located adjoining to the Brahmsarovar tank. The temple has huge image of Lord *Krishna* in the premises delighted the pilgrims those who visited at this temple. The text of Mahabharata is written on the walls of the temple. The temple's beauty is depicted by the Chariot of Lord Krishna conveying the sermon of Geeta to Arjuna.

The Lakshmi Naryana temple adds another feather in the cap of the holy city of Kurukshetra. The main deity of the temple is Lord Narayna and Goddess Luxmi. The temple is constructed in two story building. Numbers of pilgrims visited this temple to explore this magnificent temple of ancient time.

Famous museum at a small distance to this temple, Sri Krishna museum on a close distance, is preserving and presenting the multifaceted personality of Lord Krishna, the episodes of his childhood, philosophy and teaching in form of rare manuscripts, paintings, scriptures and other artistic creations as well as archaeological material unearthed in Kurukshetra and its neighbourhood. Thousands of tourists visited this museum to get close the Lord Krishna.

Adjacent to Sri Krishna museum- one comes across the nicely designed Sri Krishan Panorama & Science centre, one of its kind in the country. It has been built with the twin aim of reviving the memories of Mahabharata in an effective way and educating the mass. The ground floor of the Panorama is depicted with examples of scientific process which a person come across in routine life presented with the help of suitable instruments or devices. The first floor of the museum is exclusively developed to a light depiction of Mahabharata battle. The concerned authorities frequently put their efforts to glorify this heritage of the Kurukshetra and to attract more visitors to the Panorama.

The Bhadar Kali temple of the city is dedicated to Goddess *Satti* and claimed to be a *Shaktipeetha*. According to belief the Pandvas performed the worship of Goddess at this temple to take a boon of their victor in the battle of Mahabharata. A good number of pilgrims gathered here particularly on the occasion of Navratras, Dussehra and many more to gain the favour and blessings of Goddess.

An ancient Shiva temple named as 'Sthaneshwar Mahadev temple', situated on the outskirts of Kurukshetra is one of the most important pilgrimage site in area. Infact, it is after this temple that Kurukshetra derived its name as Sthaneshwara or Thanesar which is still popular among local peoples of the area. The holy pond situated adjoining to the temple is considered to be very sacred and it is believed that all the sins committed deliberately are naturalized as well as may attain heaven with influence of Loard Shiva and taking a dip in holy pond at this place.

Situated five kilometres west of Thanesar, identified as the birthplace of Bhagwadgita, Jyotisar is the most sacred spot reminiscent of Great Epic battle of Mahabharata. Lord Krishna is believed to have delivered his famous philosophical message of Bhagwadgita to Arjuna close to a Banyan tree at this place. The five thousand (5000) year old immortal Banyan tree is still proudly stands on a raised platform near temple complex of Jyotisar. The historic water tank, once existed here is now renovated with bathing Ghats. Pilgrims visited in thousands at this place to pay homage to Banyan tree, bathing in water tank and listen to the verses of Bhagwadgita. A tourist complex run under the aegis of Haryana Tourism Corporation is also developed here for providing pleasant staying arraignments to the tourists.

Pehowa is also an important tourist site and located about twenty five kilometres westwards Kurukshetra. According to legend, the Prajapati (Brahma) created the world and four Varnas at Prithuduka (Pehowa). After recognizing the religious importance of this place, the Kurukshetra Development Board has consistently putting its efforts to develop it as pilgrimage tourism hub in the area.

Kurukshetra is famous for the celebration of different religions and cultural events organized in the city. The most prominent among those is annual International Gita festival is a major seven days festival celebrated in Kurukshetra and the craftsmen and artesian stalls are installed in this event. Solar Eclipse is also a famous fair held here which attracts lakhs of devotees who came to take bath in the holy water of sacred tanks of Kurukshetra. The A huge crowd of pilgrims also attended the fair held on the day Amavsya falling on of Monday (Somwti Amavsya) celebrated on the banks of sacred tanks of the Kurukshetra.

A big fair is held at Sarawati tank in Pehowa on the occasion of Chaitra Amavasya also attracting lakhs of pilgrims during the occasion to perform the last rituals of their ancestors.

Along with religious tourism sites, Kurukshetra also revealed its relevance to culture-heritage attractions including Multi art cultural centre, prominent Sheikh Chellis', Dharohar Museum, Kalpna Chawla Planetarium, Mini Zoo of Pipli, are also famous among visitors.

The regional authorities and government are playing an important in development of tourism in the area. For instance, Haryana Government after recognizing the tourism potential of the city, established a high powered Kurukshetra Development Board (KDB) in August 1968 to facilities development over a vast area of about 100 miles in which 134 places of pilgrimage/tourists interest are situated. The board primarily responsible to ensure comprehensive tourism growth, to make arrangement for tourist facilities and for overall development in the area.

Empirical Findings and Result

The result regarding demographic profile of the respondents in the table 1 provides a summary profile of the respondents. The table highlights more participation of male respondents as compared to their female counterparts as revealed from its result that shows that 54.2% were male and 45.8% were female respondents. Age wise classification of respondents reported that middle age group cover highest participation in the survey. This out of total most of the respondents were belongs to 25-44 year age group (i.e. 25-34 years of 31% and 35-44 years of 31.7%). This is followed by younger age group of 18-24 years that represent 18% of total surveyed respondents. The 14.4% respondents were belongs to senior age groups thus involves who falls in the 45-59 years of age group. The remaining 4.9% of respondents were either of 60 years of more than that years of age. Though, it has been detected that highest respondents were belongs to middle age followed almost similar number of younger and senior age group respondents. The result portray variation in literacy level of the respondents and thus indicated that 26.1% respondents having postgraduate degree level education, followed by graduate degree holders covering 22.9% of total respondents. This is followed by group of respondents having senior secondary education covering 21.0% and high school education holder covering 19.7% of total respondents. The remaining 10.3% of respondents having only basic elementary school education. The education wise classification of respondents reported participation of almost all the section of the community ranging from highly educated to basic educated respondents. Furthermore, it has been noticed that just above $2/5^{th}$ of the respondents (42.1%) of respondents are working in the tourism sector including both government and private tourism related occupation. More than half of the respondents (57.69%) are related to non- tourism related occupation in the study area that involves government/private jobs and business, shopkeepers & retailers and students in the area.

Community Perception About Tourism Potential in Kurukshetra

In order to explore local views about what tourism in Kurukshetra, the results indicate that the majority (> 50%) of respondents had a positive view of the potential of tourism in providing job opportunities for locals, enhancing income, and improving their general well-being. It should be noted that lower mean score (conversely higher rank) mean strong agreement, while higher score (conversely lower rank) mean strong disagreement with the pre-determined statements presented to the respondents. The low standard deviation (SD) coefficient suggests similarities in each respondent's views. The mean scores of fourth, fifth and

Table 1. Demographic profile of respondents

Demographic Features		Percentage
Gender	Male	63.0
	Female	37.0
Age	18-24	18.0
	25-34	31.0
	35-44	31.7
	45-59	14.4
	60 +	4.9
Education	Elementary School	10.3
	High School	19.7
	Sr. Secondary	21.0
	Bachelor	22.9
	Post-Graduate	26.1
Occupation	Working in tourism (government/private)	42.31
	Working in non-tourism occupation (jobs/business/shopkeepers/ students	57.69

Source: Author's Creation.

seventh variables are below 2.91 which imply that overall participants tend to support these statements. Whilst the mean score for sixth variable was 4.22, which means that the responses to this statement was in the range between disagree and strongly disagree. It can be observed from the standard deviation scores for the first and second variables that there is similarity in participants' opinions about these statements.

In table 3, the analysis of Chi-square indicates no significant differences exist between the people (participants) who work in tourism sector and those who don't work (Chi-square= 3.401, P= 0.493) which highlights that local people in general believed that tourism can help to improve their livelihood regardless their participation in tourism industry. Regarding impact of religious tourism towards providing jobs to local community, Chi-square test shows that there is a significant differences exist (Chi-square= 9.537, P= 0.008) in the area. Furthermore, the Chi-square test fails to shows any significant differences (Chi-square= 1.337, P= 0.513) regarding income generation through tourism. This could be due to poor participation of local community by government. On statement regarding role of tourism in creation of local infrastructure, Chi-square test shows that there is a significant differences exist between the local people who work in tourism sector and those who don't work (Chi-square= 36.118, P= 0.000) in Kurukshetra. The next question from the table through Chi-square test shows that there is a significant differences exist among two groups (Chi-square= 54.189, P= 0.000) and perceived requirement of more efforts to involve private sector in setting up tourism policy and schemes. Next to this, the result of Chi-square test shows that there is a significant differences (Chi-square= 24.761, P= 0.000) was noticed between two groups of respondents and confirmed the adequate distribution of tourism schemes

Table 2. Community perception of tourism potential

Question	Percent of responses					Mean	SD
	SA	A	N	D	SD		
Tourism can offer jobs for local community	73.7	21.4	2.8	1.5	0.6	1.34	0.699
Tourism can generate income for local community	60.4	33.4	4.3	1.5	0.4	1.48	0.691
Tourism can help locals to improve their livelihood	57.8	33.4	5.1	2.1	1.5	1.56	0.812
Tourism in area help to create local infrastructure	26.6	30.8	12	19.1	11.6	2.58	1.361
Tourism development encouraged the private sector to set up tourism in Kurukshetra	8.6	44.3	7.9	25.5	13.7	2.91	1.259
Tourism policies & schemes sponsored by government are distributed fairly in area	0.2	1.3	12.4	48.4	37.7	4.22	0.724
The area deserve more tourism policy and schemes	66.6	27.8	5.1	0.2	0.2	1.40	0.618

Source: Author's Creation.

and policy to the area. At last, the Chi-square test shows that there is a slight difference exist between respondents' groups (Chi-square= 10.329, P= 0.006) regarding necessity to implementation of more policy and schemes in order to attract tourists in Kurukshetra.

Hence, from the analysis of the result of Chi-square test discussed above, the hypothesis(H_1) postulating no significant difference is not supported and thus can be concluded that two groups of respondents who are working and who are not working in tourism have different perception towards potential of religious tourism in Kurukshetra.

Community Participation in Tourism Development in Kurukshetra

The result presented in below table number 4 provides a summary of descriptive result and indicates positive agreement of local community regarding participation in tourism development in terms of all

Table 3. Chi-square test for community perception of tourism potential

Statement	Working in Tourism		Not Working in Tourism		Chi- Test	Prob.
	Mean	SD	Mean	SD		
Tourism helps locals to improve their livelihood	1.5	0.678	1.53	0.702	3.401	0.493
Tourism offer jobs for local community	1.21	0.483	1.31	0.482	9.537	0.008
Tourism generates income for local community	1.47	0.581	1.44	0.532	1.337	0.513
Tourism projects in Kurukshetra have helped create local infrastructure	2.68	1.142	2.31	1.423	36.118	0.000
The concerned tourism authority has encouraged the private sector to set up tourism plans and schemes in area	3.17	1.069	2.78	1.197	54.189	0.000
Tourism plans & schemes sponsored by government are distributed fairly in area	4.03	0.608	4.38	0.693	24.761	0.000
Kurukshetra needs more tourism policy & schemes	1.21	0.406	1.41	0.589	10.329	0.006

Source: Author's Creation.

the variable statements. The table indicates results in respect to variables having low mean scores (1.45 to 1.52) implies positive agreement among respondents which indicating a consistent pattern with these variable statements. The variable with mean scores from 1.57 to 1.63 indicating overwhelming positive affirmation and confirm that few governmental efforts have been made to involve the local community in setting up tourism plans and schemes in Kurukshetra, but that there is a desire to be involved in the decision-making processes, and awareness of challenges faced by community representatives. The highest mean score reported by statement regarding government preference in setting more tourism policy and schemes to support small scale tourism related business in study area. It means government notice influential impact of micro level earning activities from tourism.

The application of Chi-square test shows that there is a slight difference exist between the people who work in tourism sector and who don't work (Chi-square= 10.580, P= 0.032) and favourable agreement regarding participation of local community can play important role in setting up plans and schemes in Kurukshetra. The Chi-square test shows a significant differences between selected respondent groups (Chi-square= 15.682, P= 0.003) and strongly believed and agreed that local community can contribute to address tourism development challenges as they are aware of tourism needs, while some of the participants who are not involved in tourism took a neutral position as they are not interested in tourism. Furthermore, the statement regarding favourable attitude of local community towards tourism in area through Chi-square test shows that there is a significant differences exist between the people who work in tourism sector and who don't work (Chi-square= 13.309, P= 0.004). The table also covered the statement regarding views towards representatives of locals and their participation in tourism decision making and Chi-square test reported a significant difference exist between the people who work in tourism sector and who don't work (Chi-square= 14.990, P= 0.005) on this statement. The result of Chi-square test also portray existence of significant difference exist between the people who work in tourism sector and who don't work (Chi-square= 23.380, P= 0.000) on asking about their opinion regarding availability of representative of local community to contribute in making decisions about tourism development. In addition to above, the Chi-square test shows that there is a significant difference exist between two selected groups of respondents (Chi-square= 24.337, P= 0.000) which indicates the existence of community

Table 4. Community perception of participation in tourism development

Question	Percent of responses					Mean	SD
	SA	A	N	D	SD		
Local community play an important role in setting up plans and projects in Kurukshetra	61.5	33.6	3.4	1.3	0.2	1.45	0.648
Local community contributes to address tourism development challenges	59.7	34.5	3.4	1.9	0.4	1.49	0.697
Local community is in favour of schemes related to small scale tourism business in Kurukshetra	40.2	41	7.7	9.4	1.7	1.91	1.005
Local community can elect its representatives to participate in tourism decision making	51.2	42.2	4.3	1.7	0.6	1.58	0.714
Local community representatives should have a voice in making decisions about tourism development	52.9	39.8	5.1	1.7	0.4	1.57	0.709
Local community representatives face challenges in providing input to tourism decision making process in Kurukshetra	50.5	39.2	7.1	2.8	0.4	1.63	0.770
The concerned authority sets up tourism plans and schemes without consulting with the local community	53.7	35.5	5.8	4.1	0.9	1.63	0.835

Source: Author's Creation.

representatives to have voice in tourism decisions making. The statement about community representatives role in making decisions about tourism, Chi-square test indicates a significant difference between selected groups of respondents (Chi-square= 31.240, P= 0.000).

Hence, from the analysis of the result of Chi-square test discussed above, the hypothesis (H_2) postulating no significant difference is not supported and thus can be concluded that two groups of respondents who are working and who are not working in tourism have different perception towards local community involvement in religious tourism sector in Kurukshetra with each variable presented in the above table.

Role of the Government in Tourism Development

Results of the survey showed that 66.4% of the respondents were having positive agreement and remaining was having negative agreement with role of government towards tourism development in the area. This indicates that most of the respondents were having a strong perception towards government's role in tourism development. In general, there was overwhelming agreement that the role of government in tourism was critical in helping small businesses with grants and loans, prioritizing tourism - related investments, providing investment opportunities to the private sector, allocating tourism revenues for community livelihood projects, and in solving challenges within the tourism industry. With responses to all six statements below mean scores of 2.0 (strongly agree or agree), it can be deduced that most of the respondents would like to see the government playing a stronger role in tourism development with the exception of one respondent, all respondents agreed that government should offer loans and grants to local communities to help establish small businesses' as this statement had the lowest score (mean 1.18, SD 0.386; recall, low scores mean positive affirmation), implying the need for government support to tourism entrepreneurs. The same was true in case of government priority to tourism investments.

The table number 7 showcasing that the grant of loans and financial assistance for establishing tourism business was sufficient and analysis of Chi-square test indicates very slight difference among two

Table 5. Chi-square test for community perception for participation in tourism development

Statement	Working in Tourism		Not Working in Tourism		Chi-Test	Prob.
	Mean	SD	Mean	SD		
Local community can play important role in setting up plans and schemes in Kurukshetra	1.55	0.804	1.37	0.535	10.580	0.032
Local community contributes to address tourism development challenges	1.67	0.900	1.35	0.497	15.682	0.003
Local community is in favour of schemes related to small scale tourism business in Kurukshetra	1.72	0.741	1.48	0.502	13.309	0.004
Local community can elect its representatives to participate in tourism decision making	1.79	0.876	1.47	0.535	14.990	0.005
Local community representatives should have a voice in making decisions about tourism development	1.82	0.943	1.51	0.519	23.380	000
Local community representatives face challenges in providing input to tourism decision making process in Kurukshetra	1.99	0.978	1.49	0.582	24.337	000
The concerned tourism authority sets up tourism plans and projects without consulting with the local community	1.99	1.095	1.39	0.508	31.240	000

Source: Author's Creation.

Table 6. Role of the government in tourism development

Question Statement	Percent of responses					Mean	SD
	SA	A	N	D	S		
Government should offer loans and grants to local communities to businesses	82.7	17.1	0.2	0	0	1.18	0.386
Government should give priority to tourism investments	66.8	33	0.2	0	0	1.33	0.477
Government should provide investment opportunities in tourism to the private sector and local residents	57.2	41.1	1.5	0.2	0	1.45	0.539
Government should offer job opportunities for local people	55.2	38.5	4.1	1.9	0.2	1.53	0.687
Government should allocate part of the tourism revenue to local community livelihood	43	54.4	1.9	0.6	0	1.60	0.563
The concerned authorities should play important role in solving tourism-related problems	35.8	60.6	2.6	0.9	0.2	1.69	0.585

Source: Author's Creation.

groups of respondents (Chi-square= 3.896, P= 0.048). For the next statement regarding effort of government to priority and encourage tourism investments in Kurukshetra in order to attract more tourists to visit the area, the Chi-square test reported a significant difference between two groups of respondents (Chi-square= 41.269, P= 0.000). This approves that those participants who are involved in tourism have more information about tourism and aware of tourism needs in Kurukshetra due to their interactions with tourists. Next to this, the Chi-square test (Chi-square= 27.500, P= 0.000) portrayed two selected groups of respondents have different opinion and considered as significant in terms of statement that the government should provide investment opportunities in tourism to the private sector and local residents in Kurukshetra in order to enhance the tourism sector. The analysis of Chi-square test (Chi-square= 8.438, P= 0.038) shows is a slight difference in opinion of two groups of respondents regarding government effort to offer job opportunities for local people to enhance their livelihood. The statement regarding allocation of tourism revenue to local community to enhance their livelihood and help them also analysed the Chi-square test indicates a significant difference among two groups of respondents in study area (Chi-square= 16.515, P= 0.001). The Chi-square also reported no significant difference between groups selected for survey (Chi-square= 9.34, P= 0.052) on asking about role played by concerned tourism authorities in solving tourism-related problems and provide a tourism environment for tourists, investors, and local communities.

Hence, from the analysis of the result of Chi-square test discussed above, the hypothesis (H_3) postulating no significant difference is not supported and thus can be concluded that two groups of respondents who are working and who are not working in tourism have different perception towards role played by government in religious tourism sector of Kurukshetra.

SOLUTION AND RECOMMENDATIONS

The present study local community participation and role of government in development of religious tourism in Kurukshetra. The chapter also draws a picture of religious tourism attractions and major facilities in the area. The author from the study provide some suggestions to support development of tourism in an effective manner. Suggestions drawn from the study are:

Table 7. Chi-square test for the role of government in tourism development

Statement	Working in Tourism		Not Working in Tourism		Chi-Test	Prob.
	Mean	SD	Mean	SD		
Government should offers loans and grants to local communities to help establish small businesses	1.15	0.354	1.25	0.434	3.896	0.048
Government should give priority to tourism investments	1.19	0.392	1.60	0.492	41.269	0.000
Government should provide investment opportunities in tourism to the private sector and local residents	1.30	0.460	1.68	0.597	27.500	0.000
Government should offers job opportunities for local people	1.34	0.476	1.49	0.651	8.438	0.038
Government should allocates part of the tourism revenue to local community livelihood	1.55	0.500	1.47	0.689	16.515	0.001
The tourism authorities in Kurukshetra should plays important roles in solving tourism-related problems	1.62	0.486	1.66	0.733	9.384	0.052

Source: Author's Creation.

Tourism marketer and concerned authorities in the study area should take measures to develop variety of attractions and to maintain the uniqueness of culture-heritage and religious activities.

Tourism department is required to design religious tourism circuits by adjoining small temples and places situated in rural or far away territories of the area.

There is need to build more entry points at the entrance of main shrines and temples such as near the Bhadarkali temple in Kurukshetra to avoid the overcrowding situation during peak seasons.

The proper mechanisms for improvement in signage boards indicating major attractions and facilities should be placed at suitable location.

The sufficient parking space in the area would be an added advantage to increase the satisfaction scale of the visitors.

The air connectivity in the study area was not so good and not general. This is another matter to be taken by concerned government on its development agenda.

The accommodation facilities in the study area need to be increasing their efforts towards availability of luxury and star category hotels at the destination.

The government should introduce resources to provide training to tourism professionals to serve tourist in a better way.

The cleanliness at the food and eating area should be required to build good image among tourists. The food serving staff and production staff should know the importance to keep neat & clean appearance while serving to the guests.

The government should take measures to establish tourist information centres at appropriate locations to provide quick & accurate information to the tourists about destination.

The proper mechanism should be made increase the security of tourist luggage and belongings in the area during their stay at the destination.

This is the time to have a marketing strategy to utilize the merits of communication technology such as mobile network availability. So, the concerned authorities should ensure the proper mobile network availability in the area as well as to develop other information & communication facilities in the area.

The government should ensure proper regularization by keeping regular checks on development activities. There is also need to keep a close watch on price & fare charged for different services in the area such as fare of taxi/cab, room rent, prices charged by food centres.

The government and local community should take collaborative measures to maintain the natural beautification of the area by maintaining ecology, environment, using eco-friendly products and other activities to promote sustainable pilgrimage tourism in the area.

Formulation of appropriate marketing and promotion strategies should be designed to cater to different market segments of religious tourism industry.

DIRECTION FOR FUTURE RESEARCH

The present study was conducted to provide an overview of community participation and role of government in religious tourism development in Kurukshetra, thus focus only on supply side of the tourism destination. For a more comprehensive analysis, the future research can be done on demand side, by identification of tourist satisfaction and experience from its attractions and various facilities in the area. A comparative study on international and domestic tourist experience could be studied to uncover the similarities and differences in them.

CONCLUSSION

Kurukshetra is considered as most emerging destination for religious tourism in the region as it is known to be the land of Bhagwadageeta and Mahabharata war. Obliviously the place is richly dotted with many important religious tourism sites and also famous for religious events celebrated here as well as also include some culture and heritage attraction in its list. For looking after the development, government established an autonomous board to be known as, Kurukshetra Development Board (KDB). The board continuously putting its efforts for achievement of comprehensive development of the region in general and religious places in specific. The local community also participating in different employment and business related to tourism in the area, the study conclude existence of positive agreement of local community regarding impact of tourism in their region, particularly the economic contribution through employment & entrepreneurial opportunities. In addition to economic prosperity, there are also socio-cultural benefits of tourism and some benefits associated towards natural environment of area through tourism. The study also observed small sized businesses as crucial for development through involvement of lower economic section of the area and noticed as effective measure to increase local community participation. The study fails to notice any issues related to planning and management of tourism among its different tourism stakeholder. The study also notices variation in opinion of among those who working and not working in tourism towards community participation and role of government in tourism planning and management. The study determines participation of local community to a limited extent due to excessive governmental control on tourism sector in the area. Thus, suitable participation of local community supported by a collaborative and decentralized government planning is suggested to go a long way to improve tourism in area. The study also suggested to provide adequate tourism infrastructure and efforts to impart education and training to local residents, better cooperation among tourism stakeholders for better tourism planning and management.

REFERENCES

Akal, M. (2010). Economic Implications of International Tourism on the Turkish Economy. *Tourismos: An International Multidisciplinary Journal of Tourism*, *5*(1), 131–152.

Claiborne, P. (2010). Community participation in tourism development and the value of social capital. The case of Bastimentos, Bocas Del Toro, Panamá. [Master Degree Project, University of Gothenburg].

Dasgupta, D. (2011). *Tourism Marketing*. Pearson Education.

Dredge, D., & Jenkins, J. (2007). *Tourism policy and planning*. John Wiley & Sons.

FICCI. (2015). *Tourism Infrastructure: The role states play*. FICCI.

George, B. P., & Swain, S. K. (2005). *Advancements in Tourism Theory and Practices From India* (1st ed.). Abhijeet Publication.

Hall, C. M. (2000). *Tourism Planning: Policies, Processes, and Relationships*. Prentice Hall.

Hall, D., & Richards, G. (2002). *Tourism and sustainable community development*. Routledge. doi:10.4324/9780203464915

Holloway, J., & Valins, O. (2002). Pacing Religion and Spirituality in Geography. *Social & Cultural Geography*, *3*(1), 5–9. doi:10.1080/14649360120114107

Jalilvand, M. R., & Heidari, A. (2017). Comparing face-to-face and electronic word-of-mouth in destination image formation: The case of Iran. *Information Technology & People*, *30*(4), 710–735. doi:10.1108/ITP-09-2016-0204

Kandari O.P. & Ashish, C. (2004). *Tourism Development Principles and Practices.* Shree Publishers: New Delhi.

Ketmuang, S. (2005). *Local Residents and Tourists Attitudes towards Tourism Development at Patong Beach, Phuket, Thailand.* [Unpublished Ph.D. Thesis, University Putra Malaysia, Serdang].

Kumar, R. A., & Singh, L. (2014). Pilgrimage Tourism in Kurukshetra (Haryana): A Sustainable Development Approach. *African Journal of Hospitality, Tourism and Leisure*, *3*(2), 1–10.

Ministry of Tourism, Government of India. (2022). *Ministry of Tourism, Government of India, Annual Report 2022*. Ministry of Tourism, Government of India, New Delhi. www.indiatourism.com.

Mishra, R. (2000). *Pilgrimage Tourism – Case Study of Brajmandal.* [Unpublished Ph.D. Thesis, Kurukshetra University, Kurukshetra].

Mohammadi, M., Khalifah, Z., & Hosseini, H. (2010). Local People Perceptions toward Social, Economic and Environmental Impacts of Tourism in Kermanshah (Iran). *Asian Social Science*, *6*(11), 220–225. doi:10.5539/ass.v6n11p220

Obasola, K. E. (2014). Religious tourism and sustainable development: A study of eyo festival in Lagos, Nigeria. *International J. Soc. Sci. & Education*, *4*(2), 524–534.

Pendit, N. S. (2002). Tourism Science: A First Introduction (Latest Edition) with Improvements. Pradnya Paramita Publisher: Jakarta.

Samira, D., Mondal, K., & Basu, K. (2006). Dissemination of cultural heritage and impacts of pilgrim tourism at Gangasagar Island. In *Anthropological Survey of India*. Anthropologist.

Saravanan, A., & Rao, D. Y. V. (2012). Equitable Tourism Development: Need for Strategic Partnership. *International Journal of Multidisciplinary Research*, *2*(3), 344–356.

Sharma, J. K. (2000). *Tourism Planning and Development*. Kansishka Publisher.

Shinde, K. (2008). Religious tourism: exploring a new form of sacred journey in North India. In J. Cochrane (Ed.), *Asian Tourism: Growth and Change* (pp. 245–257). Elsevier Publishing Ltd. doi:10.1016/B978-0-08-045356-9.50025-4

Shinde, K. A. (2003). Residents' perceptions of pilgrimage tourism. *Sociology ofReligion*, *63*(4), 475–496.

Singh, S. (2002). Tourism in India: Policy Pitfalls. *Asia Pacific Journal of Tourism Research*, *7*(1), 45–59. doi:10.1080/10941660208722109

Singhal, V. (2009). *Perspectives and Potential of Kurukshetra as Tourist Destination*. [Unpublished Ph.D. Thesis, Kurukshetra University, Kurukshetra].

Statistical Abstract of Haryana (2022). *Data Compiled* from *Statistical Abstract of Haryana 2021-22*. Department of Economic and Statistical Analysis, Haryana.

Vijayanand, S. (2013). Pilgrimage Tourism and its Economic Dimensions in Tourism Management. *Radix International Journal of Economics & Business Management*, *2*(1), 1–31.

Wall, G., & Mathiesom, A. (2006). *Tourism: Change, Impacts and Opportunities*. Pearson.

Wilson, H., & Venes, J. (2001). *Factors for Success in Rural Promotion and Development*. Deep Publications.

World Tourism Organization. (2011). Religious tourism in Asia and the Pacific. World Tourism Organization.

World Tourism Organization. (2015).United Nation World Tourism Highlights, 2015 Edition. Published by United Nation World Tourism Organization Publication, Madrid Spain.

ADDITIONAL READING

Batra, G. S., & Dangwal, S. C. (1999). *Tourism Promotion and Development*. Deep and Deep Publications Pvt. Ltd.

Beeton, S. (2006). *Community development through tourism*. Landlinks Press. doi:10.1071/9780643093881

Goswami, B. K., & Raveendran, G. (2003). *Text Book of Indian Tourism*. Har Anand Publications.

Mathieson, A., & Wall, G. (1982). *Tourism: Economic, physical and social impacts*. Longman House.

Murphy, R. E. (1985). *Tourism: A Community Approach.* Methuen.

Sharma, S. (2008). Tourism, Technology and the Entrepreneur: A Case Study from Northeast India. *ICFAI University Journal of Business Strategy*, *5*(4), 71–80.

Singh, L. (2021). *Religious Tourism in North-West India: Current Situation and Future Prospects.* Lambert Academic Publishing.

KEY TERMS AND DEFINITIONS

Carrying Capacity: The amount of visitor activity that a site or destination can sustain.

Community: The group of people meets each other because they live nearby, not because they particularly want to be together. In a community, the members often have shared values and symbols.

Infrastructure: Construction needed to support economic development.

Marketing: Marketing is the activity, set of institutions, and processes for creating, communicating, delivering, and exchanging offerings that have value for customers, clients, partners, and society at large.

Religious Tourism: Travel to religious places, events, gatherings, rivers, caves, temples that relates to religious significance for a particular religion.

Shrine: A place, building or area which relates to some religious interest.

Sustainable Tourism: Sustainable Tourism leading to management of all resources in such a way those economic, social, and aesthetic needs can be fulfilled while maintaining cultural integrity, essential ecological processes, biological diversity, and life support systems.

Tourism Development: Development that meets the needs and aspirations of the tourism sector growth along with consequential benefits (economic, socio-cultural and environmental) to the destination.

Tourism Impacts: Effects, which may be felt because of tourism-associated activity.

Tourism Professionals: Individuals who have a vested interest in development, including community members; NGOs; planning, and government officials; hotel owners, tour operators, guides, transportation providers, and representatives from other related services in the private sector.

Tourism: Tourism, the act and process of spending time away from home in pursuit of recreation, relaxation, and pleasure, while making use of the commercial provision of services.

Chapter 12
Hedonic Experience of Visitors at a Heritage Site:
A Netnographic Study

Partho Pratim Seal
Manipal Academy of Higher Education, India

Debasish Batabyal
https://orcid.org/0000-0003-0191-9804
Amity University, Kolkata, India

ABSTRACT

Tourism is an activity that provides experience for the tourist. The experience is of high importance, and is accompanied by satisfying and pleasurable emotions. The experience allows the tourists to be actively involved in heritage consumption. The purpose of the study is to explore the hedonic experience of tourists at a UNESCO heritage site in India. The study adopts a netnographic approach to explore tourists' hedonic experiences. The results identified the three categories leading to exploring the hedonic dimension of experience while visiting a heritage site multisensory, imaginative, and emotional. The study provides appropriate understanding to the management and authorities by which they can improve the hedonic dimensions for the tourist, and improve the overall experience at the tourist destination. Further findings state that recognizing the relevant sensory clues will help the tourists to connect with the heritage monument better.

INTRODUCTION

History and heritage are intertwined as heritage (Timothy & Boyd, 2003). It is classified by the attractions which can have different classes including historic cities, archaeological sites, museums, and historic buildings (Timothy, 2011). The present-day society wishes to acquire heritage and pass it on to the future (Turnbridge & Asworth,1996). Heritage could either be a physical thing, a property, a building, or a place that could be either owned or passed to another person (Campanella, 2013).

DOI: 10.4018/978-1-6684-9957-3.ch012

Heritage tourism is one of the fastest-growing tourism in terms of visitors and attractions involving millions of people globally. For the tourist, it is a part of experiential consumption while visiting various historic sites with heritage attractions being the main components of tourist travel (Packer and Ballantyne, 2016). Heritage tourism is about the traveler's experience of place, artifacts, and activities which authenticates stories and people of the past and present. The perceptions of the past of a country help to determine its cultural heritage activities. Cultural heritage helps to share bonding with the community by giving better knowledge about our past, our identity, and our bonding with past, present, and future. The cultural heritage is about human creation which is intended to inform us about (Feather, 2006).

UNESCO provides an international brand and representation to the heritage site thereby allowing to present the site to match the demands of tourism (Salazar, 2012). This World Heritage list allows us to identify, preserve and pass on to future generations the monuments pertaining to history and art (Saipradist & Staiff, 2008). Though the prime purpose of UNESCO recognition assists in preserving and conserving the historical site and helps to increase the footfall of tourist (Adie, 2019). For a tourist, a tourism experience is of high importance which is accompanied by satisfying and pleasurable emotions. The emotional reactions of the tourist while experiencing the tourism experience lead to post-consumption behavior like satisfaction, intention to recommend, and choice (Gnoth, 1997). Tourism is an observed activity where the tourist interaction is the center of the tourist experience (Campos *et al.*, 2018) and the main benefit is hedonic, symbolic of the recognized psychological outcome (Chen and Chen, 2010). Hedonic experience is the major reason by which visitor assess the perceived value of their experience of heritage attractions (Calver & Page, 2013). The experience allows the visitors with active involvement in heritage consumption thereby involving imaginative, affective, and emotional perspectives to the historical place (McIntosh, 1999). Although hedonic dimensions have been studied in tourism the hedonic dimension of tourism experience in UNESCO heritage sites has not been studied. An in-depth understanding of the hedonic dimension of the experience of the tourist is of great importance. In India, there are a total of 40 UNESCO World Heritage sites which provides India with a lot of opportunities for travel & tourism one among them is Hampi. Hampi is a heritage, destination an architectural, archeological site in Bellary district of Karnataka and is also part of the group of monuments including the famous stone chariot. Hampi was established in fourteenth century which was later abandoned after two centuries when it was raided by the Deccan Sultans. Hampi has the remains of fortifications, palaces, and places of worship which were part of the Hindu Vijayanagara kingdom from 1336- 1646 CE. Hampi is also known as Vijayanagara by the ruler of the city. During the rule of Krishna Deva Raya between 1509 and 1529 Hampi reached its prime with the flourishing of international trade which reached great heights with progressive trading practices along with other international trade agreements.

However it is still an important pilgrimage destination for the Virupaksha Temple dedicated to Lord Shiva. Hampi is also associated with religious and mythological episodes of the epic Ramayana. It is listed as World heritage site by UNESCO in the year 1986 in the cultural criteria. Therefore the aim of the study is to explore the hedonic experience of the tourist at a Hampi on their visit of the stone chariot. A qualitative netnographical approach was considered to collect information about the experience of the tourists by the reviews posted by them on the social media. The data were analysed through content analysis by the NVivo software.

LITERATURE REVIEW

Cultural heritage is the inheritance of physical artifacts along with intangible attributes of a society that are inherited from past generations maintained by the present and bequeathed for the benefit of future generations. The study and interpretation of history motivate humans and allow them to communicate with other eras, cultures, and religions. Cultural heritage includes movable and immovable artifacts old and important and therefore necessitate conservation measures. The cultural heritage of indigenous people assists to connect the country with the locality and to the place. The heritage includes buildings, monuments, and artifacts that are in earnest need of preservation in the future and are pertinent to the history of the country. The preservation of the heritage is a recognition of the necessity of our past and the story it tells about. It also helps to create a base for cultural heritage. Understanding the value of the past is not just important for individuals related to cultural heritage but also for humankind in total. The architectural monuments of the past are of prime importance as they not only provide valuable information about architecture but also contribute essential documents to explain and illustrate all the features of the history of peoples through time (Jokiletho, 1986). Cultural heritage is important for quality of life with values and pride for all civilizations. Understanding the determinants of cultural heritage experience is of interest to both the tourist and the tourism service providers, The knowledge about these factors contributes towards tourist satisfaction and is of importance for tourists, historians, and all stakeholders. Though past studies have been on various heritage destinations minimalistic research on hedonic experience at the heritage site has been conducted. The hedonic experience is an intangible benefit and a vital component of the cultural heritage experience.

The word Hedonism is derived from the ancient Greek meaning pleasure (Veenhoven, 2003). Hedonism is stated to be part of 10 human value types (Schiffman et al., 2011) where the hedonic value type considers both pleasure and sensuous gratification. In psychology, it is stated that pleasure-seeking is the key motivator of human behavior. Pleasure includes all pleasant feelings or experiences including delight, enjoyment, elation, gladness, gratitude, and satisfaction (Veenhoven, 2003). Hedonism is applied to consumer behavior from the behavioural and cognitive psychology (Jantzen et al., 2012). Researchers Zhang (2013) and Moore and Lee (2012) have defined hedonism as the consumption of pleasure and gratification obtained from enjoyable experiences. Hedonism is also related to emotional desire and people prefer pleasurable life (O'Shaughnessy & O'Shaughnessy, 2007). Hedonism is a reflective form of pleasure and fun (Oliver & Raney, 2011), and offers possessed subjective and intangible attributes thereby creating matching hedonic reactions among the consumers (Hanzaee & Khonsari, 2011). Hedonic experience is identified by a set of determinants that are part of the experience which include a sense of achievement, a sense of satisfaction: feeling good which includes a sense of restful, a sense of nourishment, or a sense of feeling good about yourself or others, and novelty which incorporates novel activities and environment. The researchers Otto & Ritchie (1996) while preparing the experience quality scale stated four dimensions mainly hedonics, peace of mind, recognition, and involvement and found hedonics to be the most evaluative dimension.

The researchers Hirschman and Holbrook (1982) state that hedonic motivation leads to selecting a tourist destination. It is also stated that the push and pull factors of a destination along with a hedonic response motivate the tourist to opt for a trip (Goossens, 2000). Researchers in tourism have stated that tourism products always have a hedonic part (Chen et al., 2016). The heritage sites are said to be experiential spaces that are filled with emotions and mindfulness (Wilson & McIntosh, 2007) providing experiential consumption to the tourist. The heritage attractions are personal and everyone does try to find specific attributes and characteristics (Petr, 2015). Considering it the heritage attractions are currently focusing on the hedonic dimension of experiences along with visitors' emotional responses (Marshall et al, 2016).

Researchers Husyeynov et al. (2020) have stated that currently managing the customer experience and making it a hedonic experience is of prime importance and so further studies must be conducted to understand how a heritage visitor could be provided with a memorable experience. A tourist visits a heritage site with interest and motivation to learn along with a desire to entertain and inform children and friends of an experience that is both stimulating and enjoyable. Researchers Poria, Reichel, and Biran (2006) have also stated that motivation to visit heritage includes learning about heritage, leisure, and emotional involvement. The heritage attractions bring in hedonic meaning which includes enjoyment, fantasy, and fun. A heritage location has both historic and aesthetic narratives which must be conserved. The heritage attractions are infused with the cultural importance of the local community with imaginations and aspirations extending national boundaries. The hedonic tourism experience considers how vivid memories are created. Tourism experience creates lasting memories for a tourist which they will ponder and share on social media (Andrades and Dimanche, 2014) to a larger population.

METHODOLOGY

The present study is taken to consider the various components which lead to the tourist hedonic experience. The study considers netnography (Kozinets, 2002) also stated as " online ethnography," which is based upon reflexive narratives which are published online. Researchers Kozinet (2002) has defined netnography as a process or research methodology which adapts ethnographic techniques to study the cultures and communities which develop through computer-mediated communication. Data for these studies are collected from travel blogs, tourist reviews, and social networking sites which are analysed thematically (Catterall & Maclaran, 2001). Kozinets (2015) research work states that a netnographic study has the data being collected and manifested which is available freely through Internet. Netnography is a preferred method to study the customer experience as they write their reviews and observation after visiting a location (Kozinets, 2002) as usage of traditional methods of data collection with the presence of a researcher do affect the normal practices of everyday life (Hammersley & Atkinson, 1995). Kozinets (2010) provides six criteria when selecting sites for netnographic research which includes relevance, active, interactive, substantial, heterogenous and data rich.

Netnography is also preferred as it is simple, quick and an inexpensive method to access realistic, voluntary experiences of tourism information available on Internet (Mkono & Markwell, 2014; Wu & Pearce, 2014). Netnography is a non-participant observation without influencing the respondent (Elliot & Jankel-Elliot, 2003) that are based upon the reviews which are published online. The data was collected from travel site TripAdvisor, which do provide information about tourists' experience at Stone Chariot, Hampi, India. The choice of TripAdvisor was considered as it is a largest travel community site with millions of reviews and assists in collecting tourists travel and destination information.

The next step was data collection. The data collection and analysis happened between February- May 2023. The search was made on the TripAdvisor forum on stone chariot to avoid additional data, irrelevant reviews which are not related to the topic. The study follows a passive approach as the researchers did not interfere with the naturally occurring or ongoing discussions or had influenced the reviewers comments. The netnographic approach of the study is personal and maintains the distance between the reviewers and the researchers (Arsal et al., 2010). As there is a large set of data the first step of it to code it. Coding is to segregate the data into different units as per the meaning. As many codes are generated by initial coding, Charmaz (2006) advises selecting the useful analytical codes. The raw data is extracted from the TripAdvisor then initial codes is done followed by line- by -line analysis. The data analysis then leads to axial coding (Strauss & Corbin, 1998) which includes the connections between the codes to form categories (Matteucci & Gnoth, 2017). The axial coding process narrows the database to smaller themes or categories which states about process under study (Creswell, 2007). The axial coding helps to understand the hedonic experience of the tourists. Codes and categories are further obtained by rereading the coded statement. In the data analysis process, the concepts and relationships obtained from the coding process are compared with the concepts which are obtained from the existing literature.

RESULTS

The study followed Hirschman and Holbrook's (1982) process on hedonic consumption and based on data into three categories to explore hedonic dimension of experience while visiting a heritage site multisensory experience, imaginative and emotional.

Multi sensory experience- Multisensory experience is stated to be an impression created by specific events for whom the sensory elements are created by someone (Velasco & Obrist, 2020). The senses and their matching functions are the core of the formation of an impression of the object even if the object is not present (Velasco & Obrist, 2021). Senses act as a juncture between the outside world and the brain, and the interaction between the multiple sensory modalities influences the understanding of the environment (Chalmers, 2017). Researchers Barnes et.al, (2014) state that a visitor's outcome is mainly influenced by sensory experiences which suffice the hedonic needs. Multisensory images could also be by remembering an event that has occurred or from a previous experience (Hirschman & Holbrook, 1982). The multi-sensory consumption includes various psychological senses like smell, sight, hearing, taste, and touch. All these senses combined together create a strong, logical, and reliable understanding of the environment (Fulkerson, 2009). The visual effect of the chariot and its location impacts the tourist as was emphasized by reviews obtained on which basis a mind map is created as in Figure 1.

"The architecture, sculptures and the work in the temple is amazing".

"Pillars with two sided carvings and musical stone pillars are all amazing"

"Fairly well maintained and highly ornamental with exquisite statues".

"A stunning piece of architecture the stone chariot".

"The entire town (earlier Vijayanagara Samrajya) has architectural brilliance mostly destroyed by Ghazni and Ghori".

"Amazing place".

Figure 1. Mind map of Hedonic experience

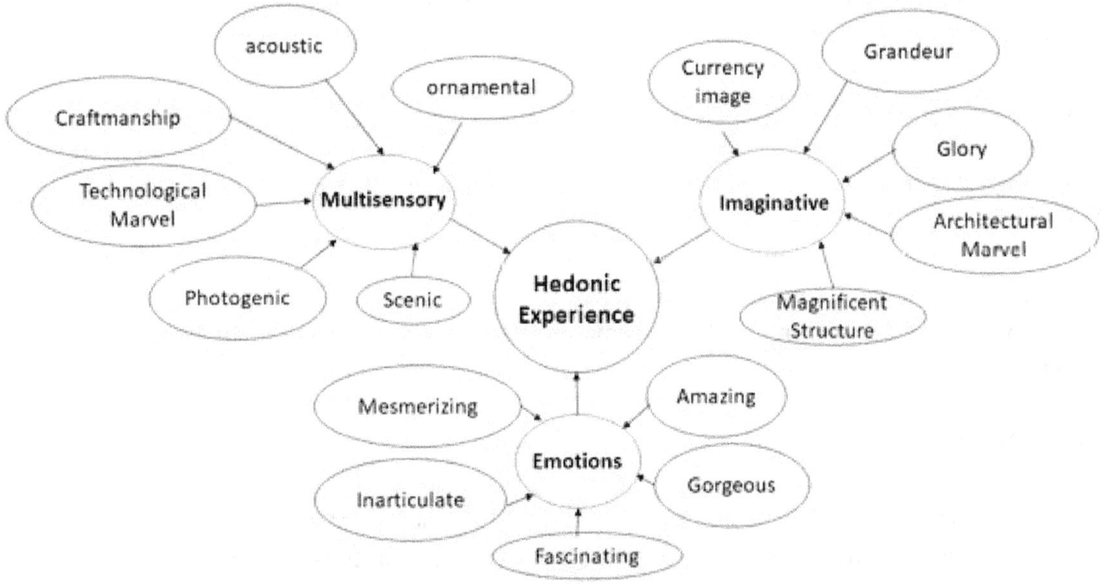

The tourists are amazed by the architecture of the temple including the sculptures around the heritage site, the musical pillars which were made during that era, and how they are still intact.

"Monolithic stone chariot of rough quartz embellished with carving of figures. The wheel carvings shows high level of craftmanship."

"Stone chariot and mantapas are good and photogenic. Complex demonstrates marvelous craftsmanship, you are left with an awe when you visit here."

"The route to temple is scenic though even stark in places".

"The temple is unique as its pillars make different sounds when tapped, it is an engineering marvel".

"Complex with temples and mantapas including the stone chariot is fascinating which requires several hours to appreciate them a lifetime experience".

The tourist is also enchanted with the craftmanship and the carvings made during the era. The visual impact of tourist while approaching the heritage site along with the beauty of the temple is also striking and memorable for the tourist.

"Most magnificent temple complex, a magnificent monument of architectural and technological marvel. Expertise in acoustics and finesse in technology reflecting architectural showpiece."

"A technological marvel, a monolithic stone-cut chariot has wheels that can rotate."

"An incredible piece of architecture as it has borne the brunt of weather, looters and zealots, and still remained to remind us of the glorious history".

"Awesome architecture."

"Stone chariot is an amazing architecture made from a single stone, an example of Indian architecture."

The construction of the temple, the architecture of the complex, the technology used for the construction during the period especially the stone-cut chariot that has wheels that could be rotated, and the quality of the construction which still remains intact though it had been attacked by many rulers, neglected and affected by weather.

Imaginative

Hedonic consumption is also related to imaginative construction of reality (Singer,1966). Hedonic consumption just does not consider what is real but also what reality is also about. So the consumer's internal construction of reality may not be matching with the verifiable world. It states about the mental imagination the consumer has around the product as they are mostly considered as a subjective reality beyond objective context. The tourists who visited the heritage monument could have imagined how the place had been in the past when people use to be habited.

"The famous stone chariot which is on the currency notes, couple of sabha grah, and a huge mandapa".

"The same chariot is on the reverse side of INR 50."

In Indian currency of Rs 50 denomination, the chariot image is featured, and the tourists who visit the location immediately recollect the image with the currency and try to relate with the same.

"Mind-boggling precision leaves you spellbound. Cannot imagine a full orchestra made with stone pillars".

"We can imagine the forgotten empire grandness"

"It has musical pillars where ancient dancers used to beat the pillars to create music while dancing".

"Musical pillars which are now not functional"

"Famous for 56 "musical" pillars used during musical/dance performances as each pillar resounded with a different note when struck".

"To identify the sound produced by a particular pillar observe the sculpture on the pillar's base, there is a person holding or playing that instrument".

The tourist could imagine by observing the pillars now, how the pillars were made sound precision to an instrument that was depicted in the base of the pillar. The tourist could also visualize how the ancient dancers used to dance around while striking the pillar with different notes matching their dance steps.

"So evident that we were using wheels before the intention in the west and using stones".

"One will wonder what on earth motivated one to build such a magnificent structure with no modern tools"

"Some times we can have a subtle smell of sandal wood which is evidently nice from the forest."

"One can visualize the glory and grandeur of the great Vijaynagar Kings."

The tourist is also awe-struck by how the heritage structure and the sculpture work are done on the stone without having appropriate tools. While some tourists could envisage the heritage structure and state about the Vijayanagara Kingdom during the 13th Century. A tourist also felt that a delicate smell of sandalwood surrounded the heritage location.

Emotions

The emotions assists in reciting a consumers experience of consumption (Babin, Darden, & Babin, 1998). Emotions are physiological processes that are mostly expressed physically. An important feature of emotion is it mediates between cognitions and key outcomes like customer satisfaction and retention (Oliver & Westbrook,1993). Past studies do state the role of emotions in affecting consumers behavior. Researchers Arora and Singer (2006) have stated that effective elements of emotions do influence hedonism by the customer. Although tourist have difficulty while expressing how they tourist were able to share about the visit which aroused positive feelings and emotions.

"Musical pillars are highlights of beautiful and mesmerizing stone artwork place of the 13th century".

"Stone chariot monument is mesmerizing".

"The temple is just beautiful and a thousand words is not sufficient to describe how I felt when I got here. I can spend hours appreciating the amazing architecture".

"Even though the place is in ruins like the rest of Hampi, it still retains the grandeur and charm"

"The stone structures are eye-catching and very gorgeous".

The tourist is enchanted with the art work of the monument and the beauty and the structure of the heritage site. A tourist didn't even have words to express their feelings after seeing the structure. A tourist in comments stated that though the stone chariot is part of ruins it continues to hold its charm and magnificence though it is a few centuries old.

"Shows a glimpse of King Krishnadevray's, mesmerizing art love. Very fine art & very keen work".

"Amazing art, the idea behind the architecture is mind-blowing, we have to save this for our future generations".

"Behind the complex is also fascinating".

"Spectacular beautiful illumination of the chariot happens in the evening from the floodlights installed in the Vittala complex. The spectacular view of the chariot and its detailed designing in the glow of the lights from the complex makes for a mesmerizing experience."

Tourists' comments also highlight how they feel about the artwork and the keenness of work the King would have possessed for which he created this heritage work. Another tourist had an opinion that the architecture is incredible and efforts must be put in to save the site for generations ahead. A tourist viewing the chariot during the evening also states how the comprehensive design appears when illuminated which is a cherished experience. A tourist adds on to state that the entire complex is captivating even behind the chariot also.

The study considered the three hedonic experience components that the tourist experienced: the imaginative, multisensory and imaginative (Figure 2).

Figure 2. The conceptual framework of Hedonic Experience at tourist destination

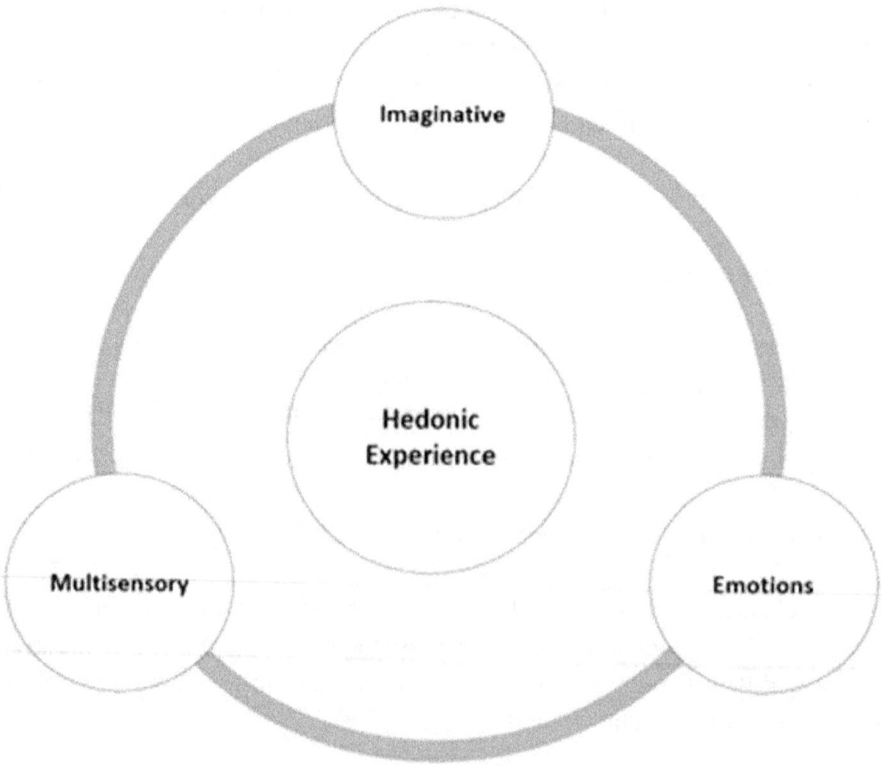

DISCUSSION AND CONCLUSION

As tourists are eager for new experiences they do get engaged when they are involved with history and culture which allow them to escape from their regular routine and learning something new motivate tourists to visit heritage attraction along with hedonics (Timothy, 2011; Medeiros *et al.,*2020). The findings of the study validate the meaning of heritage sites as the tourists are visiting the destination for hedonic pleasure.

Past research studies by Bonn et al. (2007) also investigated the effect of environmental factors on visitors' historical and cultural attractions. The emotional influence of the "atmospheric elements" is stated to be the most important construct for the visitor's experience (Slatten and Mehmetoglu, 2009).

In the present scenario the social media platforms allows the tourists to share their experience and emotions (Jacobsen & Munar, 2012). The present study considers online reviews which are abundantly available in the public domain to identify the categories of hedonic experience.

It can be stated from the research that multisensory experience, imaginative, and emotions direct tourists to the hedonic experience. The findings indicate that tourists obtain a multi-sensory experience at the location as they are amazed by the architecture at the heritage location while traveling to the destination and the musical pillars. The tourist's imaginativeness is by how they could immediately relate to the stone chariot whose image is on the back of an INR 50 denomination currency note. The tourist could also visualize the musical pillars and how the dancers use to strike them with different notes during their performance. The tourists expressed their burst of emotions on their visit as they were thrilled with the monument's magnificence, and charisma although the place is now in ruins. The feelings of the tourist is as if they had been taking them to the kingdom of Vijayanagara.

Considering the experience of visiting UNESCO heritage- the stone chariot site the results of the study states the three hedonic dimension of multi-sensory experience mainly by the tourist stimuli of the physical features of the historic monument, imaginative what they feel, and what they relate to and emotional response which is stimulated by observation and curiosity. These three main dimensions are explained by the specific features which highlight the importance of the historical site to enhance the hedonic dimension of experience.

IMPLICATIONS OF THE STUDY

For a better experience, tourist heritage places should provide more information about the place and its history (Richards,1999). The historical places along with their resources should provide opportunities for hedonic consumption by the tourist for a memorable experience (Calver & Page, 2013). The study provides appropriate understandings to the management & authorities by which they can improve the hedonic dimensions of the tourist and improve the overall experience at the tourist destination. Further findings do state that recognizing the relevant sensory clues will help the tourists to connect with the heritage monument like the stone chariot with the currency denomination. Although the symbolic representation of a heritage site helps in stimulating the tourist's senses and imagination (Lv et al.2020) the management and the tourism promoters should think of a more hedonic approach of the destination. They can also use various communicating and engaging methods which can integrate the stakeholders, resources, places and environments (Agapito and Chan, 2019).

LIMITATIONS

The study considers the hedonic dimension of a heritage location which provides pleasure to the tuorist visiting the location. Future research studies should be more diversified especially at heritage attractions as this will help to understand the tourist hedonic behavior. Another limitation of the study is it is destination specific and is restricted to only a single UNESCO heritage site. Similar studies could be conducted in other destinations to study the hedonic experience of tourists. As the study considered the content of comments posted online on the travel website TripAdvisor only further future studies could also include other travel websites about hedonic experiences at the heritage destination. The study adopted netnography an online ethnography method, as compared to ethnographic studies where researchers have prolonged engagement with the participant. Considering the non-participative role of netnography future studies must ponder upon before conducting similar studies.

REFERENCE

Adie, B. A. (2019). *World Heritage and tourism: Marketing and management*. Routledge. doi:10.4324/9780429506659

Agapito, D., & Chan, C. S. (2019). A multisensory approach to responsible management in community-based tourism: A case study in Hong Kong. *Journal of Tourism Quarterly*, *1*(1/2).

Andares, L., & Dimanche, F. (2014). Co-Creation of experience value: A tourist behavior approach. In N. Prebensen, J. Chen, & M. Uysal (Eds.), *Creating experience value in tourism Bloomington* (pp. 95–112). CABI.

Arora, R., & Singer, J. (2006). Cognitive and affective service marketing strategies for fine dining restaurant managers. *Journal of Small Business Strategy*, *17*, 51–61.

Asral, I., Woosnam, K. M., Baldwin, E. D., & Backman, S. J. (2010). Residents as travel destination information providers: An online community perspective. *Journal of Travel Research*, *49*(4), 400–413. doi:10.1177/0047287509346856

Babin, B. J., Darden, W. R., & Babin, L. A. (1998). Negative emotions in marketing research: Affect or artifact? *Journal of Business Research*, *42*(3), 271–286. doi:10.1016/S0148-2963(97)00124-0

Barnes, S. J., Mattsson, J., & Sorensen, F. (2014). Destination brand experience and visitor behavior: Testing a scale in the tourism context. *Annals of Tourism Research*, *48*, Vol-88. doi:10.1016/j.annals.2014.06.002

Calver, S. J., & Page, S. J. P. (2013). Enlightened hedonism: Exploring the relationship of service value, visitor knowledge and interest, to visitor enjoyment at heritage attractions. *Tourism Management*, *39*, 23–36. doi:10.1016/j.tourman.2013.03.008

Campanella, M. (2013). *Establishing tourist places: between social constructivism and phenomenology*. [Doctoral dissertation, Tartu Ülikoo].

Campos, A. C., Mendes, J., Valle, P. O., & Scott, S. (2018). Co-creation of tourist experiences: A literature review. *Current Issues in Tourism*, *21*(4), 369–400. doi:10.1080/13683500.2015.1081158

Catterall, M., & Maclaran, P. (2001). Body talk: Questioning the assumptions in cognitive age. *Psychology and Marketing*, *18*(10), 1117–1133. doi:10.1002/mar.1046

Chalmers, A. (2017). Experiencing the multisensory past. In M. Loannides, N. Magnenat-Thalmann, & G. Papagiannakis (Eds.), *Mixed Reality and Gamification for Cultural Heritage*. Springer. doi:10.1007/978-3-319-49607-8_14

Chen, C.-F., & Chen, F.-S. (2010). Experience quality, perceived value, satisfaction and behavioral intentions for heritage tourists. *Tourism Management*, *31*(1), 29–35. doi:10.1016/j.tourman.2009.02.008

Feather, J. (2006) Managing the documentary heritage: issues for the present and future. In: Gorman, G.E. and Sydney J.Shep. (eds.), Preservation management for libraries, archives and museums. London: Facet, pp. 1-18. doi:10.29085/9781856049214.002

Fulkerson, M. (2013). Explaining Multisensory Experience. Consciousness Inside and Out: *Phenomenology, Neuroscience, and the Nature of Experience*, 365–373. doi:10.1007/978-94-007-6001-1_25

Gnoth, J. (1997). Tourism Motivation and Expectation Formation. *Annals of Tourism Research*, *24*(2), 283–304. doi:10.1016/S0160-7383(97)80002-3

Goossens, C. (2000). Tourism information and pleasure motivation. *Annals of Tourism Research*, *27*(2), 301–321. doi:10.1016/S0160-7383(99)00067-5

Hammersley, M., & Atkinson, P. (2007). *Ethnography: Principles in practice*. Tavistock.

Hanzaee, K. H., & Khonsari, Y. (2011). A review of the role of hedonic and utilitarian values on customer's satisfaction and behavioural intentions. *Interdisciplinary Journal of Research in Business.*, *1*(5), 34–35.

Hirschman, E. C., & Holbrook, M. B. (1982). Hedonic consumption: Emerging concepts, methods and propositions. *Journal of Marketing*, *46*(3), 92–101. doi:10.1177/002224298204600314

Jacobsen, J. K. S., & Munar, A. M. (2012). Tourist informationsearch and destination choice in a digital age. *Tourism Management Perspectives*, *1*(1), 39–47. doi:10.1016/j.tmp.2011.12.005

Jantzen, C., Firchett, J., Ostergaard, P., & Vetner, M. (2012). Just for fun? The emotional regime of experiential consumption. *Marketing Theory*, *12*(2), 137–15. doi:10.1177/1470593112441565

Jokilehto, J. (1986). *A History of Architectural Conservation: The Contribution Of English, French, German And Italian Thought Towards An International Approach To The Conservation Of Cultural Property* [D.Phil. Thesis, The University Of York]. https://vk.com/doc277688559_437435201?hash=204ac4f1d4c699116f&dl=39b65a8b182586a0cd

Kozinets, R. (2010). *Netnography: Doing ethnographic research online*. Sage.

Kozinets, R. (2015). *Netnography: Redefined*. Sage.

Kozinets, R. V. (2002). The field behind the screen: Using netnography for marketing research in online communities. *JMR, Journal of Marketing Research*, *39*(1), 61–72. doi:10.1509/jmkr.39.1.61.18935

Lv, X., Li, S. C., & McCabe, S. (2020). Expanding theory of tourists' destination loyalty: The role of sensory impressions. *Tourism Management*, *77*, 77. doi:10.1016/j.tourman.2019.104026

McIntosh, A. J. (1999). Into the tourist's mind: Understanding the value of the heritage experience. *Journal of Travel & Tourism Marketing*, *8*(1), 41–64. doi:10.1300/J073v08n01_03

Mkono, M., & Markwell, K. (2014). The application of netnography in tourism studies. *Annals of Tourism Research*, *48*, 266–291. doi:10.1016/j.annals.2014.07.005

Moore, D.J., & Lee, S. P. 9 2012). How advertising influences consumption impulses. *Journal of Advertising, 41*(3), 107-120

O'Shaughnessy, J., & O'Shaughnessy, J. N. (2007). Reply to criticisms of marketing, the consumer society and hedonism. *European Journal of Marketing*, *41*(1/2), 7–16. doi:10.1108/03090560710718076

Oliver, M. B., & Raney, A. A. (2011). Entertainment as pleasurable and meaningful: Identifying hedonic and eudaimonic motivations for entertainment consumption. *Journal of Communication*, *61*(5), 984–1004. doi:10.1111/j.1460-2466.2011.01585.x

Oliver, R. L., & Westbrook, R. A. (1993). Profiles of consumer emotions and satisfaction in ownership and usage. *Journal of Consumer Satisfaction, Dissatisfaction & Complaining Behavior*, *6*, 12–27.

Packer, J., & Ballantyne, R. (2016). Conceptualizing the visitor experience: A review of literature and development of a multifaceted model. *Visitor Studies*, *19*(2), 128–143. doi:10.1080/10645578.2016.1144023

Poria, Y., Reichel, A., & Biran, A. (2006). Heritage site perceptions and motivations to visit. *Journal of Travel Research*, *44*(3), 10. doi:10.1177/0047287505279004

Richards, G. (1999). Heritage visitor attractions in Europe: A visitor profile. *Interpretation (Tulsa)*, *4*(3).

Saipradist, A., & Staiff, R. (2008). Crossing the cultural divide: Western visitors and interpretation at Ayutthaya world heritage site, Thailand. *Journal of Heritage Tourism*, *2*(3), 211–224. doi:10.2167/jht061.0

Salazar, N. B. (2012). Community-based cultural tourism: Issues, threats and opportunities. *Journal of Sustainable Tourism*, *20*(1), 9–22. doi:10.1080/09669582.2011.596279

Singer, J. L. (1966). *Daydreaming: An Introduction to the Experimental Study of InnerExperimental Study of Inner Experience*. Random House.

Timothy, D. J., & Boyd, S. W. (2003). *Heritage Tourism*. Prentice Hall.

Tunbridge, J. E., & Ashworth, G. J. (1996). *Dissonant heritage: the management of the past as a resource in conflict*. Wiley.

Veenhoven, R. (2003). Hedonism and Happiness. *Journal of Happiness Studies*, *4*(4), 437–457. doi:10.1023/B:JOHS.0000005719.56211.fd

Velasco, C., & Obrist, M. (2020). *Multisensory experiences: Where the senses meet technology*. Oxford University Press. doi:10.1093/oso/9780198849629.001.0001

Velasco, C., & Obrist, M. (2021). Multisensory Experiences: A Primer. *Frontiers of Computer Science*, *3*, 614524. doi:10.3389/fcomp.2021.614524

Wu, M. Y., & Pearce, P. L. (2014). Appraising netnography: Towards insights about new markets in the digital tourist era. *Current Issues in Tourism*, *17*(5), 463–474. doi:10.1080/13683500.2013.833179

Chapter 13
Cultural Tourism and Community Development:
Dhofar Governorate as a Case

Samskrati Gulvady
University of Technology and Applied Sciences, Salalah, Oman

Mohammad Soliman
https://orcid.org/0000-0002-9359-763X
University of Technology and Applied Sciences, Salalah, Oman & Fayoum University, Fayoum, Egypt

Hamed Almuhrzi
Oman Tourism College, Muscat, Oman & Sultan Qaboos University, Muscat, Oman

Musallam S. Hawas Al-Aamri
University of Technology and Applied Sciences, Salalah, Oman

ABSTRACT

The purpose of the current chapter is to discuss the impact of cultural tourism on the development of communities, particularly intangible cultural heritage (ICH). In addition, it highlights the ICH's property in Oman's Dhofar region. It also provides recommendations and practical guidance on how local communities can gain the most from the expansion and development of ICH tourism. It is evident that the development of cultural heritage tourism benefits the local community, but there are also some difficulties that need to be considered. In the Dhofar governorate of Oman, cultural heritage tourism, especially its intangible component, plays a vital role in the growth of the local economy. This study significantly advances the theoretical understanding of concerned academics and scholars by shedding light on the intricate connections between cultural heritage tourism and community development. Additionally, it makes contributions to practice that could affect how public policy, historical preservation, and community involvement are handled.

DOI: 10.4018/978-1-6684-9957-3.ch013

1. INTRODUCTION

The concept of cultural tourism is commonly used interchangeably with heritage tourism and cultural heritage tourism (Almuhrzi, 2020). Cultural tourism can be approached from two standpoints: supply and demand sides. According to Apostolakis (2003), the demand side (or descriptive side in his words) tends to reflect the tangible heritage resources and any intangible practices associated with such resources. The supply side also defines intangible resources such as dances, languages, and rituals. Furthermore, Apostolakis (2003) suggests that cultural tourism can also be defined from a demand side. In approaching cultural tourism from the visitor's eye, the focus is more on visitors' motivations, perceptions, and meanings of cultural resources.

Regardless of which standpoint one approaches the meaning of cultural tourism, literature agrees on the vital role of a community in shaping, driving, and developing cultural tourism and vice versa. While culture is an integral part of any tourism destination offerings, in some destinations a specific aspect of a culture is the core visitors' pulling factor (McKercher & du Cros, 2002). The fact that cultural tourism is solely built around the existence of a community, its traditions, language, rituals, way of living, and all elements of tangible and intangible resources make the relationship between tourism and community a two-way relation. On one hand, the community offers and opens its 'doors' to tourists to experience their culture; on the other, tourism contributes directly to the development of the community.

As an emerging destination, Sultanate of Oman is recognised for its wealth of history, culture, and heritage resources. Since its effort to promote itself as a destination, Oman places its unique cultural heritage resources at the centre of its tourism offerings. Several resources were commodified into tourism products at the early stage of tourism development in Oman. The Omani tourism industry has benefited from well-reserved and conserved tangible and intangible cultural heritage resources. For example, from the early 2000s, Nizwa Fort was central in different tour itineraries. Additionally, Muttrah Souq was also a must-visit cultural attraction for tourists visiting Muscat.

This relationship between the Omani tourism industry and Oman's cultural heritage resources is underpinned by several factors, including stakeholders' involvement and support. One of the success enablers is the community itself. Omani communities are known for their hospitality and welcoming outlook toward outsiders. Furthermore, tourism has contributed directly and indirectly to the revival and sustainability of different tangible and intangible Omani cultural resources (Al Riyami, Almuhrzi & Scott, 2022).

This chapter discusses cultural tourism and community development in the case of Dhofar. The chapter starts by discussing the concept of cultural tourism and its different classifications while trying to answer the question of who a cultural tourist is. Subsequently, it explores the nexus between cultural tourism and community development by discussing several cases and how cultural tourism serves as a development catalyst. It then moves to discuss positive and challenging factors facing communities using cultural tourism as a development tool. The last section discusses cultural tourism and community development from the Oman perspective, in the Governorate of Dhofar. It specifically discusses several unique cultural resources and how the community uses/promotes them as tourism products and experiences. Finally, this chapter puts forward an interesting perspective on how cultural tourism can contribute towards the development of local communities and put forward a future research trajectory that several researchers will certainly find interesting to pursue.

2. LITERATURE REVIEW

2.1 Cultural Tourism: Concept and Classification

According to the UNWTO (2018), a definition for cultural tourism was adopted during the twenty-second session of the UNWTO General Assembly in China (in 2017) which articulates that visitors' main motivation is to learn about, discover, experience, and consume the tangible and intangible cultural attractions at tourism destinations. These attractions include works of art and architecture, historical and cultural heritage, culinary heritage, literature, music, creative industries, and living cultures with their way of life, value systems, beliefs, and traditions.

The UNWTO considers culture as the main predictor for tourists to visit a destination and nature is the subsequent motivation. Cultural tourism is a flourishing tourism wherein the tourists pay a visit to places involving activities that encourage them to experience the heritage while getting involved with the local communities (Abouelmagd, 2023). A wide variety of activities that are a part of cultural tourism use cultural heritage assets (e.g., art, archaeological sites, castles, ethnic crafts, communities, famous buildings, historical galleries, buildings, museums, palaces, religious buildings, ruins, sculpture) and other things that represent people and their cultures. Broadly speaking, cultural tourism includes activities with a cultural perspective in the form of trips and visits with a variety of pursuits (Yolal & Negrusa, 2012). According to Cravidão et al. (2018), tourism provides an opportunity to discover varied cultural customs, existences, and a global perspective for a tourist to appreciate culture leading to self-development. It also encourages reciprocal support, reinforcing belongingness to cultural traditions by allowing people from various social classes to come together. Such a kind of tourism allows exposure to other cultures and familiarizing with guest cultures helps in empathizing with others. It also encourages economic growth at a local and global level. There is also scope for local industries to thrive by involving themselves in creative and innovative entertainment activities while assessing their tangible and intangible properties. Cultural tourism improves the quality of life for locals by generating jobs and business opportunities, preserving resources, and many other benefits. It supports the role of tourism as a significant part of the global tourism economy which can help ensure historical preservation, promote intercultural understanding, and boost local communities' sense of pride (Soliman, 2015).

According to Park (2013), cultural tourism is particularly well-liked in countries that are attempting to expand their offerings. Singh (2021) presupposes that in the post-pandemic years, tourism may be focused on UNESCO World Heritage sites that promote safer tourism while directly helping the local people. He calls it 'quixotic' (quick & exotic) tourism that may bourgeon under proper management. Tourism, in general, and culture tourism per se, provide a significant developmental impetus that offers benefits for the tourists, as well as the local community. Developing community-led tourism using participatory planning methods is based on the idea that it is in the community's best interest (Mair & Reid, 2007). It becomes essential to implement a participatory approach that can help communities to have greater control over their lives and resources – participation as a means of achieving improved social, economic, or environmental objectives (Lyon, 2013). As national and international tourism has been growing over the past few decades, visitors have begun understanding the importance of this 'good' mode of tourism that boosts the economy as well as aids the conservation of culture (Richards, 2018).

Cultural tourism brings together people living in various backgrounds, enabling them to contact people from the local communities and learn about their culture. Cultural tourism can be considered a complex and multidimensional phenomenon that is associated with education and cognitive journeys.

The primary motive of cultural tourism is to explore areas and facilities that are laden with historic and artistic value and participate in cultural events. Matlegiewicz (2020) considers cultural tourism to allow tourists to visit native populations who lead their lives in the traditional way while keeping the tradition alive. Such a kind of tourism involves economic and cultural benefits for the tourists and the local community residents. As a result, tolerance and mutual understanding develop among both parties. The author categorizes cultural tourism as follows:

i) intellectual culture tourism (e.g., museum tourism, cultural heritage tourism).
ii) education tourism (e.g., language seminar journeys, study, specialist).
iii) general culture tourism (e.g., culinary, exotic, industrial, regional, pilgrimage, military, urban).

A tourist who visits or intends to visit a cultural tourism attraction (for instance: an art gallery, museum, or any historic place, attending a performance or festival, and/or participating) is called a culture tourist (Gortan-Carlin, Krajnovič, & Raguž, 2022). A cultural tourist is further classified as a) a purposeful cultural tourist; b) a sightseeing cultural tourist; c) a casual cultural tourist; d) an incidental cultural tourist; and e) a serendipitous cultural tourist. Delving further into culture tourism, sustainability gains prominence (McKercher & du Cros, 2002).

Cultural heritage has two main components, which UNESCO categorizes as (1) Tangible cultural heritage, which includes buildings and historic places, monuments, artefacts, etc., that are considered worthy of preservation for the future; and (2) Intangible cultural heritage, that includes oral traditions or living expressions, performing arts, rituals social practices, festive events, knowledge, and practices. An understanding of the intangible cultural heritage helps with intercultural dialogue and encourages mutual respect for other ways of life (Gortan-Carlin, Krajnovič & Raguž. (2022). Richards (2018) observes the growth taking place in cultural tourism as it gets fragmented into niche areas like heritage tourism, arts tourism, gastronomic tourism, film tourism, and creative tourism.

2.2 Cultural Tourism and Community Development

Several studies showcase the use of cultural tourism in community development. For instance, Abouelmagd (2023) studied that cultural tourism can motivate tourists to experience the heritage and relish the opportunity to interact with the local community and suggests that we should build a sustainable relationship between the sites and the surrounding communities. Giampiccoli & Kalis (2012) see tourism as a tool for community development that can also empower people living in poverty. Community-based tourism is considered a potential strategy for poverty elimination. In that context, local culture is not seen as a tourist attraction, but rather, a resource for which community-based tourism can develop. Knowledge transfer from one generation to another creates an advantage with more opportunities for learning. Travel motivation and emotional experience have a direct effect on eudaimonic well-being, and that eudaimonic well-being has a direct effect on tourist satisfaction and destination loyalty (Al-okaily et al., 2023).

According to Lyon (2013), a community's ethnicity and culture can improve visitors' experiences. The local community will benefit from being reminded to develop local skills. The community members' usage of their traditional attire to distinguish themselves from outsiders has another good effect. Local culture is seen as a resource on which community-based tourism development may be built. Giampiccoli & Kalis (2012) indicate that community-based tourism is an additional means to meet household needs and aids in diversifying rural livelihoods. Along with that, local culture also becomes a tourism resource by using local foods, arts, and crafts as tourist attractions. Gastronomy takes the lead as a local cultural resource that facilitates many benefits for the community. According to a case study presented by UNESCO (2023) in a cultural heritage site Beit Eidis in Jordan, it is proposed that tourism can support the living of the local people around the sites. By rehabilitating the site, local communities have gained access to job opportunities. According to Al-Hashim (2015), local government and residents will start maintaining a site if they are aware of its cultural significance. Learning about local history will give residents a sense of pride in significant historical occurrences. While welcoming the tourists and politely sharing their history with them, the responders in the Mirbat region were deeply attached to their homes. It was seen that society accepted the tourism sector favourably. Al-Hashim (2015) emphasizes the possibility of economic gains because facilities and services will be needed. They could provide local crafts and mementos for sale. By attracting tourists with local cuisine, restaurants can be revitalized, improving their financial situation. Gaining experience and skills through initiatives like youth training, leadership, and conservation management are made possible through local capacity building, individual, and business collaborations. Youth must seize the opportunities that tourism has created, such as joining park boards, developing their negotiating abilities, and collaborating with the local administration, businesses, and citizens. They get experience in socio-economic development, infrastructure requirements, feasibility studies, and tourism marketing when these skills are put into practice (Colton & Whitney-Squire, 2010). Kaur et al., (2016) observe that community-based tourism (e.g., homestays) provides tourists with learning about the community lifestyle, local culture, and customs; and has a significant impact on community development. Tourists can interact with local communities in the cultural heritage and natural environment. The main agenda is to preserve the socio-cultural status and the environment of the local communities. Carefully planned and community-controlled tourism development may provide an enriching social learning experience for participants as opportunities for leadership, empowerment, and democracy are created (Mair & Reid, 2007). A case study presented by UNESCO (2023) about Oman's Land of Frankincense mentions a 'self-financing' scheme that ensures its conservation efforts continue to remain funded beyond state contribution to become self-reliant. Cultural heritage affects local economic activity in many ways than just through cultural tourism. Investments, the creation of jobs (Guzmán et al., 2017), increased economic activity, and socioeconomic repercussions on other aspects reliant on regional identity and culture are further direct effects of culture (Muštra et al., 2023).

Despite the positive effects of cultural tourism on the local community, there are some challenges related to its development. In this respect, management, and sustainability of cultural assets are top concerns from the views of both practitioners and academics due to rising tourist numbers and tourism-related activities. Cultural heritage tourism, like other forms of tourism, depends on the support of locals for the development of tourism in addition to its distinctiveness, attractiveness, maintainability, and heritage resources (Chen & Chen, 2010). Melubo and Lovelock (2020) discuss the difficulties locals face as a result of cultural tourism, including access restrictions to land resources, livelihoods being threatened, inadequate participation in decision-making and benefit sharing, the threat of relocation, and negative attitudes toward conservation principles. This study suggests that for World Heritage Sites to be considered successful by local communities, community participation in decision-making and in benefit sharing is essential. In this vein, Gursoy et al. (2002) found that factors like: the degree of concern, the clarity of eco-centric values, the use of resource base, perceived costs, and advantages of developing tourism - all influence the host community support. When locals perceive that tourism development will result in more costs than benefits, they are more inclined to oppose it (Chen & Chen, 2010).

2.3 Culture Tourism and Community Development in Dhofar

Since tourism fosters a certain level of homogeneity, interaction, and coordination across diverse production and service sectors, its expansion is essential for achieving Oman's economic and social development. Citizens experience higher levels of quality of life and prosperity as a result. The simplicity and variety of local customs and traditions, as well as easy access to prominent tourist attractions, allure large numbers of tourists (Al Amri et al., 2019). Intangible cultural heritage tourism can be actively developed to increase market demand, economic value, and the destination's integration into contemporary social life, which is a crucial strategy for heritage preservation and exploitation (Fu et al., 2023).

In this vein, Dhofar Governorate, located in the south of the Sultanate of Oman, is characterized by many elements of intangible cultural heritage (ICH) (e.g., Haffa heritage market, gold market, frankincense products and markets, *'Omani Halwa'*, traditional dance and music, traditional Dhofari food, local farmers market, Dhofari costumes, jewellery, etc.). This area has a significant chance to grow its cultural tourism, involving intangible cultural tourism, because there are many culturally themed attractions that can improve visitors' experiences and, as a result, ensure public support for cultural preservation (Al-Hashim, 2015). For instance, Dhofar offers a wide variety of traditional foods, including the well-known sweet; *'Omani Halwa'*. There are several recipes for this delectable dessert. Omani Halwa is frequently offered at markets for consumers. In addition, the cuisine of the Dhofari people is diverse. Many of these dishes are not available in other parts of the nation. '*Madhbi*', a dish of stone-grilled meat and chickens, and '*Ma'ajeen*', one of the region's most well-known meals, are two examples of well-liked dishes. Camel milk is a well-known traditional beverage in Dhofar, that is also utilized for therapeutic purposes. Thus, it has become a source of income for camel herders especially in Dhofar because of the establishment of canning and exporting businesses. Therefore, decision-makers predict that people will adopt some contemporary food tourism practices from other places and strengthen their interest in the culinary legacy of Dhofar. For the stakeholders in the destination, this has some benefits. Owing to its climatic conditions, Dhofar governorate generates a wide range of agricultural and other culinary goods that might be leveraged to offer creative food tourism experiences and raise the region's appeal to tourists not only during the much sought-after *Khareef Season* but at other times too. The preservation of local food production, culture, and tradition, the development of new jobs, and a rise in visitor spending are additional advantages.

Furthermore, the *Frankincense* tree, which grows naturally in Dhofar, is one of the most well-known species in the world. It produces about six tons of frankincense annually, which is a significant source of income for the locals. Omani frankincense was valued as highly as gold and utilized as a medium of exchange in the past (Dhanalekshmi et al., 2022; Weismann et al., 2019). Moreover, the Dhofar region produces more than twenty well-known arts, all of which reflect the Omani people's ties to their nation, civilization, and culture. The dialect used in the area, as well as the dancers' costumes, movements, leaps, and swordplay, make these traditional dances distinctive. The name of the clothing, "*Dhofari*", refers to the Dhofar region. Women's costumes from the Dhofar region are now recognized as a distinctive international brand. Families have specialized in creating and manufacturing these outfits, which are bought in bulk by tourists and notable figures in the entertainment industry.

3. DISCUSSION AND CONCLUSION

According to UNESCO (2023), cultural heritage extends beyond monuments and archaeological sites. It also encompasses some intangible aspects such as traditions or living expressions that have been passed down from our ancestors to our descendants (e.g., performing arts, social customs, rituals, holiday celebrations, knowledge of nature and the cosmos, and the skills necessary to make traditional crafts). The present study focuses on the impacts of culture tourism, especially intangible cultural heritage (ICH), on community development, highlighting the ICH constituents and assets in the Dhofar region of Oman. It also provides recommendations and practical guidelines on how local communities could benefit most from the expansion and development of ICH tourism.

Overall, the development of the local economy in Oman's Dhofar Governorate is fostered by cultural heritage tourism, particularly its intangible component. In other words, the Omani people in Dhofar may reinvigorate traditions, develop a sense of identity, and strengthen their local economy by promoting their own cultural heritage activities. Visitors' purchases of lodging, traditional food, beverages, and souvenirs generate revenue, which in turn, support regional companies and create jobs for locals. Indeed, this demonstrates the critical role that tourism (including ICH tourism) plays in raising people's socioeconomic status and quality of life, particularly in developing countries (Soliman, 2015). Furthermore, the interaction with tourists from different backgrounds offers a chance for elders to pass on important knowledge and skills to younger generations, ensuring the survival of cultural customs. In addition, it stimulates community involvement, raises standards of education, and introduces knowledge about other cultural traditions. Communities can diversify their sources of revenue, promote intercultural dialogue, and encourage the preservation of their cultural legacy through cultural heritage tourism. In addition to drawing more attention to, maintaining, and developing strategies for culture tourism, and presenting them to the world in a better way, there are other things that can be done to draw more tourists from around the world and make sure that they do not vanish or be forgotten (Al Amri et al., 2019). In addition, in international tourism, cultural heritage stimulates respect and understanding of other cultures and, consequently, promotes peace and understanding. However, the development of cultural heritage tourism, including ICH, might have negative effects on the local communities. The loss of cultural authenticity and meaning could result from excessive commercialization and catering to visitor expectations. Instability may result from the disruption of customary ways of life, reliance on tourism-related income, and potential environmental stress. Increased visitor activity can also change social relations, result in cultural appropriation, and invade locals' privacy. Consequently, it becomes imperative to strike a balance between the advantages of intangible culture tourism, safeguarding of cultural integrity, and community well-being to limit these potentially negative impacts.

This theoretical study represents substantial contributions to concerned academics and scholars by shedding light on the intricate connections between cultural heritage tourism and community well-being. By examining factors including economic advantages, sociocultural exchanges, and potential problems like over-commercialization and cultural commodification, this conceptual work contributes to our understanding of how cultural heritage tourism may affect local identities, economies, and social structures. Researchers are encouraged to use the findings of this paper to investigate a strategy for sustainable tourism growth that enriches and preserves cultural heritage while envisioning favourable social effects. This research also provides insightful knowledge to academia about how the inflow of visitors looking for cultural experience shapes, transforms, and occasionally even challenges locals. In addition, the study increases our knowledge of how tourism affects community structures and offers a nuanced view of sustainable development methods through the analysis of economic, social, and environmental factors. By bridging the divide between the academic disciplines of tourism, cultural studies, and community development, this endeavour contributes to the conversation on ethical tourist practices and encourages the preservation of cultural assets in regional settings.

The current study also produces significant contributions to practices that can affect how community engagement, urban and heritage development, and governmental policies are approached. The study provides local authorities and stakeholders with insights to help them make proper decisions by revealing precise ways that cultural tourism affects regional economics, social connections, and environmental sustainability. The findings can serve as a roadmap for the construction of focused efforts that optimize the benefits of cultural heritage tourism, such as job generation and cultural preservation, while minimizing any potential drawbacks, such as overcrowding or displacement. In addition, the current work assists tourism stakeholders and policymakers in developing a deeper understanding of the positive and negative effects of cultural heritage tourism on the local community while considering the existing goals and potential future strategies for cultural tourism in relation to the community. This paper indicated that cultural heritage tourism could provide infrastructure development for local cultural preservation and environmental protection, economic stimulus, and employment opportunities in rural areas such as the region of Dhofar in Oman. As opportunities for empowerment and entrepreneurship are made available, carefully managed and community-driven cultural tourism development may provide the people of Dhofar with valuable social learning experiences. Thus, the involvement of stakeholders (i.e., local community) in the management of cultural resources is essential to creating a symbiotic and sustainable link between tourism and heritage (Chauhan, 2022), to diversify rural livelihoods (Giampiccoli & Kalis, 2012) and preserving the socio-cultural status and environment of the local community (Kaur et al., 2016). Moreover, an understanding of ICH tourism helps with intercultural dialogue and encourages mutual respect for other ways of life. As a result, it becomes imperative for the government to promote ICH tourism which has multidimensional benefits.

In addition to its contribution, the current work offers several insightful approaches for further investigation. Given the relative dearth of studies concentrated on cultural heritage tourism in general and ICH tourism in particular, the current study paves the path for further research on the empirical evidence concerning the effects of cultural heritage tourism, especially ICH, on the local community, particularly in Oman, in terms of its economic, social, cultural, and environmental impacts. This can be achieved by conducting a quantitative study based on a questionnaire while taking the local community's and/or visitors' perspectives into consideration. In addition, a qualitative study could be carried out through interviews with different stakeholders (e.g., destination managers, destination marketers, and members of the community). Moreover, future studies may examine the most crucial variables affecting cultural heritage destination evangelism (Soliman & Al Balushi, 2023) of Oman and other Middle Eastern tourist destinations. Additionally, future research can examine the influence of cutting-edge technology, such as gamification (see e.g., Abou-Shouk & Soliman, 2021) and digital content marketing (see e.g., Mathew & Soliman, 2021) on the branding of cultural heritage destinations and how visitors engage with those locations. Furthermore, one of the promising directions for future research could be conducting a thorough review study using bibliometric techniques (see e.g., de Bruyn et al., 2023; Soliman et al., 2021; Soliman et al., 2023) of the scientific output in the field of ICH tourism. Future research can also investigate the relationship between destination social responsibility, visitor attitudes, and behaviours (Hassan & Soliman, 2021) toward cultural tourism destinations in different countries with various cultural backgrounds.

REFERENCES

Abou-Shouk, M., & Soliman, M. (2021). The impact of gamification adoption intention on brand awareness and loyalty in tourism: The mediating effect of customer engagement. *Journal of Destination Marketing & Management, 20*, 100559. doi:10.1016/j.jdmm.2021.100559

Abouelmagd, D. (2023). Sustainable urbanism and cultural tourism, the case of the Sphinx Avenue, Luxor. *Alexandria Engineering Journal, 71*, 239–261. doi:10.1016/j.aej.2023.03.041

Al Amri, I. T., Sbai, A., & Ben Rabia, K. (2019). Tourism constituents and tourism marketing in Oman. *Dhofar Governorate as a Model, 216*(2), 1-17. https://mrk.journals.ekb.eg/article_100405_16383346 8a3d5f06b9156f88895c8596.pdf

Al-Hashim, A. (2015). Embracing sustainable tourism in Oman: case study of Mirbat settlement. *International Journal of Cultural and Digital Tourism, 2*(2). 8-16.

Al-okaily, N. S., Alzboun, N., Alrawadieh, Z., & Slehat, M. (2023). The impact of eudaimonic well-being on experience and loyalty: A tourism context. *Journal of Services Marketing, 37*(2), 216–231. doi:10.1108/JSM-08-2021-0282

Al Riyami, H., Almuhrzi, H., & Scott, N. (2022). Heritage hotel experiences in Nizwa, Oman. *International Journal of Tourism Research, 24*(1), 169–180. doi:10.1002/jtr.2492

Almuhrzi, H., Hughes, K., & Ballantyne, R. (2020). Exploring Arab and Western visitors' interpretive experiences at an Omani heritage site: Does one size fit all? *Journal of Heritage Tourism, 15*(2), 180–199. doi:10.1080/1743873X.2019.1632319

Apostolakis, A. (2003). The convergence process in heritage tourism. *Annals of Tourism Research, 30*(4), 795–812. doi:10.1016/S0160-7383(03)00057-4

Chauhan, E. (2022). Residents' motivations to participate in decision-making for cultural heritage tourism: Case study of New Delhi. *Sustainability (Basel), 14*(14), 8406. doi:10.3390u14148406

Chen, C. F., & Chen, P. C. (2010). Resident attitudes toward heritage tourism development. *Tourism Geographies, 12*(4), 525–545. doi:10.1080/14616688.2010.516398

Colton, J. W., & Whitney-Squire, K. (2010). Exploring the relationship between aboriginal tourism and community development. *Leisure/Loisir, 34*(3), 261-278. doi:10.1080/14927713.2010.521321

Cravidão, F., Nossa, P., Santos, N., Vidal, M., Crous, N., Aulet, S., & Dallari, F. (2018). Tourism at world heritage sites opportunities and challenges. *Tourism Management at UNESCO World Heritage Sites,* 23-30. UNESCO.

de Bruyn, C., Said, F. B., Meyer, N., & Soliman, M. (2023). Research in tourism sustainability: A comprehensive bibliometric analysis from 1990 to 2022. *Heliyon, 9*(8), e18874. doi:10.1016/j.heliyon.2023.e18874 PMID:37636413

Dhanalekshmi, U., Alam, T., & Khan, S. A. (2022). *The folkloric uses and economic importance of some selected edible medicinal plants native to Oman: a brief overview. Edible Plants in Health and Diseases* (Vol. 1). Cultural, Practical and Economic Value. doi:10.1007/978-981-16-4880-9_1

Fu, X., Sittithai, T., & Chankoson, T. (2023). Factors affecting the development of Lipu Yi costumes: Intangible cultural heritage tourism. *Journal of Cultural Heritage Management and Sustainable Development*. doi:10.1108/JCHMSD-05-2022-0068

Giampiccoli, A., & Kalis, J. H. (2012). Tourism, food, and culture: Community-based tourism, local food, and community development in m pondoland. *Culture, Agriculture, Food and Environment, 34*(2), 101–123. doi:10.1111/j.2153-9561.2012.01071.x

Gortan-Carlin, I. P., KRAJNOVIČ, A., & Raguž, I. V. (2022). Study programs in culture and tourism in higher education in the Republic of Croatia – A critical analysis. In *Proceedings of FEB Zagreb International Odyssey Conference on Economics and Business* (Vol. 4, No. 1, pp. 462-484). University of Zagreb, Faculty of Economics and Business.

Gursoy, D., Jurowski, C., & Uysal, M. (2002). Resident attitudes: A structural modeling approach. *Annals of Tourism Research, 29*(1), 79–105. doi:10.1016/S0160-7383(01)00028-7

Guzmán, P. C., Roders, A. P., & Colenbrander, B. J. F. (2017). Measuring links between cultural heritage management and sustainable urban development: An overview of global monitoring tools. *Cities (London, England), 60*, 192–201. doi:10.1016/j.cities.2016.09.005

Hassan, S. B., & Soliman, M. (2021). COVID-19 and repeat visitation: Assessing the role of destination social responsibility, destination reputation, holidaymakers' trust and fear arousal. *Journal of Destination Marketing & Management, 19*, 100495. doi:10.1016/j.jdmm.2020.100495

Kaur, P., Jawaid, A., & Othman, N. B. A. (2016). The impact of community-based tourism on community development in Sarawak. *Journal of Borneo Kalimantan, 2*(1), 15–26.

Lyon, S. (2013). Coffee tourism and community development in Guatemala. *Human Organization, 72*(3), 188–198. doi:10.17730/humo.72.3.0011g4xj1m83x68j

Mair, H., & Reid, D. G. (2007). Tourism and community development vs. tourism for community development: Conceptualizing planning as power, knowledge, and control. *Leisure/Loisir, 31*(2), 403-425.

Mathew, V., & Soliman, M. (2021). Does digital content marketing affect tourism consumer behavior? An extension of technology acceptance model. *Journal of Consumer Behaviour, 20*(1), 61–75. doi:10.1002/cb.1854

Matlegiewicz, M. (2020). Motivation in Cultural Tourism: The Case of the Traditional and Regional Food Products of West Pomerania. *Zarządzanie w kulturze, 21*(3), 219-237.

McKercher, B., & du Cros, H. D. (2002). *Cultural tourism: the partnership between tourism and cultural heritage management*. Haworth Hospitality Press.

Melubo, K., & Lovelock, B. (2020). Living inside a UNESCO world heritage site: The perspective of the Maasai community in Tanzania. In *Sustainable Tourism Policy and Planning in Africa* (pp. 89–108). Routledge., doi:10.4324/9781003038832-6

Muštra, V., Perić, B. Š., & Pivčević, S. (2023). Cultural heritage sites, tourism and regional economic resilience. *Papers in Regional Science, 102*(3), 465–482. doi:10.1111/pirs.12731

Park, H. Y. (2013). *Heritage Tourism*. Routledge. doi:10.4324/9781315882093

Richards, G. (2018). Cultural tourism: A review of recent research and trends. *Journal of Hospitality and Tourism Management, 36*, 12–21. doi:10.1016/j.jhtm.2018.03.005

Singh, S. (2021). 'Quixotic' tourism? Safety, ease, and heritage in post-COVID world tourism. *Journal of Heritage Tourism, 16*(6), 716–721. doi:10.1080/1743873X.2020.1835924

Soliman, M., & Al Balushi, M. (2023). Unveiling destination evangelism through generative AI tools. ROBONOMICS. *The Journal of the Automated Economy*, *4*(54), 1.

Soliman, M., Cardoso, L., de Almeida, G. F., Araújo, A., & Vila, N. (2021). Mapping smart experiences in tourism: A bibliometric approach. *European Journal of Tourism Research, 28*, 2809–2809. doi:10.54055/ejtr.v28i.2254

Soliman, M., Gulvady, S., Lyulyov, O., & Pimonenko, T. (2023). Research Trends and Themes in the Top-Tier Tourism, Leisure and Hospitality Journals: A Bibliometric and Network Analysis Before and During the COVID-19 Era. *International Journal of Hospitality and Tourism Systems, 16*(1).

Soliman, M. S. (2015). Pro-poor tourism in protected areas–opportunities and challenges: "The case of Fayoum, Egypt". *Anatolia, 26*(1), 61–72. doi:10.1080/13032917.2014.906353

UNESCO. (2023). *What is Intangible Cultural Heritage?* UNESCO. https://ich.unesco.org/en/what-is-intangible-heritage-00003

UNWTO. (2017). *Tourism and culture*. UNESCO. https://www.unwto.org/tourism-and-culture.

UNWTO: World Tourism Organization (2018). *Tourism and Culture Synergies*. UNWTO. doi:10.18111/9789284418978

Weismann, N., Dziamski, P., & Haar, L. (2019). The kambārī in the museum of the frankincense land, Salalah, Oman. *International Journal of Nautical Archaeology, 48*(2), 342–359. doi:10.1111/1095-9270.12372

Yolal, M., & Negrusa, A. L. (2012). Analysis of cultural tourism motivation: the case of Turkish students. *Chinese Business Review, 11*(3). ISSN 1537-1506

Chapter 14
Rhythms of Living Cultural Heritage:
Co-Creating Memorable Tourism Experiences

Minna-Maarit E. Jaskari
https://orcid.org/0000-0002-5412-7176
University of Vaasa, Finland

ABSTRACT

Cultural heritage tourism involves traveling to a destination to experience and appreciate its cultural heritage. This type of tourism focuses on exploring historical sites, traditions, festivals, arts, crafts, and other aspects that reflect the local culture and heritage. Its aim is to promote sustainable tourism practices while preserving and enhancing the cultural identity of the region. Folk music, on the other hand, refers to the traditional music that is created and transmitted within a particular region reflecting its cultural and social heritage. Folk music is typically performed using traditional instruments and may involve storytelling, dance, and communal participation. It plays a significant role in preserving and promoting living cultural heritage. In this chapter, the author will be discussing the Kaustinen Folk Music Festival and how it co-creates memorable tourism experiences for visitors while promoting and preserving regional cultural heritage.

IINTRODUCTION

The concept of intangible or living cultural heritage has gained significant recognition and appreciation, particularly when viewed through the lenses of cultural diversity and destination marketing. The concept of Intangible Cultural Heritage (ICH) encompasses more than just immaterial aspects. In addition to practices, representations, expressions, knowledge, and skills, intangible cultural heritage encompasses tangible elements such as cultural spaces, buildings and architecture, instruments, and other artifacts (UNESCO, 2023a; Lenzerini, 2011). These intangible components play a significant role in the preserva-

DOI: 10.4018/978-1-6684-9957-3.ch014

tion and transmission of cultural heritage. By recognizing the importance of both intangible and tangible aspects, intangible cultural heritage aims to safeguard and promote the diverse cultural expressions and traditions that contribute to the richness of our global heritage (UNESCO, 2023a).

Folk music is an example of intangible or living cultural heritage that plays an important role in preserving and promoting cultural heritage. Folk music refers to traditional music that is created and passed onwards orally within a specific region. It reflects the history, practices, traditions, beliefs, and values of that region. Folk music is performed using traditional instruments, such as violins or bagpipes, and may involve storytelling, dance, and communal engagement (Kõmmus, 2019). Examples of worldwide-known folk music include Flamenco from Spain, Samba from Brazil, Bluegrass from the USA, Rajasthani Folk music from India, Irish Traditional Music from Ireland, and Scottish bagpipe music. In Finland, Kaustinen style fiddle music played by violin and other traditional string instruments represents living cultural heritage preserved also by The United Nations Educational, Scientific and Cultural Organization (UNESCO). All of these examples demonstrate how folk music is a vital and enduring component of the region's cultural identity and destination branding. Therefore, it also has implications for the tourism industry.

Cultural heritage tourism can be explored from both the perspectives of visitors and tourism management (Campos et al., 2018). From the tourism management perspective, the objective of cultural heritage tourism is to facilitate the implementation of sustainable tourism strategies, thereby preserving and enhancing the cultural distinctiveness and identity of the local community. The triple bottom line framework of sustainability, encompassing economic, social, and environmental dimensions, is commonly employed in discussions surrounding sustainability (Elkington 1994, 1998). However, cultural heritage tourism places particular emphasis on the cultural dimension of sustainability (Dessein et al., 2015).

From the tourist perspective, cultural heritage tourism involves an extended customer journey from before visiting the location, including traveling to a specific location, visiting the location in order to engage with and gain an appreciation for its cultural heritage as well as the phase after the visit (Lemon & Verhoef, 2016). In cultural heritage tourism, the emphasis is on exploring historical sites, traditions, festivals, arts, crafts, and other aspects that reflect the local culture and heritage, as well as establishing connections with these spaces (Minkiewicz et al., 2014; Seyfi et al., 2020). Many of the earlier studies on cultural heritage tourism focus on the middle part of the tourism journey, that is the experiences while visiting the site, neglecting the pre- and post-visit phase.

A memorable tourism experience is a consumer's subjective memory attached to a specific tourism experience and recalled later in life (Kim et al., 2012; Kauppinen-Räisänen et al., 2013; Hosany et al., 2022). While tourism experiences are always subjective and context-dependent, they are also co-created through visitors active or passive participation in tourism activities and social interactions with other people in different phases of the customer journey (Campos et al., 2018; Tung & Ritchie, 2011; Seyfi et al., 2020). While there has been extensive research conducted on memorable tourism experiences in a broad sense, its specific application within the context of cultural heritage tourism has received limited attention. Moreover, earlier research on music in consumer research has recognized how music and sounds have a great impact on consumer experience and behaviour (e.g. Gustafsson, 2015; Jain & Badgare, 2011; Bitner, 1992), still, the context of music in not very well studied in tourism research.

This chapter aims to fill these research gaps by analyzing how the living cultural heritage has successfully been promoted and preserved in a sustainable way while co-creating memorable tourism experiences. The chapter utilizes an illustrative and unique case of cultural heritage tourism, namely the Kaustinen fiddle playing and an annual festival event that takes place in the small town of Kaustinen. The living cultural heritage of a particular and distinctive style and playing technique of Kaustinen folk music has existed for more than 250 years. The Kaustinen Folk Music Festival has been running for over 50 years and it has impacted the people in the region, the domestic and international visitors as well as tourism, the economy, and the on-going development of the region. The illustrative case study presented in this chapter provides implications for the field of tourism management, particularly in relation to living cultural heritage and for the broader context of sustainable tourism.

LIVING CULTURAL HERITAGE

According to UNESCO (2023a), the term "cultural heritage" has changed in recent decades. Cultural heritage extends beyond concrete buildings, monuments, artefacts, or object collections. To illustrate, performing arts, social practices, rituals, festive events, knowledge, skills, and practices concerning nature or producing traditional crafts are all examples of living expressions passed down from our forefathers to our descendants. Indeed, these elements that represent the living culture of human communities are the essence of cultural heritage (Lenzerini, 2011). Furthermore, cultural heritage is an evolving entity that continues to shape and influence contemporary societies rather than being static or confined to the past (UNESCO, 2023).

UNESCO has initiated the process of creating a comprehensive inventory of intangible cultural heritage with the aim of safeguarding, promoting, and fostering respect for cultural diversity on a global scale. UNESCO anticipates that compiling this list, it will raise awareness about the importance of intangible cultural heritage and promote its preservation and appreciation. By recognizing and valuing the diverse cultural practices and traditions that shape our shared human experience, this initiative aims to contribute to the sustainable development of our societies. In September 2023 there were 676 elements listed from 5 regions (Africa, Arab States, Asia and the Pacific, Europe and North America, and Latin America and the Caribbean) and from 140 different countries (UNESCO, 2023d).

Within the cultural heritage list, traditional or folk music is listed as an element in different ways. In some elements, the music genre or playing style is the focus. To illustrate, Jamaican reggae, which originated in marginalized communities and blended various musical influences such as Caribbean, North American, Latin, and Neo-African styles, is now popular across a wide range of societal segments, addressing issues of injustice, resistance, and humanity. Another example is the Hungarian string band tradition that represents a significant facet of folk music culture, with its basic fiddle-viola-bass line-up varying across regions and eras. Historically, these bands played at various local events, and today, they are integral to stage performances and urban folk dance parties (UNESCO, 2023d).

In some elements, the music and instrument craftsmanship are closely tied together. To illustrate, Algerian Raï is a folk music genre that conveys social realities, addressing themes like love, freedom, and societal pressures. The musicians often craft and adorn their instruments, with the genre's transmission occurring both informally, through observation, and formally, through apprenticeships (UNESCO, 2023d).

Further, in some elements the focus is on musical instruments, for example, the M'Bolon is characterized by a large calabash sound box and a bow-shaped neck with strings, and is a versatile musical instrument from southern Mali. Also, Cremonese violinmaking in Italy is well-known for its traditional methods of creating and restoring violins, violas, cellos, and contrabasses (UNESCO, 2023d).

However, in most cases, the cultural heritage element combines the instrument, music, related practices, and expressions as is the case with Uilleann piping, which uses a specific type of bagpipe, and is a musical practice that is deeply rooted in Irish tradition. It is most common in Ireland and Irish communities around the world, and it serves as a socializing medium and connects people to their heritage during life events (UNESCO, 2023d).

It is worth noting that traditional music is part of a much wider range of cultural heritage elements, for example, the music is embedded in performative arts such as The Corpus Christi festival in Panama or the Dance drama Nora from Thailand. The above examples reflect elements where traditional music is at the core of the element (UNESCO, 2023d).

The context of this chapter is Finland, a country of 5.5 million people located in Northern Europe. Finland has two elements listed on its own: the Finnish sauna culture and Kaustinen fiddle playing and related practices and expressions (UNESCO, 2023a, c). Also, together with other Nordic countries, Finland is also part of the Nordic clinker boats cultural heritage.

In this chapter the focus is in folk music, in Kaustinen fiddle playing and tourism around the living cultural heritage. In UNESCO listing (2023a) Kaustinen fiddle playing and related practices and expressions are described as follows:

"Kaustinen folk music is a Finnish tradition where the fiddle (violin), although not the only instrument, is the main melodic instrument. It is in fact the violin (with or without other instruments) that cadences the dances or the songs. Based on playing by ear, it is characterized by syncopated and accented rhythms that are easy for people to dance to. The distinctive style and playing technique of Kaustinen folk music have existed for over 250 years, and there is a repertoire of several hundred tunes from the area. Kaustinen folk music is played in many contexts: in private, in organized groups, during public celebrations and ceremonies (including wedding parties), at concerts and public jam sessions, and at the annual Kaustinen Folk Music Festival. Performers often wear traditional costumes. Most inhabitants of Kaustinen and neighbouring communities consider this music an essential aspect of their personal and community identity, even if they are not practitioners themselves. It is felt to represent a sense of belonging and viewed as a means of strengthening cross-generational ties. The significance of the practice is evident in the names of public spaces and in symbols such as the presence of the fiddle in the Kaustinen coat of arms."

As shown in the extract above, Kaustinen folk music is typically played with violin and it has a distinctive style that has existed for over 250 years. While it is played in many contexts, the annual folk music festival is a focal event where both domestic and international music lovers celebrate the cultural practices.

SUSTAINABILITY IN CULTURAL HERITAGE TOURISM

According to UNESCO, the Kaustinen folk music strongly connects to sustainable development goals of "Peace, Justice and Strong institutions" that aims to promote peaceful and inclusive societies for sustainable development, providing access to justice for all and building effective, accountable and inclusive institutions at all levels and "Quality Education" that aims to ensure inclusive and equitable quality education and promoting lifelong learning opportunities for all (UNESCO, 2023b). Indeed, cultural heritage elements have an important role in reaching towards United Nations' 17 sustainability goals described in The 2030 Agenda for Sustainable Development that are adopted by UN member states (UN, 2023).

In order to manage sustainability in practice, many researchers refer to Elkington (1994). He developed in the 1990s a "triple bottom line" - framework of sustainability, encompassing economic, social, and environmental dimensions, as a philosophical orientation for companies to build their strategy in a sustainable way (Elkington, 1994, 1998). In the framework the economic dimension refers to the economic value that organizations provide to their surroundings, the social dimension refers to conducting fair business practices towards labor and human capital and the ecological dimension refers to practices that are not harmful to future generations' access to natural resources. The framework is also suggested for tourism management in order to develop sustainable tourism (Stoddard et al., 2012). For example, Landorf (2009) assessed to which extent World Heritage Site management planning processes confirm the key principles of sustainable tourism. Based on the content analysis of six World Heritage Site management plans, she highlights how the sites are mainly focusing on managing the environmental sustainability dimensions and lack the economic and social sustainability dimensions in their plans.

The triple bottom line framework has its origins in corporate social responsibility and accounting (Elkington, 1994). While it gives a usable framework for many cases, it lacks the cultural dimension (Dessein et al., 2015), namely preserving cultural diversity and living heritage. The concept of cultural sustainability holds significant importance, particularly in relation to the preservation of living cultural heritage. Puro (2021) illustrates how the cultural dimension of sustainability in tourism involves preserving the cultural heritage by ensuring that tourism product offerings do not harm the cultural heritage, making sure that the product offering and cultural heritage have an authentic connection, keeping the preservation of cultural heritage as a focal point in product offerings, and developing tourism products in collaboration with people practicing that specific culture as well as locals and other stakeholders.

To sum up, the four-dimensional sustainability framework (Picture 1) is suggested to be used for cultural heritage tourism. The framework seeks to minimize negative impacts on the environment, preserve cultural heritage, and promote social well-being, while simultaneously fostering economic growth and development.

CO-CREATION OF MEMORABLE TOURISM EXPERIENCES

Memorable Tourism Experience

The unique nature of the tourism experience is often related to the interplay between everyday life and alternative realities. This interplay becomes most apparent when there are shifts in the familiar temporal and spatial structures that individuals are used to (Binkhorst & Den Dekker, 2009). Moreover, earlier research points out how "tourists seek authentic, rewarding, meaningful, multisensory and transformative experiences when visiting places" (Hosany et al., 2022). Indeed, tourism management aims for tourism experiences to be memorable, unique, extraordinary, and exceptional (Pine & Gilmore, 1999) compared to ordinary, mundane, every-day experiences that are familiar and utilitarian (Heinonen & Lipkin, 2023).

The majority of research conducted on the subject of tourism experience integrates cognitive, emotional, behavioral, sensorial, and social aspects (Schmitt, 1999; Becker & Jaakkola, 2020), linking them with the various stages of the overall experience evolving from the customer journey, namely the pre-experience phase, the co-creation of the actual experience, and the post-experience phase (Björk et al., 2021; Lemon & Verhoef, 2016).

Figure 1. A four-dimensional framework for sustainable cultural heritage tourism

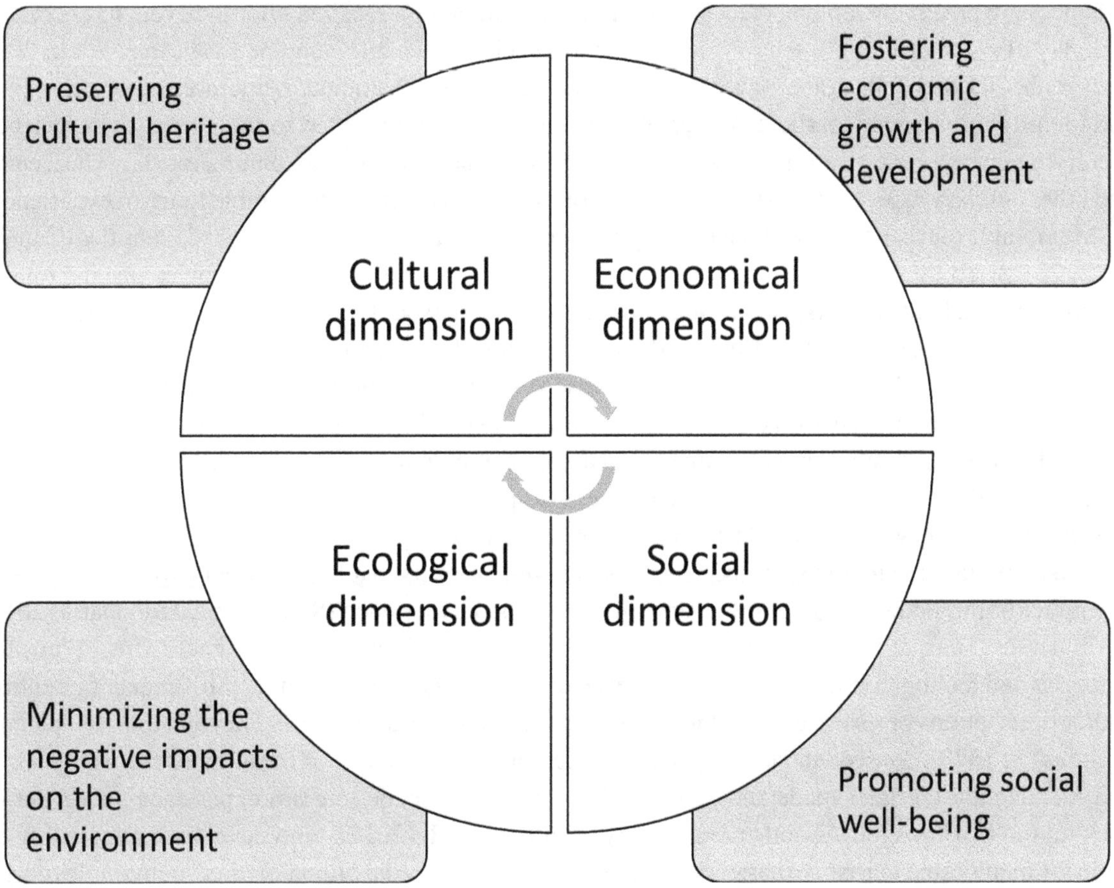

Tourism experiences are multifaceted and multidimensional (Kauppinen-Räisänen et al., 2013). Tourism experiences are subjective and personal in nature (e.g. Caru & Cova, 2003). The diverse interests and backgrounds of tourists contribute to a wide range of interpretations of an identical tourism product. This variation arises from the unique preferences, cultural backgrounds, and prior experiences of individual tourists engaging with the same tourist offering. As tourists bring their own set of values, beliefs, and expectations to their travel experiences, they inevitably perceive and interpret the same offering in distinct ways. To illustrate, tourists' individual moods and personal emotions at a given moment can significantly influence their subjective interpretations of these experiences. This phenomenon highlights the subjective nature of tourism experiences and emphasizes the importance of recognizing and accommodating tourists' diverse needs and interests (Ooi, 2005; Kauppinen-Räisänen et al., 2013; Becker & Jaakkola, 2020).

Tourism experiences are also situational and context-specific (e.g., Kauppinen-Räisänen et al., 2013), which means that the place itself has an essential impact on how an event or activity is experienced and how the experience is perceived. Moreover, the same tourism activity may be interpreted differently at different times. To illustrate, weather conditions impact the exploration in the open-air museum, or screaming children can have an impact on how the atmospherics is experienced.

At their best tourism experiences may become memorable experiences, which means that these experiences are positively remembered, stored in travelers' memory, recalled after the event has occurred, and may also affect future intentions and behavior, such as revisit (Kauppinen-Räisänen et al., 2013; Kim et al., 2012). To illustrate, Kauppinen-Räisänen et al. (2013) studied memorable food experiences and identified how many participants recalled memories of eating linked to their tourism experiences specifying places, contexts, times, types of foods, and personal responses to their memories. Thus, eating and other culinary aspects of tourism play an important role in creating memorable tourism experiences.

Memorable tourism experiences have the potential to shape and define individuals' identities (Campos et al., 2018). To illustrate, through exposure to diverse cultures, traditions, and environments, tourists are able to broaden their perspectives and gain a deeper understanding of the world around them. This exposure allows individuals to reflect on their own values, beliefs, and interests, leading to a more refined sense of self. Moreover, by immersing themselves in new and unfamiliar settings, tourists are able to explore different aspects of their personalities and discover hidden passions or talents. This process of self-discovery and self-expression may contribute to an individual's overall well-being and satisfaction. According to Tung & and Ritchie (2011), consumers view memorable experiences as the peak of satisfaction, and they work toward achieving those experiences.

The memorable tourism experience consists of several aspects that can be either explicitly expressed or implicitly implied throughout the course of the tourism experience (Sthapit & Coudounaris, 2018). To illustrate, Tung & Ritchie (2011) identify four aspects of memorable experience: affect ("positive emotions and feelings associated with the experiences" p. 1377), expectations ("fulfillment of intentions and/or descriptions of surprises encountered during the trip that was above and beyond tourists' planned agendas" p. 1378), consequentiality ("personally perceived importance or meaning of the trip" p.1379) and recollection ("efforts made and actions taken to remember the tourism experience and/or reflect back on the trip" p. 1380). Another example is from Kim et al. (2012), who developed a measurement scale for memorable tourism experience including seven experiential dimensions of memorable tourism experience, namely hedonism (e.g. thrilled about having a new experience), novelty (e.g. once-in-a-lifetime experience), local culture (e.g. closely experienced local culture), refreshment (e.g. liberating), meaningfulness (e.g. I did something important), involvement (e.g. I enjoyed activities which I really wanted to do), and knowledge (e.g. exploratory).

In the context of cultural heritage tourism, Seyfi et al. (2020) studied memorable tourism experiences in Paris. Based on 29 qualitative in-depth interviews they identified authenticity, quality of service, engagement, cultural exchange, and culinary attractions as configuring elements of memorable cultural tourism experiences. They also emphasize how pre-experience perceptions and expectations, and post-experience recollection of the significance of the experience and evaluation of its memorability play a significant role in the memorability of the experience.

In their systematic literature review, Hosany et al. (2022) suggest to be cautious when using generic scales in measuring memorable tourism experience, since those scales fail to consider the contextual and situational aspects of memorable tourism experience (Sthapit & Coudounaris, 2018). Indeed, in order to accurately capture an experience, it is important to include dimensions that are specific to the context in which the experience occurs (Hosany et al., 2022). Moreover, it can be questioned whether and to what extent ready-made measurement scales can actually capture the subjective and contextual experience. Thus, in this chapter, the memorable experience is approached in a more conceptual manner and the focus is on how the memorable experience is fostered by different stakeholders. This perspective leads us to co-creation.

Co-Creation of Tourism Experiences

The concept of co-creation is a widely and variously adapted concept in tourism research, encompassing perspectives from both tourism management and tourists (Campos et al., 2018). In their systematic literature review, four dimensions of tourism experience co-creation are identified; 1) tourist contributing to the co-creation, 2) tourist actively participating in on-site activities, 3) tourist interacting with others, and 4) the subjectivity of tourism experience (Campos et al., 2018).

First, co-creation in tourism involves a tourist contributing to some aspect, or phase of the overall tourism experience. The on-site tourism experience is affected by the pre-phase of the travel (earlier knowledge, expectations, planning, bookings, travel to the destination) and the overall experience is impacted by the post-phase of the travel (traveling back from the destination, recollecting the memories and sharing stories with other people) (Binkhorst & Den Dekker, 2009; Lemon & Verhoef, 2016). Indeed, the memorability of the tourism experience is enhanced or de-enhanced over time when the visitor is recalling the experience and evaluating its significance (Seyfi et al., 2020). This implies that all the actors involved in co-creation need to be present in different phases of the tourist journey – from planning to visiting to recollection of memories. Still, as the co-creation involves interaction where tourist is contributing, the service providers cannot co-create for tourists but suggest touchpoints where co-creation can emerge (Lemon & Verhoef, 2016).

Second, the tourist actively participates in on-site experience activities that are different from mundane, every-day routines (Campos et al., 2018; Heinonen & Lipkin, 2023). The participation may be passive, where tourists mainly gaze at attractions through visual perception as a spectator, or active, where the touristic experience is described through active participation and multisensory exploration through physical, emotional, and mental engagement (Pine & Gilmore, 1999). In the context of living cultural heritage, listening to the folk music performance can be passive participation while attending folk music jamming session is active and co-creative.

Third, social interactions are an integral part of the co-created experience. The visitors interact with others – travel companions, other tourists, staff members, locals, and so on - during the on-site experience. These interactions between people may be planned or unplanned, formal or informal, one-time occasions, or even long-term relationships (Campos et al., 2018). The significance of social interactions is recognized in impacting experiences becoming memorable (Tung & Ritchie, 2011).

Fourth, the co-creation of experience is impacted by the tourist's psychological state and processes (such as perceptions, thoughts, expectations, emotions, and involvement) making each experience subjective and context-dependent, and thus service providers can impact it only partly (Campos et al., 2018).

Campos et al. (2018) also suggest that the on-site tourism experience is co-created in an "experiencescape" that incorporates the influencers (physical, social, and organizational), tourists' on-site physical and mental participation, and social interaction with other people and the environment (Bitner, 1992; Ooi, 2005; Cutler & Carmichael, 2020).

Earlier studies on the co-creation of experiences in the context of cultural heritage tourism experiences are not many. One exception is Minkiewicz et al. (2014), who explored the co-creation of experiences in two heritage organizations, namely a museum and a gallery, using consumer critic analysis. Their findings reveal that consumers actively choose to co-create their heritage consumption experiences through co-production, engagement, and personalization. The co-production refers to an "active participation in one or more activities performed throughout the consumption experience, focusing typically on physical elements of the experience". Engagement refers to "how cognitively and emotionally immersed one is in the experience". Their findings illustrate how visitors' cognitive immersion is revealed through their reflections on what they experienced, a desire to learn more, and challenging of their assumptions, beliefs, and attitudes. They highlight how emotional immersion requires a sense of personal relevance and connection to the experience and can be achieved through social interaction with other people, such as travel companions, other visitors, locals, or staff members. They emphasize that personalization is a critical component of co-creation referring to "individuals tailoring their experiences to meet their needs through self-directed customization of the experience, interaction with a service representative, and technology." Their findings illustrate how consumers personalize their experiences by selecting a self-directed path according to their preferences, using experience spaces in their own unique way, and utilizing technology and employee assistance. They also point out how visitors' interactions within the museum/gallery were influenced by visitors' familiarity and knowledge of the type of experience and previous exposure to the experience space.

KAUSTINEN FOLK MUSIC FESTIVAL

In this section, a unique case of living cultural heritage, the Kaustinen Folk Music Festival (KMFF) is explored. The case is used in an explorative manner, it provides insight on how living cultural heritage is preserved in a sustainable way while aiming at co-creating memorable experiences with visitors.

The qualitative data used for this case analysis is holistic and multiple, consisting of participatory observation, public records from the event organization, public statistics from Visit Finland, news articles, festival programs, social media publications (@kaustinenfmf, #kaustinenfolkmusicfestival, #kaustinen-fmf, #kansanmusiikki), archival material as well as interviews with festival organizers and visitors. The data is analysed in a holistic manner from a tourism perspective focusing on events and practices that exemplify how living cultural heritage is produced and reproduced during the festival and beyond it.

The location of Kaustinen is in Finland. Kaustinen is a small town established in 1868 with a population of around 4200. In spite of its tiny size, Kaustinen and the areas surrounding it have a rich tradition in folk music, and the traditional way of playing the fiddle (violin) is inscribed in UNESCO's representative list of intangible cultural heritage of humanity. The town's brand identity is strongly communicated with folk music, which is at the heart of the place brand. For both locals and visitors, it is visible since for example, the town roundabout has a large violin sculpture. Indeed, the town is renowned for its strong connection to folk music, with numerous musicians, bands, and ensembles actively participating in the local music scene.

The annual Kaustinen Folk Music Festival is one of the most important cultural events in Finland. It was organized for the first time in 1968 and has since then emerged as a prominent folk music event in the Nordic countries and Europe. The festival is organized in July and it lasts for 7 days. It brings together musicians, dancers, and enthusiasts from around the world, providing a platform to celebrate and experience the diverse expressions of folk music. Also, it attracts tourists and local people who all share a love for folk music. In 2022 KFMF organized 1575 musical events, out of which 235 were free of charge. In 2022, 43,350 visitors attended the festival, which makes it one of the largest music festivals in Finland (Finland Festivals, 2023).

In 2023 over 5000 performers participated in the festival. The festival has a unique feature in that if you are accepted to perform as a musician, you are given free admission for all days. This results in a large number of performers performing and participating in activities on all days. While the main performers are paid, most of the musicians perform for free. This arrangement makes it possible for a small town to recruit thousands of musicians. This form of co-creation enhances collaboration and active participation from all actors. Indeed, it is considered an honor to participate as a musician to the event and many musicians stay several days at the event.

The Festival park is the main location of the event. However, the festival extends outside the park to the town center, parks, parking lots, and local shops. While attending Festival Parks requires tickets, events outside of it are free of charge. This is an example of how living cultural heritage is reproduced in local region to all locals and visitors. This is also aligned with sustainability goal of peace, justice and strong institutions (UN, 2023).

The music festival has a strong emphasis on authentic fiddle music. The region has produced many accomplished fiddlers who have made significant contributions to a more contemporary Finnish folk music. The fiddle is often the primary instrument in performances and is known for its lively and energetic playing style. To celebrate the heritage, each year during the opening ceremony of KFMF a "Fiddler(s) of the Year" is chosen. This prize is highly appreciated and reported in national news throughout the country.

Despite the focus on authentic fiddle music, the Kaustinen Folk Music Festival follows the principle of "something for everyone". Besides the different kinds of folk music concerts, also more pop-oriented concerts are offered to visitors. In this sense, Kaustinen is living the tradition and future together in collaboration with different generations, understanding how the traditional heritage impacts contemporary music. Different genres of music are treated as complimentary, not competing with each other. Indeed, in generally, folk music festivals have an active role in generating new musical genres.

Folk music in Kaustinen is often characterized by its collaborative and participatory nature. Musicians frequently come together in jam sessions, workshops, and informal gatherings to share tunes, techniques, and improvisations. Moreover, all the visitors are invited to join the workshops, jamming sessions or just to take part in sing-along events. This participatory approach fosters creativity, learning, and a sense of community among folk musicians and also invites new people to join in carrying the folk music practices. Also, from the co-creation point of view, active participation invites visitors to immerse themselves in the experience.

At Kaustinen, cultural heritage tourism impacts the local communities and region in general in different ways. The Kaustinen folk music scene and folk music-based tourism play a crucial role in driving the local economy, making a substantial contribution to its overall growth and development. Since the festival attracts a significant number of visitors – both musicians and tourists – revenue for various businesses in the area increases. These businesses include accommodations, such as hotels and guesthouses, that are fully booked during the festival week. Additionally, restaurants and cafes benefit from visitors seeking for dining options during their stay. Many of the visitors drive to the town and sleep in travel vans, thus buying gas and using grocery stores. Indeed, in order to co-create the tourism experience, not only festival organizers or tourism businesses, but a whole ecosystem of actors need to be orchestrated. The local community has succeeded well in incorporating the actors before, during, and after the event.

Tourism-related activities have been found to generate employment opportunities, both directly and indirectly, thereby contributing to the sustenance of local communities. This is no exception in Kaustinen. Even though there are hundreds of volunteers organizing the smooth festival, there is also direct employment within the tourism sector such as hotel staff, restaurant and café workers. The festival's impact does not restrict itself to the town, but it contributes to regional development. For example, due to the accommodation shortage, visitors are booking hotels and guest houses from neighbouring towns and larger cities.

Although the Festival gathers around tens of thousands of participants every year, the cultural heritage itself is preserved throughout the year on a much larger scale. For example, the Finnish Folk Music Institute which aims to advance Finnish Folk music and dance is located in Kaustinen. One of the music-oriented high schools is located in the region. Moreover, a method of music education based on the Kaustinen tradition of Finnish folk music is one example of preserving the living culture beyond the festival and the region. The "Näppäri" method is primarily aimed at children and young people, and the name refers to a common method of playing string instruments. The method is used in courses and workshops throughout Finland, not just in the region. These activities continue to educate new musicians and play an important role in the preservation, promotion, and protection of the folk music tradition. These activities support the sustainability goal of Quality education that aims to ensure inclusive education and life-long learning opportunities for all (UN, 2023).

Näppärit, which is a name for young performers using the method, can vary from a small group to several hundreds of young players. They are also specifically addressed during the Kaustinen Folk Music Festival. There is a music bootcamp preceding the festival and the final concert of the music bootcamp is one of the main events during the family day of the festival where hundreds of children and teenagers play together.

All of the above-mentioned illustrations ensure that the well-known brand image of the town and region as a folk music region attracts visitors not only during the festival period, but around the year. Moreover, the fact that Kaustinen fiddle music was listed on UNESCO's representative list of the intangible cultural heritage of humanity has impacted the region in many ways. There has been a lot of interest, not only from local journalists but also from journalists from other countries to report on the specific cultural heritage as well as the festival. The inclusion on the list has also had an impact on the ways in which various elements of cultural heritage collaborate together. For instance, in 2023 approximately ten UNESCO-listed destinations participated in the festival representing their living culture.

To sum up, the case study illustrates how the Kaustinen Folk Music Festival plays a vital role in preserving and promoting the living cultural heritage of Kaustinen folk music. This particular folk music is distinguished by its strong connection to local traditions, the prevalence of fiddle music, lively dance forms, and its collaborative nature. These distinct features contribute to the region's unique musical identity and make Kaustinen and the region a vibrant hub for folk music enthusiasts and performers.

The annual Kaustinen Folk Music Festival serves as a means of passing down traditional tunes, melodies, and styles from one generation to another. The authentic, engaging, and participatory nature of the festival is apt to co-create experiences and impact the memorability of those experiences. The music reflects the cultural identity and local traditions of the Kaustinen region, contributing to the overall cultural heritage of Finland.

DISCUSSION

Based on the illustrative case of the Kaustinen Folk Music Festival four implications are drawn to illustrate the successful case of living cultural heritage tourism. These are authenticity, collaboration and co-creation, the four dimensions of sustainability, and the significance of UNESCO's list of representative intangible cultural heritage elements.

First, the illustrative case shows how authenticity is a key component of living cultural heritage, and it should apply to all aspects of cultural heritage tourism. People involved in the culture must be able to identify and recognize artifacts, events, communication, and branding. The tourism industry must interact respectfully with cultural sites, artifacts, practices, and locals. Authenticity, however, does not imply that the culture does not change; on the contrary, living cultural heritage lives and changes in times and places, connecting people of our generation to generations before and after.

Second, the case study shows how the foundation of living cultural heritage tourism is collaboration and co-creation. Collaboration among various stakeholders, including municipalities, regional, national, and international development organizations, as well as various industries, lays the meta-level groundwork for long-term development. Participatory collaboration with various actors in a living cultural heritage regime promotes cultural authenticity and engages actors in the preservation and safeguarding of the culture. Collaboration with non-tourism event participants allows for the creation of events that lay the groundwork for the co-creation of memorable experiences for all participants - visitors, performers, staff, and locals.

The memorable tourism experience is both subjective and contextual. The Kaustinen Folk Music Festival has been successful in providing its visitors with a wide range of events that allow each visitor to create their own memory path before, during, and after the festival. This festival world of possibilities serves as an experiencescape where visitors can participate in whatever way they want and find the best for themselves.

The potential negative effects of tourism on tradition and its practitioners can be mitigated by adopting sustainable tourism practices that prioritize cooperation among stakeholders. By actively engaging in responsible activities, tourism can be transformed into a force that supports and preserves local traditions, rather than eroding them. This requires a commitment from all involved parties to work together towards sustainable tourism development that respects and safeguards the living cultural heritage.

Third, the triple bottom line framework of sustainability—economic, social, and environmental—is often suggested as a framework to be used in sustainable tourism management (Stoddard et al., 2012). While the three dimensions are significant in sustainable tourism, the framework lacks the cultural dimension of preserving and safeguarding living cultural heritage (Dessein et al., 2015). While cultural heritage researchers and tourism practitioners have recognized the need for the cultural dimension, there is a need to upgrade the triple bottom line framework to a four-dimensional framework also in tourism research. Economic growth, social well-being, environmental protection, and cultural heritage are all promoted by the four-dimensional sustainable tourism framework.

Fourth, the significance of the UNESCO Representative list of intangible cultural heritage has been overlooked in previous research on cultural heritage tourism. The preliminary efforts undertaken for the purpose of listing enhance the sense of community within the area and emphasize the significance of cultural heritage in the region. Both local and global publicity have the potential to enhance awareness of cultural heritage and attract visitors to gain a deeper understanding of the tradition, on-site and online. The appearance of a particular cultural heritage element in the representative list often facilitates the acquisition of funding for the advancement and safeguarding of cultural heritage. Hence, the inclusion of a cultural destination on the UNESCO list holds significant importance, making it highly advisable to consider applying for such recognition.

CONCLUSION

This chapter has discussed the co-creation of memorable tourism experiences in the context of living cultural heritage. The case study of Kaustinen fiddle folk music and the focal event of the Kaustinen Folk Music Festival was analysed using earlier literature from sustainable cultural heritage tourism and co-creation of experiences.

Kaustinen has a rich folk music heritage, with traditional fiddle playing being listed as intangible cultural heritage by UNESCO. This musical legacy is vividly displayed at the annual Kaustinen Folk Music Festival, which takes place every July for seven days. The festival, which is well-known not only in the Nordic countries but also throughout Europe, attracts tens of thousands of visitors and offers a diverse range of performances, workshops, and collaborative sessions, with a particular emphasis on fiddle music.

Beyond its cultural significance, the festival profoundly impacts the local economy, driving local business revenue and employment. Additionally, institutions like the Finnish Folk Music Institute and the Näppäri music education method, ensure that the regional folk music heritage remains vibrant year-round.

The Kaustinen Folk Music Festival forms an experiencescape where memorable tourism experiences are co-created. While the experience itself is subjective and context-dependent, the festival provides a great amount of possibilities for forming memorable experiences. The communication about the annual event begins right after the preceding one has ended through web pages and social media sites. Returning visitors can book their hotel or guest house while they are still attending the previous festivals. Performers are invited to submit their demo presentations during the winter. During the festivals, there are hundreds of small events around the festival area ranging from musical performances to instrument workshops, jamming sessions, and lectures. These small events build on authenticity, they call for both active and passive participation and engagement from visitors and enable visitors to personalize their festival stay.

The Kaustinen Folk Music Festival highlights four essential aspects of living cultural heritage tourism. Authenticity, which is central to the experience, ensures that evolving cultural narratives are honored and accurately communicated. Collaboration and co-creation place an emphasis on close and active collaboration between local communities, development organizations, and tourists. In terms of theoretical implications, the traditional framework of sustainability, which is often limited to economic, social, and environmental dimensions, needs to be expanded to include cultural dimensions. Finally, the UNESCO Representative list provides direct advantages such as global recognition and funding opportunities, demonstrating its importance not only in maintaining and safeguarding cultural heritage but also in cultural heritage tourism.

Living cultural heritage tourism provides several interesting possibilities for future research. In this chapter, the goal was to provide a holistic understanding by exploring both tourism management and tourist perspectives. Future research could study these separately. For example, the tourism management perspective could look more into the tourism ecosystem and collaboration between different actors in co-creation. Then again, the tourist perspective could study festival visitors in more detail in order to understand how they co-create memorable experiences throughout the tourism journey. Another research path is cultural sustainability. Since the earlier tourism research has mostly neglected the cultural sustainability dimension in sustainability, more studies are needed on how to incorporate it in business management, both in tourism and other businesses as well. Also, this illustrative case study affirms that UNESCO's list of cultural heritage for humanity is not only a recognition or preservation tool but also important for tourism. Future research could study the effects of listing in different places and societies.

REFERENCES

Becker, L., & Jaakkola, E. (2020). Customer experience: Fundamental premises and implications for research. *Journal of the Academy of Marketing Science, 48*(4), 630 648. doi:10.100711747-019-00718-x

Binkhorst, E., & Den Dekker, T. (2009). Agenda for co-creation tourism experience research. *Journal of Hospitality Marketing & Management, 18*(2), 311–327. doi:10.1080/19368620802594193

Bitner, M. J. (1992). Servicescapes: Impact of physical surroundings on customers and employees. *Journal of Marketing, 56*(2), 57–71. doi:10.1177/002224299205600205

Björk, P., Prebensen, N., Räikkönen, J., & Sundbo, J. (2021). 20 years of Nordic tourism experience research: A review and future research agenda. *Scandinavian Journal of Hospitality and Tourism, 21*(1), 26–36. doi:10.1080/15022250.2020.1857302

Campos, A. C., Mendes, J., Valle, P. O. D., & Scott, N. (2018). Co-creation of tourist experiences: A literature review. *Current Issues in Tourism, 21*(4), 369–400. doi:10.1080/13683500.2015.1081158

Carù, A., & Cova, B. (2003). Revisiting consumption experience: A more humble but complete view of the concept. *Marketing Theory, 3*(2), 267–286. doi:10.1177/14705931030032004

Cutler, S. Q., & Carmichael, B. A. (2010). The dimensions of the tourist experience. In M. Morgan, P. Lugosi, & J. R. B. Ritchie (Eds.), *The tourism and leisure experience: Consumer and managerial perspectives* (pp. 3–26). Channel View. doi:10.21832/9781845411503-004

Dessein, J., Soini, K., Fairclough, G., & Horlings, L. (Eds.). (2015). *Culture in, for and as Sustainable Development. Conclusions from the COST Action IS1007 Investigating Cultural Sustainability*. University of Jyväskylä.

Elkington, J. (1994). Towards the sustainable corporation: Win-win-win business strategies for sustainable development. *California Management Review, 36*(2), 90–100. doi:10.2307/41165746

Elkington, J. (1998). Cannibals with forks: The triple bottom line of 21st century business. Gabriola Island, BC; Stony Creek, CT: New Society Publishers.

Gustafsson, C. (2015). Sonic branding: A consumer-oriented literature review. *Journal of Brand Management, 22*(1), 20–37. doi:10.1057/bm.2015.5

Hosany, S., Sthapit, E., & Björk, P. (2022). Memorable tourism experience: A review and research agenda. *Psychology and Marketing, 39*(8), 1467–1486. doi:10.1002/mar.21665

Jain, R., & Bagdare, S. (2011). Music and consumption experience: A review. *International Journal of Retail & Distribution Management, 39*(4), 289–302. doi:10.1108/09590551111117554

Kauppinen-Räisänen, H., Gummerus, J., & Lehtola, K. (2013). Remembered eating experiences described by the self, place, food, context and time. *British Food Journal, 115*(5), 666–685. doi:10.1108/00070701311331571

Kõmmus, H. (2019). How the Concepts of Folk Music Emerge: The Terminology of Folk Music Festivals. *Etnomusikologian vuosikirja [Etnomusicology Yearbook], 31,* 112-142.

Landorf, C. (2009). Managing for sustainable tourism: A review of six cultural World Heritage Sites. *Journal of Sustainable Tourism*, *17*(1), 53–70. doi:10.1080/09669580802159719

Leipämaa-Leskinen, H., Syrjälä, H., Kauppinen-Räisänen, H., & Jaskari, M. M. (2020). Saunascape in the Nexus of Socio-Material Sauna Bathing Practices. In *Research in consumer culture theory*. Consumer culture theory consortium.

Lemon, K. N., & Verhoef, P. C. (2016). Understanding customer experience throughout the customer journey. *Journal of Marketing*, *80*(6), 69–96. doi:10.1509/jm.15.0420

Lenzerini, F. (2011). Intangible cultural heritage: The living culture of peoples. *European Journal of International Law*, *22*(1), 101–120. doi:10.1093/ejil/chr006

Minkiewicz, J., Evans, J., & Bridson, K. (2013). How do consumers co-create their experiences? An exploration in the heritage sector. *Journal of Marketing Management*, *30*, 1–2, 30–59. doi:10.1080/0267257X.2013.800899

Ooi, C. S. (2005). A Theory of Tourism Experiences: The Management of Attention. In T. O'Dell & P. Billing (Eds.), *Experiencescapes: Tourism, Culture, and Economy* (pp. 51–68). Copenhagen Business School Press.

Pine, B. J., & Gilmore, D. M. (1999). *The experience economy: Work is theater and every business a stage*. Harvard Business School Press.

Puro. (2021). *Yhteinen polku. Opas elävän perinnön vastuulliseen matkailulliseen tuotteistamiseen*. Visit Finland ja Museovirasto.

Rasoolimanesh, S. M., Ramakrishna, S., Hall, C. M., Esfandiar, K., & Seyfi, S. (2023). A systematic scoping review of sustainable tourism indicators in relation to the sustainable development goals. *Journal of Sustainable Tourism*, *31*(7), 1497–1517. doi:10.1080/09669582.2020.1775621

Schmitt, B. (1999). *Experiential marketing: How to get customers to SENSE, FEEL, THINK, ACT and RELATE to our company and brands*. The Free Press.

Seyfi, S., Hall, C. M., & Rasoolimanesh, S. M. (2020). Exploring memorable cultural tourism experiences. *Journal of Heritage Tourism*, *15*(3), 341–357. doi:10.1080/1743873X.2019.1639717

Stoddard, J. E., Pollard, C. E., & Evans, M. R. (2012). The triple bottom line: A framework for sustainable tourism development. *International Journal of Hospitality & Tourism Administration*, *13*(3), 233–258. doi:10.1080/15256480.2012.698173

Tung, V. W. S., & Ritchie, J. B. (2011). Exploring the essence of memorable tourism experiences. *Annals of Tourism Research*, *38*(4), 1367–1386. doi:10.1016/j.annals.2011.03.009

UN. (2023). *The 17 goals*. United Nations. https://sdgs.un.org/goals

UNESCO. (2023a). *What is Intangible Cultural Heritage?* UNESCO. https://ich.unesco.org/en/what-is-intangible-heritage-00003

UNESCO. (2023a). *Kaustinen fiddle playing and related practices and expressions*. UNESCO. https://ich.unesco.org/en/RL/kaustinen-fiddle-playing-and-related-practices-and-expressions-01683

UNESCO. (2023b). *Sustainable development and living heritage*. UNESCO. https://ich.unesco.org/en/sustainable-development-and-living-heritage

UNESCO. (2023c). *Browse the Lists of Intangible Cultural Heritage and the Register of good safeguarding practices, Finland.* UNESCO. https://ich.unesco.org/en/lists?text=&country[]=00076&multinational=3#tabs

UNESCO. (2023d). *Browse the Lists of Intangible Cultural Heritage and the Register of good safeguarding practices.* UNESCO. https://ich.unesco.org/en/lists

Chapter 15

Unlocking the Power of Stakeholder Perception:

Enhancing Competitive Heritage Planning and Place-Making

Aditi Nag
https://orcid.org/0000-0002-0604-6945
Birla Institute of Technology, Mesra, India

Smriti Mishra
https://orcid.org/0000-0001-7594-4960
Birla Institute of Technology, Mesra, India

ABSTRACT

This research aimed to examine the Travel and Tourism Competitiveness Index (TTCI) as an evaluation tool for heritage sites with limited tourism development. The authors aimed to assess how well the TTCI aligns with stakeholders' perceptions when applied to sites facing conservation challenges. The research revealed that stakeholder perceptions significantly impact heritage planning strategies. The study highlighted the potential of stakeholder engagement in enhancing a destination's competitiveness by addressing visitor satisfaction, overall tourism appeal, and the likelihood of repeat visits. The findings emphasized the importance of a collaborative governance model and suggested potential modifications to the TTCI to better suit underserved heritage sites. The research aimed to transform heritage planning into an engaging and profitable endeavor, emphasizing the need for conservation improvements at sites lacking proper attention and promoting successful heritage planning for other sites by making them more appealing to tourists and visitors.

INTRODUCTION

Sjöholm (2013) states, "Incredibly selective material artefacts, mythologies, memories, and customs become resources for the present." Contemporary needs make the selection. Cultural heritage and the

DOI: 10.4018/978-1-6684-9957-3.ch015

changing regional landscape are intricately linked (Flores de León, Babere & Swai, 2020). In emerging nations, cities and towns are compelled to balance urban development and conservation due to rising pressure from urbanization and expansion (Puren and Jordaan, 2014). A more recent advancement with the aim of inclusive and all-encompassing urban rehabilitation is the significance of intangible cultural heritage as a rudimentary element that influences the existence and character of communities (Flores de León, Babere & Swai, 2020). The apex of communities are historical sites because they are essential to creating an exciting and parsimoniously engaging environment for residents. The people who live there, the urban spaces, and the historic buildings all contribute to raising the standard of living there (Nag and Mishra, 2023a; El Menchawy, Aly & Hakim, 2011). As a result, the tourist sector has unique potential and formidable hurdles, as correctly stated by UNWTO and UNESCO, as visitors' interest in cultural activities grows and tourism and culture collide. Because of this, heritage tourism is not only an exclusive sector of the heritage business, but its economic and numerical importance has also drawn additional political, management, and scientific attention. As a result, heritage planning as a concept was created in response. Heritage planning is a comprehensive discipline that safeguards and promotes a community's cultural, historical, and natural heritage. It entails diverse strategies and initiatives designed to preserve and celebrate these invaluable aspects of a region or society. Cultural heritage encompasses tangible items such as historic buildings and artworks and intangible elements like languages and traditions. Natural heritage pertains to the conservation of natural landscapes and resources. Heritage planning ensures that these significant facets of a place's identity are protected, maintained, and made accessible for present and future generations to appreciate and learn from. It often involves a combination of historic preservation, urban development, tourism management, and community engagement to balance preservation and progress. Planning for heritage promotes cultural tourism, one of the proliferating global tourist industry segments (Yuen, 2005). In its broadest sense, the cultural heritage management sector manages the assets that make up cultural tourism (Živanović, 2016). Several European towns have turned to historic planning to increase their expansion chances. So, protecting cultural landscapes may help communities feel more empowered and proud of themselves. On the other hand, cultural heritage refers to a community's capacity to reflect on itself on a societal level. The cultural and heritage sector has helped the tourism industry expand into a sizeable economic sector for wealth expansion (Gannon, Lynch & Harrington, 2010). Heritage assets do not, however, ensure local or tourist growth (Ferri, 2017).

Urban planning studies and practices have long addressed human connections to foster and encourage dynamic public spaces and locations (Black and Street, 2014). However, research that aims to better articulate or understand stakeholders' perspectives at these places is seriously absent. This is true even if effective urban planning may contribute to creating more welcoming and inclusive historical tourist environments (Živanović, 2016; Gannon, Lynch & Harrington, 2010). Accordingly, the study suggests that more may and should be done to include "stakeholder perceptions" in urban design talks, i.e., a profound grasp of and sensitivity to the diverse demands of historic planning. In countries like India, Australia, or China, where history is deeply embedded or where stakeholders' perspectives are heavily weighted in the development of planning procedures, this reimagines "human connections" as something that is substantially more dynamic and inclusive.

Gibson (Gannon, Lynch & Harrington, 2010; Rapoport and Hawkes, 1970) was the first to propose a hierarchy of levels through which perception might occur. Rapoport and Hawkes contend that perception is fundamentally "concrete" and that objects are perceived as different from their surroundings and as being inside the perceptual world. In contrast, perception is most "symbolic" when it links a country's flag, a religion's cross, and freedom's landscape. Studying landscapes and preferences for certain landscapes has used the idea of an associational universe of meanings. As we progress to higher levels of meaning, the perception of an object becomes more culturally determined, with attitudes and expectations playing a more significant role (Rapoport and Hawkes, 1970). Between these two extremes, there is a continual spectrum of intermediate levels.

Protecting the public's interests is a crucial duty of government in skirmish resolution, even if unambiguous consideration of the public interest in a tourist growth problem is uncommon (Dredge, 2010). The perceptions of destination competitiveness, tourist trip return intentions, visitor satisfaction, and tourism impressions have not been adequately researched for forthcoming heritage site strategic development despite the tourism industry's rapid expansion (Nag and Mishra, 2023a; Zainuddin, Radzi & Zahari, 2016). It is widely known that excellent collaboration in managing cultural resources and inter- and intra-regional development supports efficient planning interventions (Flores de León, Babere & Swai, 2020). Integrating image and reputation with identity and history can increase a site's visitor appeal and help it remain competitive (Foroudi et al., 2020). The perspectives of tourists and other stakeholders are essential to this link. The problems found include a lack of integration in the perception studies sector, heritage management and conservation, and competition in the tourist, transportation, and economic generation industries (Nag and Mishra, 2023b).

The relevance of perception in heritage tourism is highlighted by visualizing heritage as a supply chain and the interfaces amid important components to form a composite tourist product (Nag and Mishra, 2023a; McCamley, 2016). Because of the sector's economic importance, international bodies have prioritized the competitiveness indicators (Sliem, Hegazi & El-Alfy, 2019). Numerous investigations on the variables influencing residents' attitudes towards tourist development in urban and rural W.H.S.s have shown promising heritage development challenges (Rasoolimanesh, 2017).

HISTORICAL CONTEXT OF HERITAGE SITES

Heritage sites, throughout history, have held profound cultural and historical significance for various civilizations. In the ancient world, iconic structures like the Pyramids of Egypt, built over 4,000 years ago, served as tombs for pharaohs and showcased the architectural prowess of the Egyptian civilization. Similarly, Stonehenge in England, erected around 2500 BCE, remains a testament to the technological achievements and astronomical knowledge of its time. These early heritage sites exemplify the enduring human fascination with monumental constructions and their ability to convey a society's cultural and historical legacy.

Religion has also played a pivotal role in the creation of heritage sites. Across the globe, sacred places such as the Acropolis in Athens, the Vatican City in Rome, and the Angkor Wat temple complex in Cambodia bear witness to their respective religious communities' spiritual and artistic achievements. These sites were conceived not only as places of worship but also as architectural marvels that embodied their societies' religious beliefs and values. In addition to religious significance, heritage sites often have deep cultural roots (Smith, 2006). The historic city of Petra in Jordan, known for its intricate rock-cut architecture, and the city of Venice in Italy, renowned for its intricate canal system and Gothic architecture, are prime examples of cultural heritage sites that have captivated the world with their architectural and artistic excellence. These sites reflect the unique cultural expressions and creativity of their societies.

Furthermore, heritage sites can be associated with pivotal historical events or figures (Timothy and Boyd, 2003). The Palace of Versailles in France, with its opulent architecture and significance in the history of the French monarchy, stands as a symbol of political power and cultural grandeur. The Hiroshima Peace Memorial in Japan, a somber reminder of the atomic bombing during World War II, is a testament to the enduring pursuit of peace and reconciliation. Independence Hall in the United States, where the Declaration of Independence and the Constitution were debated and adopted, holds a special place in American history. As societies evolved, so did the recognition of the need to preserve and protect heritage sites. In the 20th century, organizations like UNESCO were established to designate and support the preservation of World Heritage Sites around the globe. These efforts underscored the universal importance of heritage sites as repositories of human history, culture, and identity.

In contemporary times, heritage sites have become historical and cultural treasures and significant tourist attractions that contribute significantly to local economies and global cultural exchange (Harrison, 2013; Land, 2002). However, they also face environmental degradation, urbanization, and tourism-related pressures, highlighting the need for sustainable heritage preservation and management practices to ensure these sites continue inspiring and educating future generations (Hall, 2016; King, 2014).

The impact of stakeholder engagement, or the lack thereof, on the competitiveness of heritage sites is a critical aspect of their sustainable management and preservation. Real-world examples from iconic locations around the globe vividly illustrate how the interaction between heritage site management and various stakeholders can shape the site's attractiveness, authenticity, and long-term viability (Ashworth and Larkham, 2013; Meskell, 2002; Hall and McArthur, 1998). By examining these case studies, we can gain valuable insights into the intricate dynamics of stakeholder engagement within heritage tourism, shedding light on both successful strategies and challenges heritage sites face in maintaining their competitiveness on the global stage.

1. Machu Picchu, Peru (Stakeholder Engagement):
   Engagement: The management of Machu Picchu, a UNESCO World Heritage Site, has actively involved local communities in its preservation and tourism management. They receive economic benefits and employment and have a say in visitor numbers and infrastructure development decisions.
   Impact: This approach has led to sustained stakeholder support, minimizing social and environmental impacts and preserving the site's authenticity. Machu Picchu remains a top tourist attraction, maintaining its competitiveness and cultural significance.
2. Venice, Italy (Lack of Stakeholder Engagement):
   Lack of Engagement: Venice has faced over-tourism and insufficient stakeholder engagement challenges. Residents have protested against the adverse effects of mass tourism on their quality of life.
   Impact: The lack of adequate engagement and planning has resulted in overcrowding, environmental stress, and a decline in the city's competitiveness as a desirable destination. In recent years, Venice has focused on improving stakeholder engagement to address these issues.
3. Angkor Archaeological Park, Cambodia (Mixed Engagement):
   Mixed Engagement: Angkor Wat, a UNESCO World Heritage Site, has seen both positive and negative impacts of stakeholder engagement. While local communities benefit from tourism, issues like unregulated construction have raised concerns.
   Impact: Effective stakeholder engagement has sustained Angkor Wat's popularity and economic benefits. However, challenges stemming from inadequate planning and regulation threaten the site's long-term competitiveness.

4. Giza Plateau, Egypt (Historical Perspective):
   Historical Lack of Engagement: In the past, the management of the Giza Pyramids lacked adequate stakeholder engagement, leading to unregulated development and neglect of the site's preservation.
   Impact: This historical lack of engagement damaged the site's integrity and exclusion from the UNESCO World Heritage List. However, recent efforts to engage local communities and improve conservation practices aim to enhance its competitiveness and global recognition.
5. Borobudur, Indonesia (Community Engagement):
   Community Engagement: Borobudur, a UNESCO World Heritage Site, actively engages local communities in its management. They participate in cultural events, benefit from tourism, and play a role in site preservation.
   Impact: Strong stakeholder engagement has contributed to Borobudur's competitiveness as a cultural and spiritual destination. It remains an example of how involving local communities can enhance the visitor experience and preserve cultural heritage.

These case studies demonstrate that effective stakeholder engagement can lead to the sustainability and competitiveness of heritage sites by minimizing negative impacts, enhancing visitor experiences, and fostering support from local communities and other stakeholders. Conversely, a lack of engagement can result in challenges that undermine a site's attractiveness and long-term viability as a tourism destination.

RELATIONSHIP BETWEEN STAKEHOLDERS AND HERITAGE PLANNING

The term "stakeholder" refers to people or groups with an active role in the project or whose interests might be affected if it is successful or executed. They could also affect the goals and results. The management team must identify the stakeholders, ascertain their expectations and requirements, and minimize their effect to achieve a successful venture. Investigations should be conducted into any organizations or individuals vying for few resources or who have aims that conflict with those of the program since they might endanger it (Hajialikhani, 2008). In heritage planning, cooperation and collaboration are crucial considerations. They are associated with ideas like integration of community-based tourism, participation, and sustainable tourist development. Alnafeesi contends that planning efforts must be coordinated and collaborative because the tourist industry is fragmented and involves many other players. The effectiveness of the collaborative planning strategy depends on identifying and legitimizing all prospective stakeholders, including those participating in the planning process (Alnafeesi, 2013). In places where tourism is just starting to flourish, it might be challenging to identify stakeholders since their interests aren't well-organized (Mariotti, 2012). Another argument favoring collaboration is allowing all interested parties to participate in decision-making by letting them take on responsibility, grow self-reliance, and become more aware of the issues at hand. This results in greater consensus and shared ownership, whereas the absence of collaboration can result in a state of alarm (Abeysinghe and Ariyawansa, 2017).

The complexity is further increased by the relevance of the stakeholders' complicated depictions of the neighborhood. Part of the issue is the ambiguous and evasive meaning of the term "community" (Cibinskienea and Snieskiene, 2015; Assari & Assari, 2012). A geographical area or a group of individuals who come together because of their shared heritage or interests can be called a community. If a geographic definition is utilized, the community might be characterized as the inhabitants of a particular region.

When using the common interest method, the business sector is typically employed to represent the local community, emphasizing financial difficulties. In contrast to the benefits, expanding collaboration has several drawbacks and barriers. Among these factors are the rising cost of planning and development, the recognition of real stakeholders, and the capacity of the stakeholders to participate. The authority is usually concentrated in the hands of a well-established resident elite and/or the most "vocal," overshadowing the quiet majority and any resident minority. Expectations may be raised beyond what is feasible.

An essential problem emphasized in prior studies is the requirement for heritage professionals to interact with various stakeholders and other specialists who value heritage structures (Kumar and Dhir, 2020). Encouraging common knowledge of a place's communal values helps improve conservation results (Nag, 2020). Heritage experts must seek to mediate a resolution that is in the best interests of conservation when stakeholders' objectives and perspectives for cultural assets disagree. Therefore, legacy professionals need to be able to negotiate and be familiar with tried-and-true tactics for coming to agreements, achieving consensus, settling disputes, and ultimately creating long-term contracts. For more than 20 years, the field of public policy consensus building and conflict resolution has developed and implemented solutions to a range of transnational issues. Numerous heritage-related concerns, such as conservation and land-use problems, urban planning, international development, community associations, resource organization, and policy formulation, have been addressed using this approach. Surprisingly, however, it has only been applied sparingly in historic conservation techniques (Nag and Sarkar, 2022).

Planners, managers, and other decision-makers are frequently required to interact with various stakeholders and their often divergent beliefs, interests, and identities. They also report conflicts resulting from cultural differences in their efforts to maintain and safeguard heritage locations. As a result, legacy experts frequently work with various governmental organizations, some of whose mandates overlap (Nag, 2022).

Stakeholders in heritage sites encompass diverse groups, each with distinct interests, perspectives, and roles in preserving and managing these cultural treasures. Understanding the dynamics of these stakeholder groups is crucial for effectively managing heritage sites. Here's a breakdown of some key stakeholder groups and how their interests may differ and overlap:

1. Local Communities:
   Interests: Residents living near or within heritage sites are often concerned about their quality of life, economic opportunities, and access to infrastructure and services. They may value the preservation of cultural traditions and heritage.
   Overlap: Local communities' interests may align with tourists' interests regarding promoting and benefiting from tourism. However, they may also have concerns about overcrowding, environmental impacts, and changes to their way of life.
2. Tourists:
   Interests: Tourists seek authentic cultural experiences, historical knowledge, and recreational activities when visiting heritage sites. They want convenient and safe access, interpretive materials, and quality services.
   Overlap: Tourists' interests often align with heritage sites' goals to provide educational and enjoyable experiences. However, they may sometimes pressure infrastructure and the environment, potentially conflicting with conservation efforts.
3. Government Authorities:

Interests: Government agencies at various levels (local, regional, and national) have interests in heritage sites related to tourism development, cultural preservation, regulation, and revenue generation.

Overlap: Government interests may overlap with those of heritage site managers and local communities when it comes to ensuring the sustainable use and protection of the site. However, they must balance these interests with economic development and regulatory considerations.

4. Heritage Site Managers:

   Interests: Site managers, often employed by public or private entities, are primarily concerned with the conservation, interpretation, and presentation of the site's cultural and historical significance. They aim to provide meaningful visitor experiences while safeguarding the site's integrity.

   Overlap: Their interests align closely with tourists seeking educational and enriching experiences. They also collaborate with local communities to engage in sustainable practices.

5. Businesses and Entrepreneurs:

   Interests: Local businesses, such as hotels, restaurants, and souvenir shops, aim to benefit economically from tourism. Their interests revolve around attracting visitors, increasing revenue, and providing services.

   Overlap: Their interests can align with those of tourists, as both seek economic benefits from tourism. However, conflicts may arise if businesses prioritize short-term profits over long-term sustainability.

6. Environmental and Cultural Organizations:

   Interests: NGOs and advocacy groups often focus on environmental conservation, cultural preservation, and sustainable tourism practices. They advocate for responsible tourism and heritage protection.

   Overlap: Their interests align with heritage site managers and local communities regarding sustainability and preservation. They may also collaborate with government agencies on policy advocacy.

7. Academics and Researchers:

   Interests: Researchers aim to advance knowledge about heritage sites, contributing to their preservation and interpretation. They conduct studies on history, archaeology, ecology, and visitor behavior.

   Overlap: Their interests may overlap with site managers, government authorities, and environmental organizations in promoting research-based conservation and management strategies.

Understanding the diverse interests of these stakeholder groups and finding common ground where their goals align is crucial for achieving a balanced and sustainable approach to heritage site management. Effective communication, collaboration, and stakeholder engagement are vital in addressing potential conflicts and preserving the authenticity and significance of heritage sites.

PERCEPTION POWER

Space may be seen in many different ways, including independently and selectively. To examine the value attached to particular settings, qualitative methodologies for analyzing socio-spatial contexts, perceptual processes, and attribution are necessary (Töppel and Reichel, 2021). Stakeholders' perceptions of the significance of collaborative governance and their separate roles are quite varied. For instance, if stakeholders don't understand the roles and importance of other contributing groups, they may encounter many problems during the urban regeneration process. However, there is a dearth of information on the diverse stakeholder perspectives on collaborative governance (Zhang, Kang & Koo, 2021). Complexity is communicated through urban vision (Rapoport and Hawkes, 1970; Nasar, 1989). Hidden patterns emerge when people's mental representations of a city mix with its spatial characteristics (Yoshimura et al., 2020). Richness and diversity are well-known visual characteristics of favored situations, but apparent associational meanings and detectable underlying structures constrain them. Many people believe that modern transformational processes are mostly to blame for the substantial failure of contemporary cities to satisfy these requirements (Gjerde, 2010).

The management of tourism destinations has benefited from long-term community interactions and observation, as well as from tourism stakeholders' in-depth knowledge and understanding of the tourism management and industry, as well as their skilled connections and active involvement in tourism planning and advancement (Yoon, 2002). According to a Vengesayi (2003) model, promotion (including reputation), branding, and price all impact how competitive a location is. A location's competitiveness is also increased by a positive reputation, which increases its attraction and, as a result, tourist demand (Perles-Ribes et al., 2019). "How can tourism be linked to a sustainable development pattern by valorizing the destination's legacy?" is a crucial query in heritage tourism development (Ursache, 2015). To aid in developing a representation of value in the built environment, individual views, such as a person's relationship to their community and how it influences their activities, are documented (Black and Street, 2014). Zainuddin et al. (2020) and Amaya-Molinar et al. (2017) found a substantial correlation between perceived destination competition and tourist reaction behavior in tourism sites, including heritage and cultural sites.

While some have questioned the necessity of studying authenticity, particularly objective authenticity, academics have long been fascinated by visitors' genuineness (Wu et al., 2019). There have always been tensions between cultural preservation, authenticity, and commercialization during the growth and practice of tourism, despite numerous scholars studying tourist authenticity. The bulk of individual studies that have been discussed in the literature also involve individuals who are outsiders. Participants include those who may be visiting or have already visited the particular place under investigation and local and foreign visitors. Tourists like assessing qualities and perceptions of the destination's attractiveness to rate the relevance of the indicators and/or features considered for the study (Johnson, 2017). These characteristics encourage tourists and locals to build a sense of place connection. The term "place attachment" is used in environmental psychology to describe people's emotional connections to their environments over time. This expression is used to investigate how locals view their neighborhoods and how tourists view a particular tourist destination. Academic research has long focused on the emotional ties between individuals and places. This bond has been described by the phrases "community attachment," "destination attachment," and "sense of place" (Wang, 2020; Wu et al., 2019).

According to a prior study, a destination's image, brand, personality, and identification as a hyperlink can be seen in various ways (Della Corte, Piras & Zamparelli, 2010). While tourism development appears to be a practical strategy for historic city expansion, experts have long cautioned that striking a balance between preservation and growth is difficult without stakeholder participation and a constructive attitude (Li et al., 2016). By bridging the gap between morphological and perceptual strategies, urban planning can assist in achieving the desired goals (Gray, 1989; Goodey and Gold, 1987). Locals may be able to resist change in the face of economic and political pressures (Burgess, 1979), the effects of losing structures with special local significance (Sairinen and Kumpulainen, 2006), and discrepancies in residents' and outsiders' assessments of the region's character and quality (Samadi and Yunus, 2012) through positive perception inclusion. From a supply-side standpoint, tourism destinations may increase their level of competition by promoting good destination branding and sustainable trade ideas that take advantage of developed cultural assets (Gupta and Singh, 2019). As a result, rural communities and the local economy will prosper (Nastase, Lucaci & Chașovschi, 2020). Despite extensive research on visitor motivation and destination competitiveness, gap analysis-based empirical research that concurrently examines both topics is still lacking in the literature (Pansiri, 2014).

BUILT ENVIRONMENT AND STAKEHOLDER PERCEPTION

Place-making combines people, structures, streets, and activities (Friedmann, 2010). Buildings are essential in place-making because they serve various purposes, from providing basic shelter to hosting several symbolic and ceremonial events. According to Glass (1993), "Policy-makers and administrators can quite happily continue to consider their own subjective choices as objective, universal ones while the plurality of needs, wants, and ambitions in society is masked." This section contends that planners and designers impose their ideas on what makes a pleasant environment to varied degrees. However, it's equally crucial to remember that political and financial constraints limit them.

The degree of complexity also changes whenever a stakeholder enters or exits a construction or planning project. Clients, sponsors, suppliers, purchasers, users, contractors, developers, investors, government agencies, small and medium-sized companies (SMEs), third parties, urban or rural planners, and regulators are typical stakeholders in construction and planning projects. Primary and secondary stakeholders are generally identified in academic research. Primary stakeholders, such as housing developers, have direct economic and operational involvement in the housing development project, unlike secondary stakeholders, who play an important role but are only occasionally involved throughout the development process. Collaboration allows stakeholders to "handle a set of difficulties that neither can face alone" (Hamdan, Andersen & De Boer, 2021). According to Grey (1989), the theoretical underpinning of collaboration is Ashby's 1960 law of necessary variety. Teamwork, in other words, allows the project to gain enough internal variation (complexity) to balance the complexity or turbulence caused by disruptions and challenges in the project's perimeter (Hamdan, Andersen & De Boer, 2021; Zhang, Kang & Koo, 2021; Hajialikhani, 2008; Yoon, 2002).

Planners face difficulties in determining how to consider the opinions of several stakeholders before coming to a consensus when making decisions about policies that need socially significant decisions or agreements (Hamdan, Andersen & De Boer, 2021). Cooperative governance could make it possible to provide the planners with the answers they need. "Collaborative governance" refers to the government's governance structure to implement programs, settle disputes, and increase productivity. Locals must be included in and assisted with significant decision-making (Zhang, Kang & Koo, 2021; Wang, 2020; Hajialikhani, 2008). Even though the bulk of research focuses on stakeholders' engagement techniques, merits, and relevance (Hamdan, Andersen & De Boer, 2021; Zhang, Kang & Koo, 2021; Wang, 2020), very few studies have explored the perspectives of various participants as a foundation for collaborative governance. The efficacy of historic planning endeavors is significantly impacted by the sensitive interaction between stakeholders, which emerges from their conflicting opinions and goals (Wang, 2020; Zehrer and Hallmann, 2015; Yoon, 2002). A few management programs have proposed a power/interest grid for stakeholder analysis that deals with the differences in impact, power, attitude, and support for construction and planning projects to establish management and engagement techniques for beneficial outcomes (Hamdan, Andersen & De Boer, 2021). Grid power and interest, support, and significance A few stakeholder matrixes created due to the grid include the Stakeholder Matrix, Business Process Management Stakeholder Matrix, and Attitude and Knowledge Stakeholder Map. Even if various evaluation matrices have been developed, it has not understood how perception and planning methods interact to affect stakeholders' influence on heritage planning (Friedmann, 2010). Few matrices compare stakeholders' interests to some extent by relating their power to their support, their impact on the project to their opinion, or their significance to its success to their commitment. Since few evaluation matrices cover stakeholder influence, problem position, knowledge, or attitude, an integrated comparison assessment is challenging (Hamdan, Andersen & De Boer, 2021; Friedmann, 2010; Grey, 1989). The effectiveness of any evaluation matrix depends on how interdependent politicians, administrators, and stakeholders are. Frooman (1999) developed a "Typology of Influence Strategies" (See Figure 1) to illustrate the connection between a stakeholder's reliance on management and policy-makers and the sort of influence approach they would employ.

When a stakeholder relies on the government and policymakers rather than vice versa, they have enormous influence. Under certain situations, stakeholders may refuse to use their products, services, labor, money, or other resources. Even though use strategies are equally dependent on administration and policymakers, they might directly impact the project (Nag and Mishra, 2023b). A stakeholder can only effectively influence using an indirect strategy if the administration and policymakers do not require it.

A good framework for determining the strategy a stakeholder will most likely employ to influence politicians and the administration is provided by Frooman's study. Frooman's Typology of Effect Strategies may be used with other stakeholder analysis techniques to offer a more complete review. However, a perceptual approach is still required to adequately account for stakeholders' effect on construction and planning initiatives.

A STAKEHOLDERS' PERSPECTIVE ON HERITAGE PLANNING'S COMPETITIVENESS

The travel and tourism (T&T) industry has grown to be one of the largest in the world's most industrialized nations. Every two years, the World Economic Forum (WEF), dedicated to improving global

Figure 1. Typology of influence strategies adapted from Frooman
(Frooman, 1999)

		Is the stakeholder dependent on the policy-makers and administration?	
		No	Yes
Are the policy-makers and administration dependent on the stakeholder?	No	Indirect/Withholding (Low dependence)	Indirect/Usage (Policy-makers and administration power)
	Yes	Direct/Withholding (Stakeholder power)	Direct/Usage (High Interdependence)

standards, produces research called the T&T Competitiveness Index that assesses the elements and legal frameworks that make a nation attractive as a destination for the T&T industry. The T&T Competitiveness Report (TTCR) is recognized as a notable innovation by the Platform for Shaping the Future of Mobility, which brings together world experts to guarantee that travel and transport systems are current for the 21st century. By employing new policies and technical innovations that solve economic, sociological, and environmental problems, corporations, policymakers, and allied industries have the chance to reinvent mobility systems. These organizations now have a powerful weapon to understand and combat developing trends and problems in global T&T, change their practices and regulations, and speed up the development of new technologies thanks to the fast proliferation of innovative transportation methods and disruptive business models. This mixture draws leaders to the platform and makes it possible for stakeholders to share information and take action to hasten the transformation.

Due to their increased competitiveness and contribution to this global trend, emerging economies are becoming increasingly well-liked as T&T travel destinations. The T&T Competitiveness Report contains updates to the T&T Competitiveness Index (TTCI), which rates and compares the competitiveness of 140 economies and indicators in the T&T region. The report is a practical benchmarking tool for businesses and governments looking to grow their T&T sectors. It informs policy and investment choices relating to T&T business and industry growth by providing cross-country assessment and benchmarking of nations' progress on the competitiveness criteria. The research offers a distinctive perspective on each country's strengths and chances for growth to increase industrial competitiveness, in addition to providing a platform for multi-stakeholder interaction at the national level to formulate applicable policies and initiatives. It comprises 90 separate indicators dispersed across 14 pillars and four sub-indices (see Figure 2). These sub-indices help break down the complex concept of tourism competitiveness into specific dimensions, allowing for a deeper understanding of a country's strengths and weaknesses in the tourism sector.

Figure 2. TTCI 2019
(WEF TTCR, 2019)

The division of the TTCI into sub-indices is significant because it offers a structured and comprehensive framework for assessing and improving a country's tourism competitiveness. It allows for a detailed analysis of different dimensions, identifies specific areas for development, promotes sustainable practices, and facilitates collaboration among stakeholders. This multi-dimensional approach contributes to the overall growth and resilience of a country's tourism industry. Despite the WEF's exhaustive and inclusive index, stakeholders' perspectives, historical preservation, and management tactics have received little attention.

The competitiveness of a site as a tourist destination is influenced by a wide range of elements, including destination management, resources, demand, and situational considerations (Zehrer and Hallmann, 2015). Several authors who have written about entrepreneurship have previously emphasized the need for the combined capabilities of numerous stakeholders in a tourism destination to promote competitiveness, even though models of destination competitiveness have rarely ever focused on the role of firms (Hamdan, Andersen & De Boer, 2021; Zhang, Kang & Koo, 2021; Wang, 2020; Zehrer and Hallmann, 2015; Hajialikhani, 2008; Yoon, 2002; Frooman, 1999). The creation and management of a destination are all that these authors have concentrated on (Zehrer and Hallmann, 2015). According to the stakeholder theory, companies that act in their stakeholders' best interests outperform those that do not. Stakeholder theory is a management and business philosophy that suggests organizations should consider the interests and concerns of all stakeholders, not just shareholders or owners when making decisions and formulating strategies. Balancing their diverse needs and concerns is crucial for the destination's sustainability. This theory emphasizes long-term thinking, community engagement, and responsible practices. When residents benefit from tourism and have a say in its development, they become advocates for the destination's success. Additionally, prioritizing sustainability and corporate social responsibility aligns with stakeholder theory, enhancing a destination's competitive advantage and reputation. Applying stakeholder theory can lead to more holistic, responsible, and competitive tourist destinations. A tourist destination could benefit from this mentality. The significance of private tourism companies at heritage-type sites has been overlooked in research on destination competitiveness. Several authors have emphasized the existence of various stakeholders and hypothesized that proactive consideration of each party's interests results in significant long-term benefits for the industry as a whole (Hamdan, Andersen & De Boer, 2021; Zhang, Kang & Koo, 2021; Wang, 2020; Zehrer and Hallmann, 2015; Hajialikhani, 2008; Yoon, 2002). On the other, developing a suitable framework for tourism's sustainable growth, particularly for its sustainable competitiveness, requires considering economic, social, and ecological components and all stakeholders active in the tourism sector (Nadalipour, Imani Khoshkhoo & Eftekhari, 2019).

The study's fundamental thesis is that stakeholder support is crucial for tourism to develop, run well, and endure over time. If tourist stakeholders gain from the effects of tourism, value the environment, and care deeply about their communities, they will support tourism growth and defend destination competitive strategies. Tourism stakeholders' support for destination competitive strategies may increase destination competitiveness and increase the likelihood of booming tourism in a region. Increased competition will benefit communities in tourist areas in terms of social and economic benefits. If the tourist location and its attractions are sufficiently developed and promoted, tourists and visitors will profit more from their travel experiences (Yoon, 2002).

As we delve into the nuances of the TTCI tool and its implications for tourism competitiveness, it's crucial to recognize that stakeholders' perceptions play a pivotal role in shaping the industry's landscape. While the TTCI provides valuable quantitative data and metrics to evaluate competitiveness, it is equally essential to consider how these metrics align with and influence the subjective perspectives of stakeholders involved in the tourism ecosystem. Understanding how stakeholders perceive the competitive factors assessed by the TTCI can shed light on the practical implications of these metrics and help bridge the gap between quantitative analysis and real-world decision-making within the tourism sector.

TTCI EVALUATION

The T&T company has a significant impact on both local economies and the global economy. The sector contributed 10.4% to global GDP in 2018 and employed a comparable number of people, exhibiting remarkable steadiness over the previous ten years. This expansion and relative longevity are caused by the middle class, which continues to expand in Asia and other regions. Over the following ten years, the industry's share of the GDP is anticipated to rise by over 50%. If T&T is to benefit from this increasing market, policymakers, business leaders, and other stakeholders must focus more on the country's competitiveness. While the government and industry players must consider what competitiveness levers they may utilize to keep or increase market share, preserving tourist infrastructure, services, and assets will demand special attention. Improvements in competitiveness, as well as careful planning for tourism management and destination carrying capacity, will be required as the number of business and leisure tourists increases internationally. Policymakers, T&T businesses, and other stakeholders utilize the TTCI's results to understand better and encourage the conversations and efforts that will ensure the sector's long-term health (WEF TTCR, 2019).

When nations perform well across various pillars and begin to outperform the global average in T&T competitiveness, tourist numbers frequently grow dramatically (see Figure 3). Because T&T's competitiveness is impacted by factors including the business climate, robust labor markets, technology, and infrastructure, less developed countries behind more developed economies in overall ranking (Kovačević, 2018) since these sorts of enabling conditions have favorable consequences on the economy as a whole, T&T stakeholders, both public and private, may use this to justify investment in the sector.

The average TTCI score varies by around 38% between high-income economies and low- to lower-middle-income nations. However, the average score for these economic clusters is too much below the pillar's 11 percent cutoff. The allocation of natural resources across countries with varied economic origins is the cause of this. Many low-income economies with abundant natural resources should consider using labor to drive economic progress through a thriving travel and tourism sector since they lack the fundamental elements needed to pursue other emerging economic pathways like unconventional manufacturing. The power of natural and cultural resources, particularly to draw capital investment, is demonstrated by the close to 76% connection between T&T capital investment and Natural and Cultural Resources sub-index scores (WEF TTCR, 2019).

Figure 4 depicts the connection between the TTCI pillars and the number of foreign travelers in high- and low-income countries. According to this data, low- to lower-middle-income countries have a more significant association between arrivals and pillars, including the business climate, ICT readiness, international openness, human resources, and inclusive infrastructure, than their high-income counterparts. Therefore, developments in these fields may significantly impact the demand for T&T in low-income countries than in high-income countries. On the other hand, Figure 4 shows that advanced countries are more likely to engage in resource competition as they have less in common regarding ICT readiness, business climate, and other characteristics linked to greater levels of economic development (WEF TTCR, 2019).

Figure 3. TTCI 2019 and international tourist arrivals
(WEF TTCR, 2019; UNWTO, 2020)
Note: *Liberia is not included in international visitor arrivals (as of 2017 or the most recent data available). Each region's top achievers are recognized.*

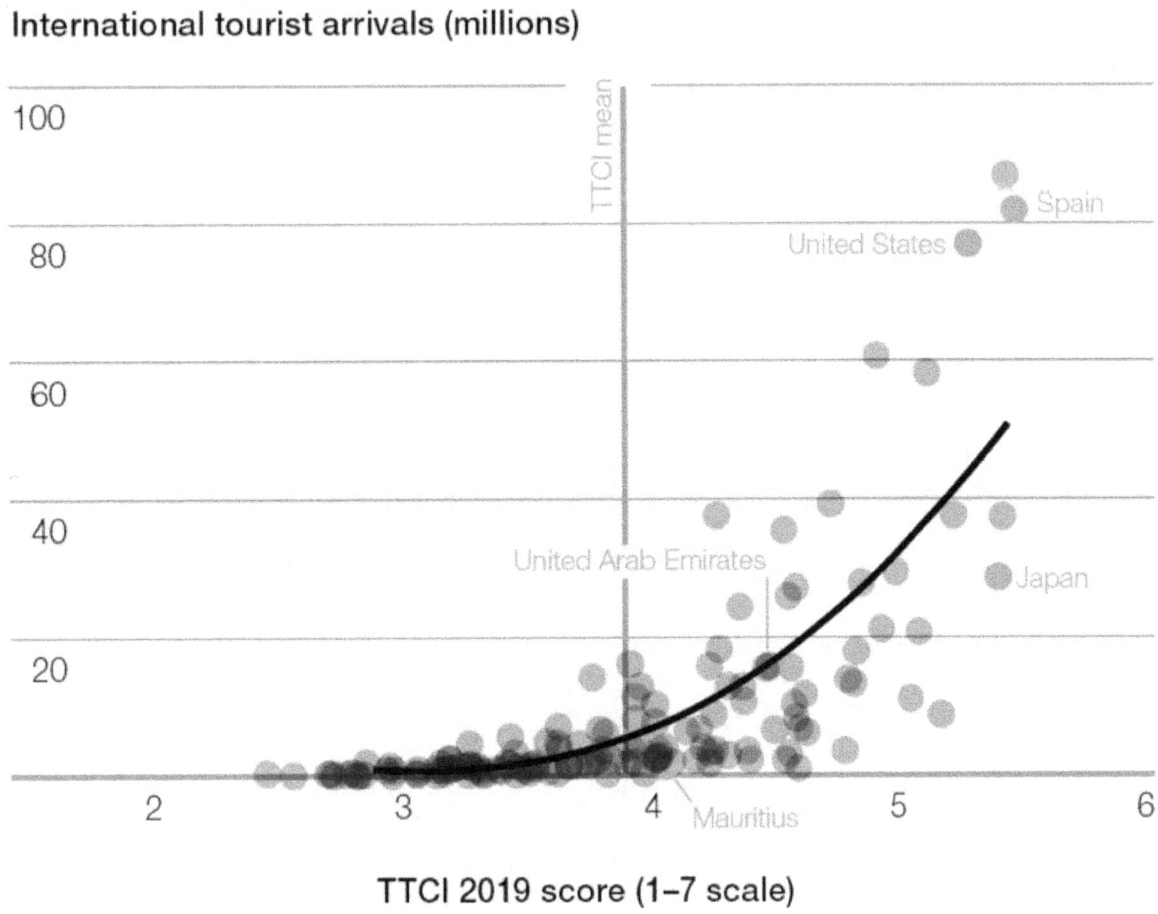

Better infrastructure and advantageous enabling conditions, which may be attained through well-managed T&T-led growth, may boost fiscal efficiency and national competitiveness. Figure 5 depicts the relationship between T&T competitiveness and the Global Competitiveness Index 4.0 (GCI 4.0), which evaluates the assortment of institutions, policies, and elements that impact an economy's production level. Since T&T and general competitiveness adhere to the same competitiveness principles, they are tightly connected. As a result, increasing T&T's competitiveness through such foundations may produce a second productive sector that helps reduce poverty, income disparity, and opportunity inequality.

A nation's capacity to compete in the global T&T market is influenced by various factors, many of which were intended to be reflected in and quantified by the TTCI. Each of the 14 pillars that make up the index has the potential to boost competitiveness and returns in visitor influxes and revenues if managed with suitable investments and policies. The deep connections between the pillars and their overall influence on the industry's long-term sustainability are less extensively studied and thoroughly researched (WEF TTCR, 2019). The creation of a new competitiveness index and associated indicators is necessary.

Figure 4. Correlation between TTCI pillars and foreign tourist arrivals by income level (WEF TTCR, 2019; UNWTO, 2020; 57. UNESCO, 2020)

Notes: *High-income economies are represented by light-colored bars, while low- and lower-middle-income countries are represented by dark-colored bars—foreign visitor arrivals (as of 2017, the most recent year for which data is available). Arrivals from Liberia are not listed.*

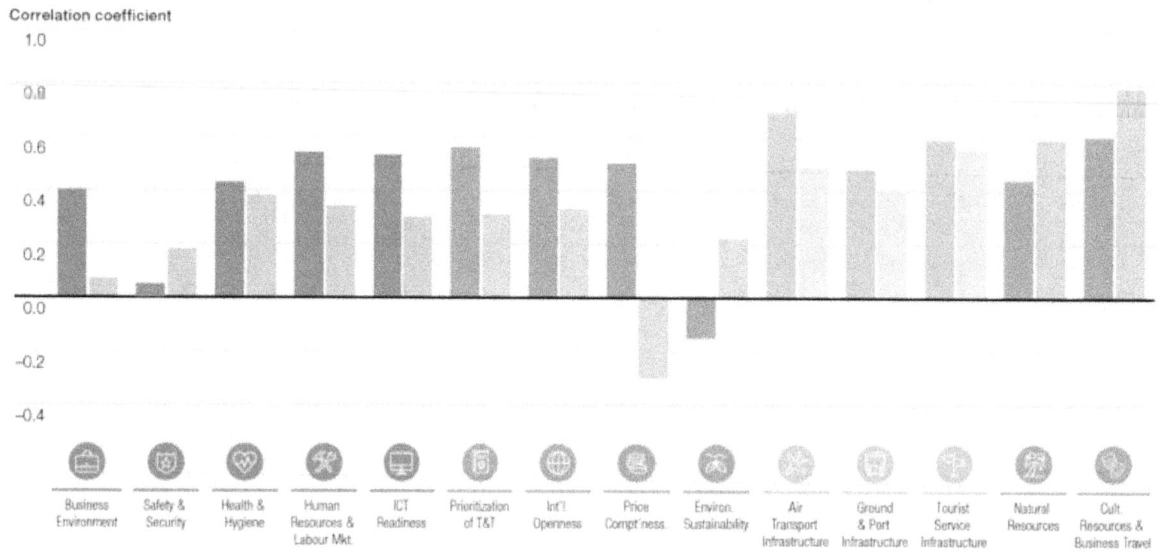

Figure 5. Correlation between T&T and overall competitiveness (WEF TTCR, 2018 and 2019)

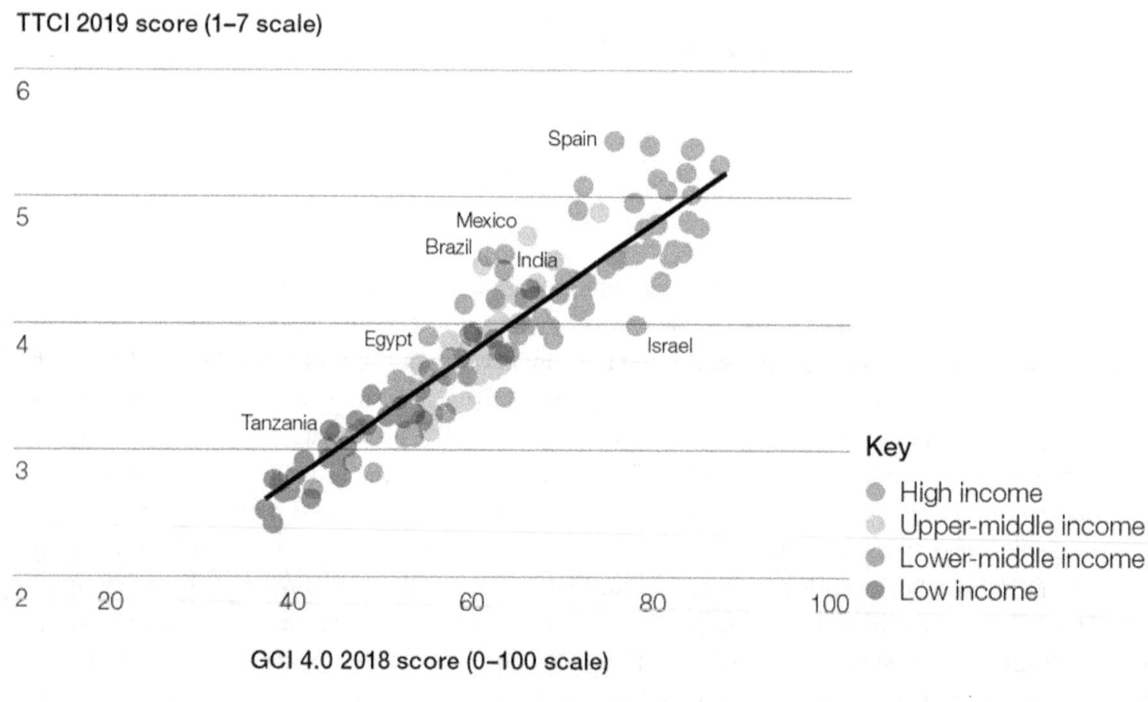

Public-private partnerships are essential, as well as cross-industry collaboration. A well-balanced mix of these ties may enable the healthy expansion of T&T's competitiveness with an emphasis on community and environmental protection in and around such regions.

GLOBAL CONTEXT REGARDING HERITAGE SITE COMPETITIVENESS AND STAKEHOLDER ENGAGEMENT

In the realm of heritage site management and tourism, the significance of global context cannot be overstated. Heritage sites span the globe, encompassing diverse cultural, historical, and environmental attributes. Understanding how competitiveness and stakeholder engagement issues play out in various regions and countries offers invaluable insights into the intricate tapestry of heritage tourism. Regional uniqueness, governance structures, and socio-cultural factors shape these multifaceted dynamics. In this exploration, we delve into the global context, shedding light on regional variations and highlighting the importance of localized approaches in preserving and promoting our world's rich cultural heritage.

Heritage Site Competitiveness

1. Europe: Europe has many heritage sites, from ancient ruins to medieval cities and cultural landmarks. The competitiveness of European heritage sites often lies in their rich history and architectural wonders. Sites like the Colosseum in Rome and the Louvre in Paris attract millions of tourists annually. Competition among European heritage sites is intense, leading to a focus on high-quality visitor experiences, advanced infrastructure, and preservation efforts. European nations often collaborate to promote heritage routes, such as the European Route of Historic Gardens, to enhance their competitiveness collectively.
2. Asia: Asia boasts a diverse range of heritage sites, including historical monuments, religious temples, and natural wonders like the Great Wall of China. The competitiveness of Asian heritage sites often stems from their unique cultural and spiritual significance. Asian countries like Japan and Thailand actively promote heritage tourism, leveraging their rich traditions and landscapes. However, challenges related to overcrowding and environmental conservation are pressing concerns, leading to increased efforts to manage visitor numbers and implement sustainable practices.
3. Latin America: Latin America is celebrated for its archaeological sites, colonial cities, and natural beauty. Machu Picchu in Peru and the historic center of Mexico City exemplify the region's competitive heritage offerings. Latin American countries often emphasize the preservation of their indigenous cultures and traditions, attracting tourists seeking authentic experiences. Collaborative efforts to promote heritage tourism within the region, such as the Mayan Route in Central America, contribute to competitiveness.
4. Middle East: The Middle East combines ancient history and modernity. Heritage sites like Petra in Jordan and the Old City of Jerusalem are prime examples of the region's unique offerings. In the Middle East, heritage tourism often intersects with religious tourism, attracting pilgrims and history enthusiasts. Stakeholder engagement here may involve collaboration between religious authorities and tourism agencies to manage the flow of visitors and preserve sacred sites.

Stakeholder Engagement

1. Europe: European heritage sites often involve local communities in preservation efforts and tourism management. Community engagement is crucial for maintaining authenticity and ensuring equitable economic benefits. In regions like Scandinavia, indigenous Sami communities actively participate in cultural heritage tourism, sharing their traditions and way of life with visitors.
2. Asia: In Asia, stakeholder engagement often centers on balancing the interests of local communities, government authorities, and religious institutions. Community-based tourism initiatives in countries like Bhutan aim to empower residents and preserve cultural heritage. In contrast, in densely populated areas like Kyoto, Japan, where overtourism has been a concern, authorities work closely with communities to manage visitor numbers and protect heritage sites.
3. Latin America: Many Latin American heritage sites are closely tied to the daily lives of local communities. Engagement here frequently focuses on economic development and cultural preservation. Indigenous communities in countries like Ecuador and Guatemala offer authentic cultural experiences to tourists, providing a source of income while preserving their heritage.
4. Middle East: Stakeholder engagement in the Middle East often involves religious authorities and government bodies working to ensure respectful and sustainable management of sacred heritage sites. Religious groups and government agencies collaborate in Israel to preserve and promote heritage sites in the Old City of Jerusalem.

In this global context, heritage site competitiveness and stakeholder engagement take on diverse forms, reflecting each region's unique attributes and challenges. While competitiveness is often driven by historical and cultural significance, stakeholder engagement strategies vary based on the involvement of local communities, government bodies, religious institutions, and international collaborations. Understanding these regional dynamics is essential for heritage site managers, policymakers, and stakeholders to navigate the complexities of heritage tourism and ensure its sustainability and benefits for all involved parties.

ECONOMIC IMPACTS OF TOURISM

Tourism is a powerful economic force that can significantly impact heritage sites and their surrounding communities. It has the potential to generate jobs, create revenue, and contribute to financial sustainability. In this exploration, we delve into the multifaceted economic impacts of tourism on heritage sites, examining the data and research findings related to job creation, revenue generation, and economic sustainability.

Job Creation: A Vital Component

Job creation is one of tourism's most immediate and visible economic benefits for heritage sites and local communities. Tourism-related activities necessitate a workforce to provide services, maintain infrastructure, and offer cultural and historical interpretations to visitors. These employment opportunities often extend to diverse sectors, including hospitality, transportation, retail, and cultural heritage preservation. According to data from the World Travel & Tourism Council (WTTC), the tourism and travel industry accounted for approximately 330 million jobs globally in 2019, directly and indirectly. This figure underscores the significant role that tourism plays in the job market. In particular, heritage sites and their surrounding communities benefit from these employment opportunities, as they often become focal points for cultural and historical experiences. For instance, the Great Wall of China, a UNESCO World Heritage Site, draws millions of tourists yearly, creating a demand for tour guides, ticketing staff, maintenance personnel, and artisans producing traditional handicrafts. Similarly, historic European cities like Rome and Florence see high employment levels in the hospitality sector, with hotels, restaurants, and tour operators hiring residents to meet the needs of tourists.

Research findings have consistently shown the link between tourism and job creation. A study by the World Tourism Organization (UNWTO) and the International Labour Organization (ILO) found that tourism supports one in every ten jobs worldwide. Moreover, this report highlighted the potential for even more significant employment growth with the right policies and strategies.

Revenue Generation: A Source of Income

Revenue generation is another critical economic impact of tourism for heritage sites and their host communities. Tourists spend money on various goods and services, from entrance fees to heritage sites, accommodations, dining, souvenirs, and local experiences. According to the United Nations World Tourism Organization (UNWTO), international tourism receipts reached approximately $1.7 trillion in 2019. This demonstrates the substantial financial contribution that tourists make to destinations globally. This revenue is often vital for heritage sites' maintenance, preservation efforts, and educational programs. Local communities surrounding heritage sites also benefit from the spending of tourists. A study by Oxford Economics and the World Travel & Tourism Council (WTTC) found that 2019 travel and tourism contributed $9.2 trillion, or 10.3% of global GDP. This figure reflects the economic significance of tourism for regions and localities. In the case of Venice, Italy, a city known for its historic heritage sites, tourism generates a significant portion of local revenue. According to data from the City of Venice, tourism accounted for about 40% of the city's income in 2019. This revenue supports the preservation of historic buildings and subsidizes local services and infrastructure. Furthermore, research conducted by Deloitte for the U.S. National Park Service found that every dollar invested in national parks (a form of heritage site) in the United States generated $10 in economic activity. This includes spending on lodging, food, transportation, and recreational activities, creating a substantial return on investment.

Economic Sustainability: Balancing Act

While the economic impacts of tourism on heritage sites and local communities are undeniable, achieving financial sustainability is a complex and often delicate balance. Sustainable tourism practices aim to maximize the positive economic effects of tourism while minimizing negative impacts on culture, environment, and society. The potential downsides of tourism, such as overcrowding, cultural commodification, and environmental degradation, require careful management. Over-tourism, in particular, can strain local resources and infrastructure. Barcelona, Spain, is a notable example where concerns over tourism's negative impact on the local quality of life led to protests and policy changes. Residents expressed grievances about rising rents, overcrowding, and the commercialization of their neighborhoods. Consequently, Barcelona implemented measures to regulate tourist accommodations, restrict cruise ship arrivals, and promote responsible tourism.

Economic sustainability also involves reinvesting tourism revenue into heritage site preservation, infrastructure development, and community well-being. The city of Venice, as mentioned earlier, has sought to achieve this balance by using tourism-generated revenue to maintain its historic buildings, improve transportation systems, and support cultural activities. Studies have shown that sustainable tourism practices can enhance economic sustainability. Research published in the Journal of Sustainable Tourism indicates that heritage sites adopting sustainable management practices tend to experience positive economic outcomes in the long run. Sustainable tourism can create a positive feedback loop where preservation efforts attract more tourists, increasing revenue, which can be reinvested in further preservation and community development.

In conclusion, the economic impacts of tourism on heritage sites and local communities are substantial and multifaceted. Job creation and revenue generation are two fundamental aspects that underscore the importance of tourism in supporting livelihoods and preserving cultural and historical treasures. However, achieving economic sustainability requires thoughtful management to mitigate potential adverse effects and reinvest in the well-being of heritage sites and the communities that cherish them. As destinations worldwide grapple with the challenges and opportunities of tourism, a balanced approach is essential to ensure that economic benefits align with the long-term preservation of our shared cultural heritage.

RESULT AND DISCUSSION

Despite many publications and studies on local views of tourism development, tourism professionals believe local perception has been a crucial research topic. The challenge is finding a way to combine preserving history with making the necessary changes to satisfy the region's expectations as a tourist attraction without taking advantage of (and consequently detracting from) the locals. Popular ideas have argued that the social exchange hypothesis is a sound theoretical basis for sustained cultural tourism. The social exchange hypothesis, rooted in sociology and anthropology, offers a valuable lens through which to examine the sustainability of cultural tourism. This theory asserts that social interactions are built on the principle of reciprocity, where individuals or groups engage with each other, expecting mutual benefits. Applied to cultural tourism, it becomes evident that both tourists and host communities seek advantages from their interactions. Tourists seek cultural experiences and knowledge, while host communities often desire economic benefits, cultural preservation, and opportunities for cultural exchange. This reciprocity and mutual exchange are at the core of cultural tourism's sustainability, fostering a sense of balance and satisfaction for all parties involved. Moreover, the hypothesis underscores the significance of cultural exchange, promoting greater cultural understanding and preservation. For cultural tourism to thrive over time, these exchanges must continue to be perceived as mutually beneficial. This framework elucidates why cultural tourism can be a lasting and mutually enriching endeavor, benefiting host communities and tourists seeking meaningful cultural experiences. The social exchange hypothesis, as applied to cultural tourism, not only elucidates the sustainability of tourism but also links to residents' perceptions. When host communities benefit economically and culturally from interactions with tourists, residents are more likely to perceive tourism positively. Favorable resident perceptions can, in turn, enhance the overall sustainability of cultural tourism by fostering community support, cooperation, and a welcoming environment for visitors. Residents' opinions regarding tourist development fall into three categories in the great majority of research in this field: economic, social, and environmental. Additionally, community participation is believed to be a key element of long-term tourist success. Furthermore, most research has noted that stakeholders are crucial to tourist growth. Although the TTCI has considered cultural resources for tourism development, the absence of stakeholder involvement in any of the pillars or indicators highlights a crucial gap needed to conduct an integrated assessment and develop a heritage site's tourism through competitive heritage planning. The TTCI 2013 pillars' construction is seen in Figure 6. The previous framework (see Figure 6) only included three primary sub-indices and a sparsely connected network, allowing space for mistakes. The TTCI 2019 categorized the sub-indices into four groups (see Figure 2).

Although the TTCI has recognized cultural resources as a critical aspect of tourist development, entirely using them has presented another challenge. Figure 7 shows the overall performance of the 14 pillars the WEF chose to evaluate competitiveness in 2019 and indicates that cultural resources scored the lowest out of the 14 in the top quartile, middle quartile, and top quartile, with the lowest global average among the 14 pillars.

Figure 6. TTCI pillar by sub-indices 2013
(Authors' interpretation of WEF TTCR, 2013)

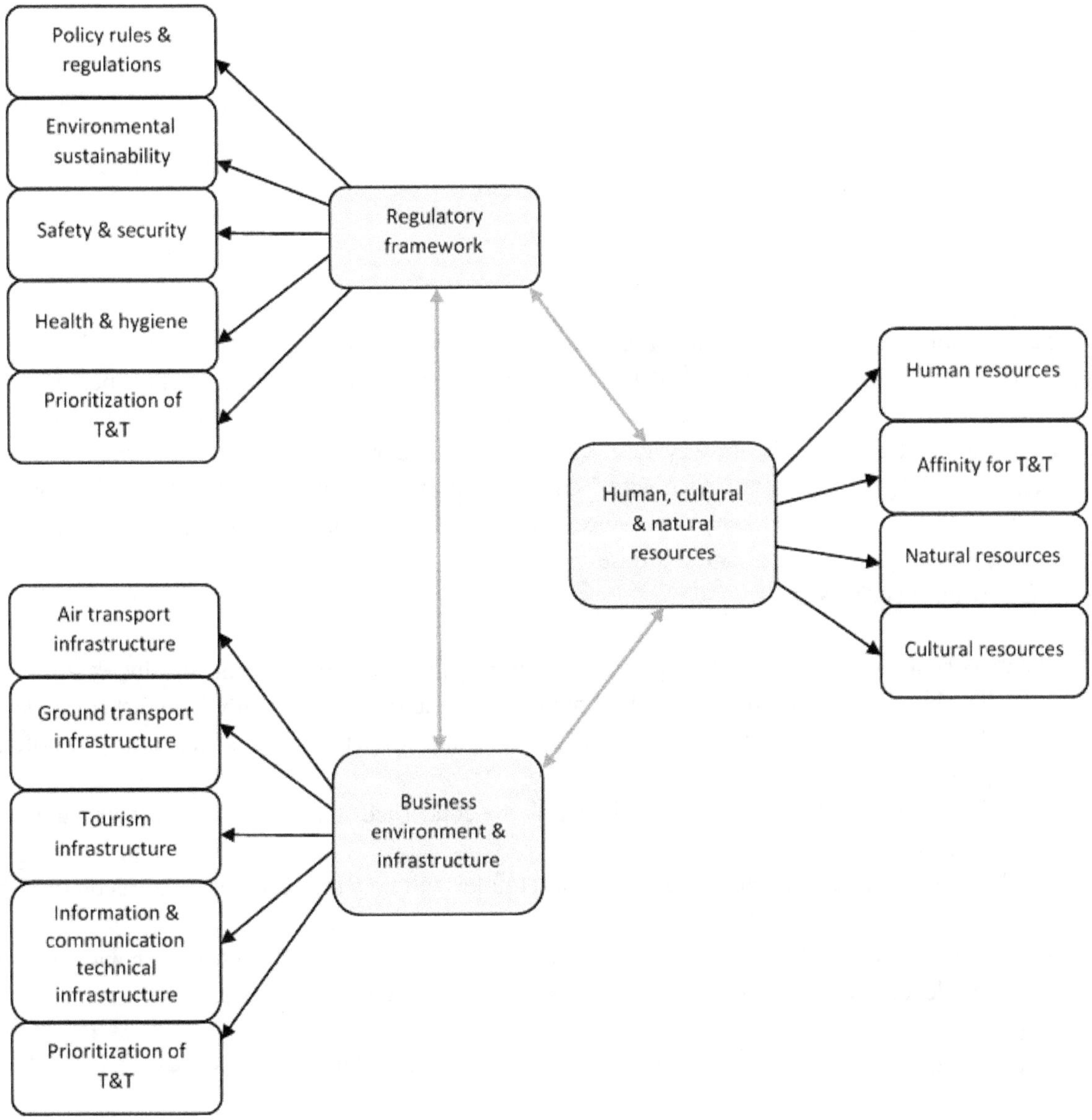

Figure 7. Pillars performance overview, 2019
(WEF TTCR, 2019)

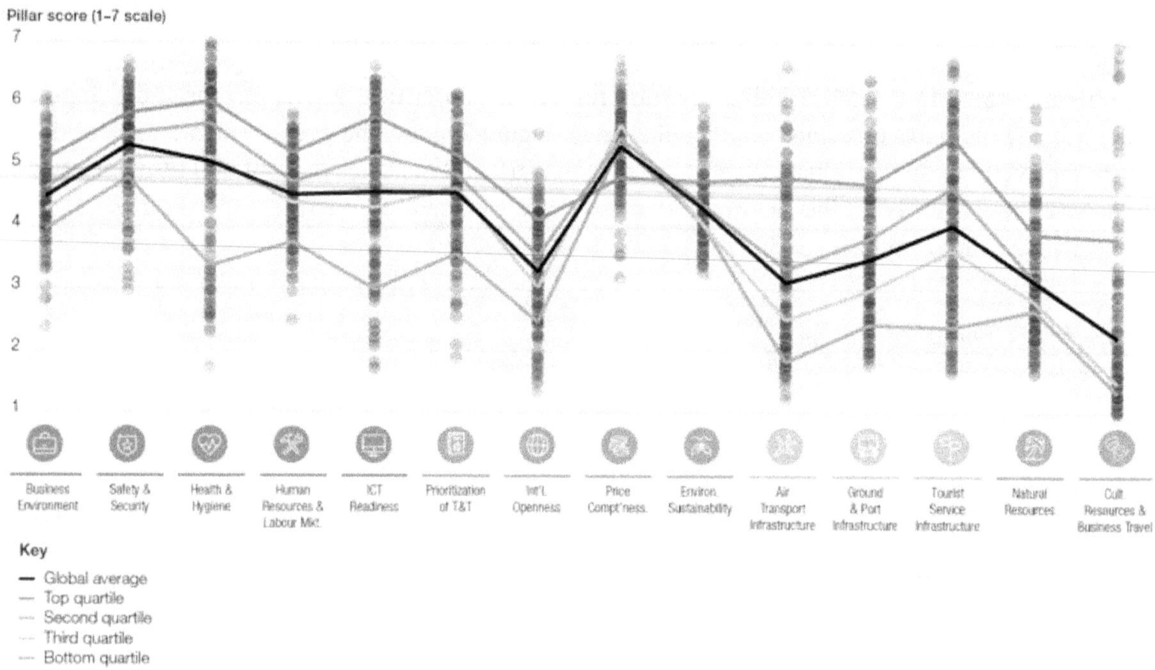

The stakeholder perspective is implicitly considered in the competitiveness evaluation pillars and indicators, which makes it challenging to determine the precise score from the stakeholders' perspective. A new TTCI that entertains stakeholder perception for each pillar and connects across sub-indices is crucial for better integration. The strategy for integrating stakeholders for each pillar is shown in Figure 8.

The proposed TTCI considers stakeholders' engagement and viewpoints when filtering the pillars to evaluate indicators based on their performance. It also expands the sub-index for natural and cultural resources by adding a new pillar called "conservation and management." The proposed TTCI will aid in the following specific ways –

1. The incorporation of stakeholders' perceptions in the 'enabling environment' sub-index –
    - The business environment pillar will help understand the economically linked stakeholders,
    - The safety and security pillar will help understand the stakeholders linked with heritage preservation,
    - The health and hygiene pillar will help link the concerned stakeholders to built environment sustainability,
    - Through human resources and labor market pillar will help the stakeholders directly or indirectly pro-actively and
    - ICT readiness will assist in determining the current condition of ICT utilization and penetration levels among medium and small businesses and their stakeholders.

The resulting output of the 'enabling environment' sub-index will be a Stage A result.

Figure 8. Stakeholder integration approach for TTCI pillars
(Authors)

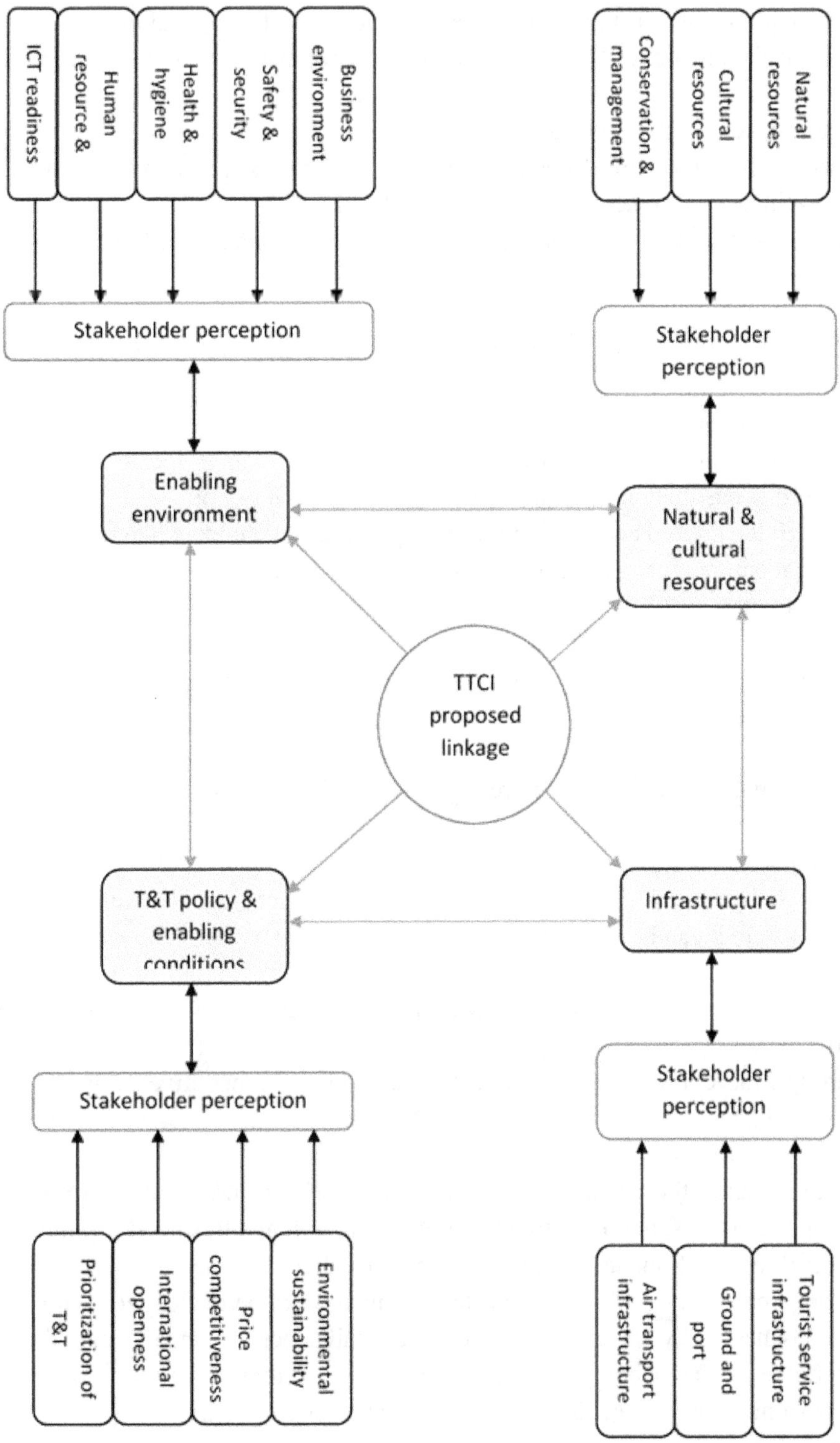

2. The incorporation of stakeholders' perceptions in the 'T&T policy and enabling conditions' sub-index –
    - The environmental sustainability pillar will aid in identifying and incorporating stakeholders who have a direct or indirect impact on the environment around heritage sites,
    - Through the price competitiveness pillar, beneficial policy planning for economically connected stakeholders will be aided,
    - Through the international openness pillar, stakeholders' roles and benefits deriving from a competitive environment can be recognized and
    - Through the prioritization of the T&T pillar, stakeholders can be directly linked with the T&T services for economy generation and community building.

The resulting output of the 'T&T policy and enabling conditions' sub-index will be a Stage B output.

3. The incorporation of stakeholders' perceptions in the 'infrastructure' sub-index –
    - Through the air transport infrastructure pillar, stakeholders like tourists can be identified who prefer destinations with an air transport approach or not,
    - Through the ground and port infrastructure pillar, stakeholders like tourists and transporters can be identified who prefer or provide transit to destinations with base or port transport approaches and
    - Through the tourist service infrastructure pillar, stakeholders like tourists and hoteliers can be identified who prefer or provide tourist services for various tourist destinations, including heritage sites.

The resulting output of the 'infrastructure' sub-index will be a Stage C output.

4. The incorporation of stakeholders' perceptions in the 'natural and cultural resources' sub-index –
    - Through the natural resource pillar, stakeholders influencing the natural environment and the benefit from it can be recognized,
    - Through the cultural resource pillar, stakeholders that are part of the cultural environment and are benefiting the community with the cultural aspect incorporation can be recognized, and
    - Through the conservation and management pillar, competitive heritage planning can be beneficial.

The resulting output of the 'natural and cultural resources' sub-index will be a Stage D output.

Cross-linking the outputs, or connecting Stage A output to Stage B, C, and D outputs and vice versa to create a multi-directional linkage, would help link the pillars together and make it easier to develop performance when one or more of the pillars start to lag and can be reenergized with the help of other pillars that are connected by stakeholders who have similar performance levels. The planned TTCI will help stakeholders create a competitive heritage strategy, either directly or indirectly, to elevate or provide an opportunity for heritage assets that need conservation but are now ignored because they are uncompetitive.

CONCLUSION

Despite the tourism sector's fast expansion, opinions on destination competitiveness, visitor satisfaction, visitor perceptions, and tourist trip return intentions have not been thoroughly examined. This is especially true for heritage sites that lack adequate conservation. Heritage sites not conserved and maintained are likely to degrade and go off the list of economic generators if they are not managed, monitored, or maintained. This jeopardizes cultural heritage as a critical element impacting the existence and identity of communities. It also tends to accelerate the natural degradation of cultural resources and a lack of interest on the part of stakeholders in developing or supporting the continuity of cultural resources. On the other hand, place-making can turn the loss of cultural resources into a boon if done correctly, as the blending of people is a critical component of the process. As a result, stakeholders' perspectives are essential in this process.

Although stakeholders' perspectives have long been acknowledged as a crucial element in ensuring that a site is both functional and appealing, they have not yet been sufficiently researched to the point where they can accurately assess a site's competitiveness for tourist evaluation. Even while the well-respected TTCI acknowledges the influence of stakeholders on a project or a site, far less effort has been made to consider these influences or opinions when developing the competitiveness index for evaluating a location. This has a significant effect on tourist destinations with little attraction for conservation. In addition to acknowledging the existence of stakeholders, the suggested TTCI sieves the competitiveness pillars according to their viewpoints, leading each site to act uniquely and in line with its stakeholders. Heritage sites that lack conservation can compete because of the new pillar of "conservation and management." More study is required to properly understand a special integrated index that can weigh each TTCI indication differently based on the kind and tier of the site.

FUTURE TRENDS

The landscape of heritage site management and tourism is continuously evolving, influenced by a myriad of factors, including technological advancements, changing visitor preferences, and global events. This exploration delves into emerging trends shaping heritage tourism's future.

1. Technology-Enhanced Experiences:
    a. Augmented Reality (AR) and Virtual Reality (VR): Technology transforms visitors' engagement with heritage sites. AR and VR applications allow tourists to experience historical events, artifacts, and architectural wonders in immersive and interactive courses. For example, visitors can use AR apps to see ancient ruins restored to their former glory or use VR headsets to explore historic sites remotely.
    b. Mobile Apps and Smart Devices: Mobile apps provide visitors with real-time information, interactive maps, audio guides, and augmented reality experiences. These apps enhance visitor engagement and offer a personalized experience. Smart devices like beacons and sensors enable location-based information and storytelling, adding depth to the visitor's understanding of the site.
    c. Digital Preservation: Heritage sites increasingly adopt digital technologies to preserve cultural artifacts and historical records. High-resolution 3D scanning and modeling techniques digi-

tally document and archive heritage sites and objects, ensuring their preservation for future generations.

2. Sustainable Tourism Practices:
    a. Regenerative Tourism: Regenerative tourism seeks to go beyond sustainability by actively contributing to restoring and enhancing natural and cultural environments. It involves initiatives such as reforestation, habitat restoration, and supporting local community development.
    b. Carbon-Neutral and Eco-Friendly Tourism: With growing concerns about climate change, heritage sites and tourism destinations are trying to reduce their carbon footprint. This includes adopting renewable energy sources, encouraging eco-friendly transportation options, and promoting responsible waste management.
    c. Community-Based Tourism: Empowering local communities by involving them in tourism decision-making and revenue-sharing is a central theme in sustainable heritage tourism. It helps preserve cultural authenticity, foster local pride, and ensure equitable economic benefits.
3. Changing Visitor Preferences:
    a. Authentic Experiences: Tourists increasingly seek authentic and meaningful experiences. They want to interact with local culture, traditions, and cuisine. Heritage sites offering hands-on experiences, cultural immersion, and local interactions are gaining popularity.
    b. Off-Peak and Niche Tourism: The desire to avoid crowds and experience sites in a more intimate setting leads to off-peak and niche tourism. Travelers are exploring heritage sites during non-peak seasons, and niche markets like heritage-focused educational and genealogical tourism are rising.
    c. Wellness Tourism: Combining heritage and wellness, tourists increasingly seek experiences that nourish the body and mind. This trend has led to the developing of wellness programs at heritage sites, where visitors can practice yoga, meditation, or wellness rituals in historical settings.
4. Global Events and Uncertainties:
    a. Pandemic Preparedness: The COVID-19 pandemic has highlighted the importance of adaptability and preparedness in tourism. Heritage sites invest in health and safety measures, including contactless ticketing, crowd control, and improved sanitation protocols, to ensure visitor safety during future crises.
    b. Resilience and Recovery: Heritage site managers are exploring strategies for resilience and recovery in the face of unexpected global events. This includes diversifying revenue streams, building a more substantial online presence, and creating contingency plans for crisis management.
    c. Geopolitical and Environmental Factors: Political tensions, natural disasters, and climate change affect heritage tourism. Sites address these challenges by implementing risk management strategies, including disaster preparedness and mitigation measures.

In conclusion, heritage site management and tourism are evolving in response to a dynamic and interconnected world. Technology enhances the visitor experience, sustainability is a growing imperative, changing preferences are reshaping tourism offerings, and global events underscore the need for adaptability and resilience. The future of heritage tourism lies in striking a balance between preserving the past and embracing the innovations and challenges of the present and future.

FUNDING

This work was supported by the Department of Science and Technology – Inspire [Grant number IF210191, 2022], Ministry of Science and Technology, Government of India.

REFERENCES

Abeysinghe, A. R. G. M. I., & Ariyawansa, R. G. (2017). Competitiveness of Kandy City as a Heritage Tourist Destination in Sri Lanka. *Sri Lankan Journal of Real Estate*. http://dr.lib.sjp.ac.lk/handle/123456789/8421

Alnafeesi, A. M. (2013). The Role & Impact of Preservation Architectural Heritage on Domestic Tourism. *Proceedings of International Conference on Tourism Development*, 93-100. https://www.semanticscholar.org/paper/The-role-%26-impact-of-preservation-architectural-on-Alnafeesi-Mohamed/ed9c1eb33a237f7b34e2654936a383055eac1520

Amaya-MolinarC. M.Sosa-FerreiraA. P.Ochoa-LlamasI.Moncada-JiménezP. (2017). The perception of destination competitiveness by tourists. *Revista Investigaciones Turísticas,* 14, 1-20. https://dx.doi.org/doi:10.14461/INTURI2017.14.01

Ashworth, G. J., & Larkham, P. J. (2013). *Building a new heritage: Tourism, culture, and identity in the new Europe*. Routledge. https://books.google.co.in/books?hl=en&lr=&id=qWS6oMt5x7MC&oi=fnd&pg=PR1&dq=64.%09Ashworth,+G.+J.,+%26+Larkham,+P.+J.+(2013).+Building+a+new+heritage:+Tourism,+culture,+and+identity+in+the+new+Europe.+Routledge.&ots=nWmKE1Fb0h&sig=OIFvGq6QwxhcQACUzvYTNbQlqHA&redir_esc=y#v=onepage&q&f=false

Assari, A., & Assari, E. (2012). Urban spirit and heritage conservation problems: Case study Isfahan city in Iran. *The Journal of American Science, 8*(1), 202–209. https://www.academia.edu/1314591/Urban_spirit_and_heritage_conservation_problems_case_study_Isfahan_city_in_Iran

Black, P., & Street, E. (2014). The power of perceptions: Exploring the role of urban design in cycling behaviours and healthy ageing. *Transportation Research Procedia, 4*, 68–79. doi:10.1016/j.trpro.2014.11.006

Burgess, J. A. (1979). Place-Making: The Contribution of Environmental Perception Studies in Planning. *Geography: Journal of the Geographical Association, 64*(4), 317-326. https://www.jstor.org/stable/40569988

Cibinskiene, A., & Snieskiene, G. (2015). Evaluation of city tourism competitiveness. *Procedia - Social and Behavioral Sciences, 20th International Scientific Conference Economics and Management - 2015 (ICEM-2015), 213*, 105-110. DOI:10.1016/j.sbspro.2015.11.411

Della Corte, V., Piras, A., & Zamparelli, G. (2010). Brand and Image: The Strategic Factors in Destination Marketing. *International Journal of Leisure and Tourism Marketing, 1*(4), 358–377. https://www.inderscienceonline.com/doi/abs/10.1504/IJLTM.2010.032064. doi:10.1504/IJLTM.2010.032064

El Menchawy, A., Aly, S. S., & Hakim, M. A. (2011). The impact of urban sprawl on the heritage areas through the urban fabric of cities. *Conference: Sustainable Development and Planning V. WIT Transactions on Ecology and the Environment, 150*, 299–314. doi:10.2495/SDP110261

Ferri, C. (2017). Competitiveness of Heritage Sites: a comparison between Scuola Grande della Misericordia in Venice and Oude Kerk in Amsterdam. *Working Papers 05*. Department of Management, Università Ca' Foscari Venezia. https://ideas.repec.org/p/vnm/wpdman/138.html

Flores de León, R.M., Babere, N.J., & Swai, O. (2020). Implications of Cultural Heritage in Urban Regeneration: The CBD of Dar es Salaam. *Revue Interventions économiques, 63*. doi:10.4000/interventionseconomiques.9171

Foroudi, P., Cuomo, M. T., Foroudi, M. M., Katsikeas, C. S., & Gupta, S. (2020). Linking identity and heritage with image and a reputation for competition. *Journal of Business Research, 113*, 317–325. doi:10.1016/j.jbusres.2019.09.042

Frooman, J. (1999). Stakeholder influence strategies. *Academy of Management Review, 24*(2), 191–205. doi:10.2307/259074

Gannon, C., Lynch, P., & Harrington, D. (2010). *Framework for Enhancing Competitiveness in Culture and Heritage Organisations in the Irish Tourism Industry*. Semantic Scholar. https://www.semanticscholar.org/paper/FRAMEWORK-FOR-ENHANCING-COMPETITIVENESS-IN-CULTURE-Gannon-Lynch/f077c904847ac8641454ea06a0a8a032396f63dc

Gjerde, M. (2010). Visual Aesthetic Perception and Judgement Of Urban Streetscapes. *Paper for Building a Better World: CIB World Congress*, 12-22. https://scholar.google.co.nz/citations?view_op=view_citation&hl=en&user=_Iw_Wf4AAAAJ&citation_for_view=_Iw_Wf4AAAAJ:u-x6o8ySG0sC

Goodey, B., & Gold, J. R. (1987). Environmental Perception: The Relationship with Urban Design. *Progress in Human Geography, 11*(1), 126–133. doi:10.1177/030913258701100106

Gray, B. (1989). Collaborating: Finding Common Ground for Multiparty Problems. *Overcoming obstacles for successful collaboration*. https://www.ojp.gov/ncjrs/virtual-library/abstracts/collaborating-finding-common-ground-multiparty-problems#:~:text=This%20book%20presents%20an%20innovative,%2C%20national%2C%20and%20international%20problems

Gupta, S., & Singh, A. (2019). Measurement Scale of Tourism Destination competitiveness: Supply Side Perspectives. *SCMS Journal of Indian Management, 16*(1), 105-116. https://www.proquest.com/scholarly-journals/measurement-scale-tourism-destination/docview/2231860053/se-2?accountid=132709

Hall, C.M., & McArthur, S. (1998). Integrated heritage management: Principles and practice. The Stationery Office.

Hall, M. (2016). Cultural heritage as civilising mission: From decay to recovery. *International Journal of Heritage Studies, 23*(2), 176–177. doi:10.1080/13527258.2016.1218917

Hamdan, H. A., Andersen, P. H., & De Boer, L. (2021). Stakeholder collaboration in sustainable neighborhood projects—A review and research agenda. *Sustainable Cities and Society, 68*, 102776. doi:10.1016/j.scs.2021.102776

Harrison, R. (2013). *Heritage: Critical approaches.* Routledge., doi:10.4324/9780203108857

Johnson, C. L. (2017). *Millennials' Perception of Destination Attractiveness.* HDL. https://hdl.handle.net/1805/15114

King, T. F. (2014). *Cultural heritage and tourism: An introduction.* Channel View Publications. https://books.google.co.in/books?hl=en&lr=&id=oADPBQAAQBAJ&oi=fnd&pg=PR5&dq=King,+T.+F.+(2014).+Cultural+heritage+and+tourism:+An+introduction.+Channel+View+Publications.&ots=uT7c721YmT&sig=WHafF3i2jRwtJ0yit56fMmYn7bg&redir_esc=y#v=onepage&q&f=false

Kovačević, N. D., Kovačević, L., Stankov, U., Dragićević, V., & Miletić, A. (2018). Applying destination competitiveness model to strategic tourism development of small destinations: The case of South Banat district. *Journal of Destination Marketing & Management, 8*, 114–124. doi:10.1016/j.jdmm.2017.01.002

Kumar, S., & Dhir, A. (2020). Associations between travel and tourism competitiveness and culture. *Journal of Destination Marketing & Management, 18*, 100501. doi:10.1016/j.jdmm.2020.100501

Land, N. (2002). *The thirst for annihilation: Georges Bataille and virulent nihilism.* Routledge. https://books.google.co.in/books?hl=en&lr=&id=d1SIAgAAQBAJ&oi=fnd&pg=PP1&dq=68.%09S%C3%B6rlin,+S.+(2009).+The+thirst+for+annihilation:+Georges+Bataille+and+virulent+nihilism.+Routledge.&ots=T5Lg6GHulz&sig=3Dmvbup2hZKvJpv--iMHqoF5rg8&redir_esc=y#v=onepage&q&f=false

Li, Y., Xiao, L., Ye, Y., Xu, W., & Law, A. (2016). Understanding tourist space at a historic site through space syntax analysis: The case of Gulangyu, China. *Tourism Management, 52*, 30–43. doi:10.1016/j.tourman.2015.06.008

MariottiA. (2012). Local System, Networks and International Competitiveness: from Cultural Heritage to Cultural Routes. *Journal of Tourism, Culture and Territorial Development, 3*(5), 81-95. doi:10.6092/issn.2036-5195/3208

McCamley, C. (2016). Achieving Competitiveness In An Emerging Heritage Region; Examining The Role Of Trust As A Critical Antecedent To Heritage Tourism Development. *Global Marketing Conference 2016: Bridging Asia and the World: Global Platform for Interface Between Marketing and Management.* HUD. https://pure.hud.ac.uk/en/activities/achieving-competitiveness-in-an-emerging-heritage-region-examinin

Meskell, L. (Ed.). (2002). *Archaeology under fire: nationalism, politics and heritage in the Eastern Mediterranean and Middle East.* Routledge. https://books.google.co.in/books?hl=en&lr=&id=QaWEAgAAQBAJ&oi=fnd&pg=PP1&dq=Archaeology+under+fire:+Nationalism,+politics,+and+heritage+in+the+Eastern+Mediterranean+and+Middle+East&ots=TlOCryURJK&sig=LheXmbPirhLtWhglV-e9vxw5w-U&redir_esc=y#v=onepage&q=Archaeology%20under%20fire%3A%20Nationalism%2C%20politics%2C%20and%20heritage%20in%20the%20Eastern%20Mediterranean%20and%20Middle%20East&f=false doi:10.4324/9780203029817

Nadalipour, Z., Imani Khoshkhoo, M. H., & Eftekhari, A. R. (2019). An integrated model of destination sustainable competitiveness. *Competitiveness Review, 29*(4), 314–335. doi:10.1108/CR-12-2017-0086

Nag, A. (2020). Health and Housing for Urban Poor in India Post Covid-19. *International Journal of Recent Technology and Engineering, 9*(3), 208–216. doi:10.35940/ijrte.C4350.099320

Nag, A. (2022). Industrial infrastructure development of cottage industries for inclusive economic growth in a sustainable manner: Case study of the urban growth centre in Bishnupur, West Bengal, India. *International Journal of Indian Culture and Business Management, 26*(2), 204–233. doi:10.1504/IJICBM.2022.123591

Nag, A., & Mishra, S. (2023a). Destination Competitiveness and Sustainability: Heritage Planning From the Perspective of the Tourism Industry Stakeholders. In I. G. I. Giuseppe Catenazzo (Ed.), *Cases on Traveler Preferences, Attitudes, and Behaviors: Impact in the Hospitality Industry* (pp. 1–32). Global. doi:10.4018/978-1-6684-6919-4.ch001

Nag, A., & Mishra, S. (2023b). Stakeholders' perception and competitiveness of heritage towns: A systematic literature review. *Tourism Management Perspectives, 48*(101156), 101156. doi:10.1016/j.tmp.2023.101156

Nag, A., & Sarkar, S. (2022). Developing an integrated model for assessment of tourism, transportation and economy in small heritage sites. *International Journal of Technological Learning. Innovation and Development, 14*(3), 272–296. doi:10.1504/IJTLID.2022.125694

Nasar, J. L. (1989). Perception, Cognition, and Evaluation of Urban Places. *Public Places and Spaces* [Advances in Theory and Research]. *Human Behavior and Environment, 10*, 31–56. doi:10.1007/978-1-4684-5601-1_3

Nastase, C. E., Lucaci, A., & Chaşovschi, C. (2020). Circular Economy: A Perspective for the Transformation of Rural Built Heritage in Europe. *Competitiveness and sustainable development, Economic International Conference*. UTM. http://repository.utm.md/handle/5014/17575

Pansiri, J. (2014). Tourist Motives and Destination Competitiveness: A Gap Analysis Perspective. *International Journal of Hospitality & Tourism Administration, 15*(3), 217–247. doi:10.1080/15256480.2014.925718

Perles-Ribes, J. F., Ramón-Rodríguez, A. B., Moreno-Izquierdo, L., & Such-Devesa, M. J. (2019). Online Reputation and Destination Competitiveness: The Case Of Spain. *Tourism Analysis, 24*(2), 161–176. doi:10.3727/108354219X15525055915518

Puren, K., & Jordaan, T. (2014). Towards integrating built heritage resources in urban development through spatial planning. *Conference: The Sustainable City IX. WIT Transactions on Ecology and the Environment, 191*, 209–220. doi:10.2495/SC140181

Rapoport, A., & Hawkes, R. (1970). The Perception of Urban Complexity. *Journal of the American Institute of Planners, 36*(2), 106–111. doi:10.1080/01944367008977291

Rasoolimanesh, M., Roldán, J. L., Jaafar, M., & Ramayah, T. (2017). Factors Influencing Residents' Perceptions toward Tourism Development: Differences across Rural and Urban World Heritage Sites. *Journal of Travel Research, 56*(6), 760–775. doi:10.1177/0047287516662354

Sairinen, R., & Kumpulainen, S. (2006). Assessing social impacts in urban waterfront regeneration. *Environmental Impact Assessment Review, 26*(1), 120–135. doi:10.1016/j.eiar.2005.05.003

Samadi, Z., & Yunus, R. M. (2012). Physical and Spiritual Attributes of Urban Heritage Street's Revitalization. *Procedia: Social and Behavioral Sciences, 36*, 342–349. doi:10.1016/j.sbspro.2012.03.038

Sjöholm, J. (2013). *Built Cultural Heritage in an Urban Planning Context: Literature Review*. Semantic Scholar. https://www.semanticscholar.org/paper/Built-cultural-heritage-in-an-urban-planning-%3A-Sj%C3%B6holm/1849c5ada310878db95d966014d23cc1b3520468

Sliem, M. A. E., Hegazi, Y. S., & El-Alfy, M. M. F. (2019). Smart Heritage Buildings As A Tool To Achieve Tourism Competitiveness. *International Journal Of Eco-Cultural Tourism, Hospitality Planning And Development, 3*(1), 1-9. https://ijecth.journals.ekb.eg/article_181298_b17ef9bab59cbe1477ca2f-1b08c4be38.pdf

Smith, L. (2006). *Uses of heritage*. Routledge. https://books.google.co.in/books?hl=en&lr=&id=bYx_AgAAQBAJ&oi=fnd&pg=PP1&dq=Smith,+L.+(2006).+Uses+of+heritage.+Routledge.%5C&ots=a0uN22RtEk&sig=KOMynKkGgRoIpTm3Mrl_W9J1KUY&redir_esc=y#v=onepage&q=Smith%2C%20L.%20(2006).%20Uses%20of%20heritage.%20Routledge.%5C&f=false

Timothy, D. J., & Boyd, S. W. (2003). Heritage tourism in the 21st century: Valued traditions and new perspectives. *Journal of Heritage Tourism, 1*(1), 1–16. doi:10.1080/17438730608668462

Töppel, M., & Reichel, C. (2021). Qualitative Methods and Hybrid Maps for Spatial Perception with an Example of Security Perception. *Urban Planning, 6*(1), 105–119. Advance online publication. doi:10.17645/up.v6i1.3614

UNESCO. (2020). *Heritage in Urban Contexts: Impacts of Development Projects on World Heritage properties in Cities.* UNESCO. https://whc.unesco.org/en/events/1516/#:~:text=It%20proposes%20new%20policies%2C%20standards,national%20and%20urban%20development%20policies

UNWTO. (2020). Tourism and Culture. UNWTO. https://www.unwto.org/tourism-and-culture

Ursache, M. (2015). Tourism – significant driver shaping a destinations heritage. *Procedia: Social and Behavioral Sciences, 188*, 130–137. doi:10.1016/j.sbspro.2015.03.348

Vengesayi, S. (2003). A conceptual model of tourism destination competitiveness and attractiveness. *ANZMAC 2003 Conference Proceedings Adelaide*. ANZMAC. https://anzmac.org/conference_archive/2003/

Wang, C. H. (2020). The necessity of taking a community approach in a historical cultural landscape conservation: A case of the Jianan Irrigation System in Taiwan. *GeoJournal, 85*(1), 107–126. doi:10.100710708-018-9950-8

WEF TTCR. (2013). *The Travel & Tourism Competitiveness Report 2013*. We Forum. https://reports.weforum.org/travel-and-tourism-competitiveness-report-2013/

WEF TTCR. (2018). *The Travel & Tourism Competitiveness Report 2018*. We Forum. https://reports.weforum.org/global-competitiveness-report-2018/downloads/

WEF TTCR. (2019). *The Travel & Tourism Competitiveness Report 2019*. We Forum. https://www.weforum.org/reports/the-travel-tourism-competitiveness-report-2019

Wu, D., Shen, C., Wang, E., Hou, Y., & Yang, J. (2019). Impact of the Perceived Authenticity of Heritage Sites on Subjective Well-Being: A Study of the Mediating Role of Place Attachment and Satisfaction. *Sustainability, 11*(21). https://doi.org/ doi:10.3390/su11216148

Yoon, Y. (2002). Development of a Structural Model for Tourism Destination Competitiveness from Stakeholders' Perspectives. *ETDs: Virginia Tech Electronic Theses and Dissertations, Theses and Dissertations, Hospitality and Tourism Management (HTM)*. http://hdl.handle.net/10919/29174

Yoshimura, Y., He, S., Hack, G., Nagakura, T., & Ratti, C. (2020). Quantifying memories: Mapping urban perception. *Mobile Networks and Applications, 25*(4), 1275–1286. doi:10.100711036-020-01536-0

Yuen, B. (2005). Strengthening Urban Heritage In Singapore: Building Economic Competitiveness And Civic Identity. *Global Urban Development, 1*(1). https://docplayer.net/41368762-Strengthening-urban-heritage-in-singapore-building-economic-competitiveness-and-civic-identity.html

Zainuddin, Z., Mohd Radzi, S., Mohd Zahari, M. S., & Azman Ong, M. H. (2020). Perceived destination competitiveness and response behavior: A test of mediation. *Journal of Tourism, Hospitality & Culinary Arts, 12*(1), 279-288. https://fhtm.uitm.edu.my/images/jthca/Vol12Issue1/Chap_19.pdf

Zainuddin, Z., Radzi, S. M., & Zahari, M. S. M. (2016). Perceived Destination Competitiveness of Langkawi Island, Malaysia. *Procedia: Social and Behavioral Sciences, 222*, 390–397. doi:10.1016/j.sbspro.2016.05.190

Zehrer, A., & Hallmann, K. (2015). A stakeholder perspective on policy indicators of destination competitiveness. *Journal of Destination Marketing & Management, 4*(2), 120–126. doi:10.1016/j.jdmm.2015.03.003

Zhang, Y., Kang, S., & Koo, J. H. (2021). Perception Difference and Conflicts of Stakeholders in the Urban Regeneration Project: A Case Study of Nanluoguxiang. *Sustainability, 13*(5). doi:10.3390/su13052904

Živanović, M.L. (2016). Culture and heritage tourism – destination competitiveness and regional clusters review. *Horizon A*, 69-77. doi:10.20544/HORIZONS.A.20.1.17.P05

Chapter 16
Nurturing the Human Capital: Exploring the Role of Essential Skills and Competencies for the Sustenance of Intangible Cultural Heritage (ICH) Tourism – A Systematic Literature Review

Sarani Bhaumik
School of Management Sciences, Maulana Abul Kalam Azad University of Technology, West Bengal, India

Sreeparna Bhattacharyya
Bharatiya Vidya Bhavan, Institute of Management Science, Maulana Abul Kalam Azad University of Technology, West Bengal, India

Sarbari Bhowmick
School of Management Sciences, Maulana Abul Kalam Azad University of Technology, West Bengal, India

ABSTRACT

Tourism, as well as ICH, boosts the country's economy. Humans are the most significant aspect of intangible cultural assets and tourism, because other resources cannot meet goals and many factors influence institution development. The purposes of the current study are twofold, i.e., to explore the role of human resources in the preservation of intangible cultural heritage and to examine the process of maintaining and developing skilled and efficient human resources in the sustenance of ICH. To collect data, the articles from databases were searched using keywords and ultimately sixteen articles were reviewed and analyzed. The study findings suggested that unlike tangible heritage, which can be physically touched, intangible cultural heritage is more difficult to preserve. It is being neglected due to a lack of government initiative as well as other social or financial factors. Due to the host community's lack of skill, as a result of inadequate training, the potential for human resource development is also being overlooked. Training guidelines can be prepared for maintaining ICH.

DOI: 10.4018/978-1-6684-9957-3.ch016

INTRODUCTION

Scope of the Study

Over the past few decades, globalisation has complicated cultural heritage and governance. Tourism has diversified as economic development accelerated urbanisation. Cultural goods and services and business operations based on intangible cultural heritage benefit practitioners. Communities that practice and preserve this history could benefit from this income. Job creation, foreign currency receipts, increases in household income and government revenue, improvements to the balance of payments, and an upsurge in the number of government initiatives supported by the expansion of the tourism industry are just a few of the many benefits tourism has for the economy. (Ivanova et al., 2022)

Tourism boosts Gross Domestic Production (GDP), and ICH is vital to the industry. Furthermore, tourism globalisation has also increased the demand for authentic cultural experiences, making preserving and promoting intangible cultural heritage even more critical. However, balancing economic development and cultural heritage protection is crucial to ensure its sustainability for future generations. However, the influx of tourists and the commercialisation of cultural heritage also pose challenges. Although it generates significant economic benefits, mass tourism negatively impacts the environment and society. (Fang et al., 2021)

Commodification and exploitation are risks, as traditional practices may be modified or exaggerated to cater to tourist expectations. Additionally, the increased demand for cultural experiences can lead to overcrowding and degradation of heritage sites, threatening their authenticity and long-term sustainability. Therefore, it is crucial to strike a balance between economic development through tourism and the preservation of cultural heritage for future generations. An integrative idea called "slow tourism" encourages sustainability in all facets of travel. In contrast to mass tourism, slow travel promotes environmentally friendly and sustainable travel. (Oak et al., 2023)

Humans are the most significant aspect of organisational systems, notably intangible cultural assets and tourism, because other resources cannot meet goals, and many factors influence institution development. Metaphysical cultural heritage-oriented tourism growth requires sustainable HRD initiatives for tourism sector workers, host communities, and tourists. These initiatives should focus on enhancing the knowledge and skills of tourism sector workers, fostering a sense of pride and ownership within host communities, and providing enriching experiences for tourists. Continuous innovation and active public participation are essential to the scientific management and preservation of intangible cultural heritage (ICH). (Li et al., 2023)

Effective HRD programs should also address the conservation and promotion of intangible cultural assets, ensuring their continued relevance and value in the tourism industry. These initiatives should focus on enhancing the knowledge and skills of tourism sector workers, fostering a sense of pride and ownership within host communities, and creating meaningful experiences for tourists. Collaboration between stakeholders, such as government bodies, local communities, and private organisations, is crucial for successfully implementing these HRD initiatives. These initiatives should focus on enhancing the knowledge and skills of tourism sector workers, fostering a sense of pride and ownership within host communities, and creating memorable experiences for tourists. Additionally, collaboration between stakeholders such as government bodies, non-profit organisations, and local communities is crucial to ensure the long-term success of these initiatives.

Intangible Cultural Heritage (Ich)

The Evolution of Terminology

The Convention for the 'Safeguarding of Intangible Cultural Heritage' 2003 introduced the term "intangible cultural heritage" (ICH). It has since been accepted globally. Before that, the concept of ICH was known as 'traditional culture' or 'folklore'. It has been put forward by the United Nations Educational, Scientific, and Cultural Organizations (UNESCO) in 1989 recommendation on the Safeguarding of Traditional Culture and Folklore. 'Non-physical culture,' 'traditional culture,' and 'folklore' were only tangentially related to the concept of "ICH" and unable to define it accurately. Both physical and non-physical aspects of cultural heritage needed to be protected, and identifying the subject matter was critical. It was time to focus on the crucial part—the conservation of "intangible" cultural heritage—once "ICH" was acknowledged as a word capable of describing the cultural "traits" and "customs" of distinct cultures.

Diverse Aspects of ICH and Their Significance

The diversity and complexities of different cultures persist. It is critical to recognise, acknowledge, protect, and raise awareness about various people, their cultures, heritage, customs, skills, and practices, among other things. The domains wherein "ICH" was displayed in the 2003 UNESCO Convention were oral traditions and expressions, performing arts, social practices, rituals, festive events, knowledge and practices respecting nature and the universe, and traditional handicrafts. ICH supports the development of various communities by fostering their growth through their ancestors' ethnic arts and crafts. Their traditional arts and crafts have evolved into a way of life and a source of income. Artisans do not receive formal education since their craft is inherited from generation to generation. The protection and preservation of ICH are crucial to recognising these artisans as the ones who have kept their traditional skills alive and to raise awareness of them worldwide.

The Role of ICH In Supporting Communities and Their Traditional Arts and Crafts

Intangible Cultural Heritage (ICH) is vital for preserving and promoting traditional arts and crafts within communities. It safeguards and enables these cultural expressions, ensuring their continuity for future generations. ICH initiatives provide opportunities for communities to preserve, transmit, and revitalise their knowledge, skills, and practices, fostering social cohesion, economic development, and cultural diversity. These efforts empower communities to share their knowledge, skills, and traditions with future generations, ensuring the continuity of their cultural heritage. Since ICH was born in a particular human-natural setting, it must be nurtured by local cultural backgrounds that exhibit typical cultural ecological traits. The environment in which natural disasters, modern culture, and a lack of inheritors can significantly impact ICH lives. (Zhang et al., 2023)

ICH initiatives also provide economic opportunities for artisans and craftspeople, contributing to the sustainable development of communities and promoting global cultural diversity. Overall, ICH plays a vital role in fostering a sense of identity and pride among its members.

Impact of Technological Advancement and Changing on Traditional Craftsmanship - Lifestyle Challenges and Threats to ICH Preservation

The impact of technological advancements and changing lifestyles on traditional craftsmanship has been significant. As technology advances, traditional craftsmanship faces the challenge of adapting to new methods and techniques and losing originality. Additionally, changing lifestyles and consumer preferences have led to a shift in demand for handmade products, posing both opportunities and challenges for traditional artisans has been significant. To make it possible for classic folk art to advance to a particular stage of development, it is essential to innovate based on traditional folk art and incorporate contemporary advances to strengthen the innovation of conventional craftsmanship and materials from a fresh viewpoint. (Li et al., 2023)

With the rise of automation and mass production, traditional craftsmanship has faced challenges regarding demand and relevance. The shift towards a fast-paced consumer culture has also led to a decline in appreciation for the time and skill required in traditional craftsmanship. Traditional craftsmanship, which once thrived to produce unique and high-quality goods, has faced mass production and automation challenges. Additionally, the shift towards digitalisation and online shopping has further affected the demand for traditional crafts, as consumers now have access to a wide range of products from around the world at their fingertips.

The pace of life is altering as a result of technological advancement. Artificial intelligence is paving the way for a more relaxed and comfortable lifestyle. Human beings have created artificial intelligence so that they can complete tasks more quickly than usual, working at the convenience of the individual. People's productivity and efficiency have been affected by this in both positive and negative ways. The significance of human factors in organisations has also diminished as a result. However, the quality of handcrafted goods, art, and craft cannot be easily replaced, but their appeal is gradually waning. Whether AI image-generation technology can outperform its creative human counterparts is still debatable. (Rasyad et al., 2023)

Additionally, those who are continuing the legacy of their ancestors are not being paid enough to support their families, which has led them to abandon their craft in favour of new lines of work that will allow them to keep themselves. This is one of the reasons why protecting and preserving ICH at all costs is crucial.

Therefore, ICH is a testament to the timeless allure of human history, artistry, and tradition. It encompasses exploring and preserving diverse cultural expressions and historic landmarks, fostering an intricate interplay between the past and the present. In this modern age of globalisation and technological advancement, the significance of safeguarding and promoting Intangible Cultural Heritage (ICH) through tourism has gained remarkable prominence. However, the effectiveness and sustainability of such tourism endeavours profoundly rely on the human element – the individuals and communities that serve as stewards, guides, and ambassadors of their cultural legacies.

Intersection among ICH, Tourism, and Human Capital

Nurturing the human capital within ICH tourism emerges as a pivotal concern. The seamless orchestration of cultural and heritage experiences demands a multidimensional skill set and a profound understanding of both tangible artefacts and intangible narratives. Tour guides, artisans, curators, local communities, and other stakeholders constitute the intricate tapestry forming ICH tourism's backbone. As such, identifying, cultivating, and enhancing these individuals' essential skills and competencies become imperative for the sustained growth and enrichment of the sector.

Importance of protecting and preserving ICH

The importance of protecting and preserving ICH. ICH, or Intangible Cultural Heritage, is crucial for maintaining cultural diversity and identity. It represents the knowledge, practices, and traditions passed down from generation to generation, shaping our collective memory and sense of belonging. By safeguarding ICH, we ensure that these unique expressions of human creativity continue to thrive and contribute to the richness of our global heritage. If living heritage is neglected in the community, it could be lost forever or become a thing of the past. Strengthening living heritage enables safeguarding this history, its inheritance to succeeding generations, and its capacity to change and adapt to any situation. (Abdul Aziz et al., 2023)

Protecting ICH fosters intercultural dialogue and understanding, promoting respect for different cultures and strengthening community social cohesion. ICH, or Intangible Cultural Heritage, is crucial for maintaining the diversity and richness of our global cultural heritage. It encompasses practices, traditions, knowledge, and skills passed down from generation to generation, shaping our identities and fostering social cohesion. By safeguarding ICH, we ensure its continuity and contribute to the sustainable development of communities while promoting mutual understanding and respect among different cultures. Additionally, protecting ICH helps us appreciate the intangible aspects of our heritage that cannot be physically touched but hold immense.

SIGNIFICANCE OF THE STUDY

The purposes of the current study are twofold. The first objective is to explore the role of human resources in preserving intangible cultural heritage, and the second is to examine the process of maintaining and developing skilled and efficient human resources in the sustenance of ICH. By investigating human resources' role in preserving intangible cultural heritage, this study aims to shed light on the crucial link between people and the safeguarding of cultural traditions. Additionally, understanding the process of maintaining and developing skilled and efficient human resources in this context can provide valuable insights for organisations and policymakers striving to ensure the long-term sustainability of ICH. This study highlights human resources' significance in conserving and promoting intangible cultural heritage. Additionally, it seeks to provide insights into practical strategies for nurturing and retaining a competent workforce that can contribute to the sustainable safeguarding of ICH. This study aims to shed light on the crucial role that human resources play in preserving intangible cultural heritage. By investigating their involvement, we can better understand how to safeguard and promote these valuable traditions for future generations. Additionally, understanding the process of maintaining and developing skilled and efficient human resources in this context is essential for ensuring the long-term sustainability of intangible cultural heritage practices.

Amidst the burgeoning research on cultural and heritage tourism, a comprehensive exploration of the human factors driving the success of ICH tourism remains relatively understudied. This research tries to bridge the gap by synthesising existing knowledge to investigate both the crucial skills and competencies essential for the cultivation of a vibrant and enduring ICH tourism industry as well as the process of preserving and developing skilled and efficient human resources by collating insights from a diverse array of scholarly works. This review offers a holistic perspective on the multifaceted dimensions of human capital in ICH tourism.

LITERATURE REVIEW

Role of Human Resources in the Preservation of Intangible Cultural Heritage

According to Francis-Lindsay (2010), within the rhetoric of sustainable tourist development, words like 'local participation' and 'local communities' have become increasingly popular in the last two decades when researchers addressed the issue. These phrases emphasised a 'bottom-up' approach over a traditional 'top-down' one, intending to empower communities and ensure their active involvement. However, despite its conceptual appeal, this approach has often remained theoretical, with limited practical implementation. This research aimed to contribute to reshaping the importance of communities in tourism by emphasising cultural heritage tourism, a burgeoning socio-economic force. Cultural heritage encompasses tangible and intangible characteristics, positioning local people as significant stakeholders in the tourism industry, constituting a crucial gap in the existing literature. In regions such as Asia and the Caribbean, many cultural custodians and heritage sites are rooted in local enclaves, which hold significant allure for exotic travellers. This paper sheds light on participatory models and management frameworks that are driving the engagement of local communities in emerging forms of tourism development. By examining cases from Asian and Caribbean contexts, the research underscores the evolving dynamics of community involvement in 'new' tourism paradigms. It highlights various participatory typologies and management strategies that have facilitated the integration of local communities into the tourism value chain. Ultimately, the paper concludes that cultural heritage, with its unique appeal and intrinsic ties to local identities, has the potential to foster increased participation of communities in tourism activities. This increased involvement contributes to the economic aspects of their empowerment and catalyses socio-cultural growth. It emphasises the need to move beyond romantic notions of 'local participation' and 'bottom-up' development in sustainable tourism. It examined how the cultural heritage sector, with its rich assets, has reshaped the significance of local communities in tourism by analysing case studies from different regions. The paper underscores the transformative potential of cultural heritage tourism in enhancing local communities' participation, leading to economic advancement and socio-cultural empowerment.

Lenzerini (2011) discoursed the Intangible Cultural Heritage (ICH) as a repository of intangible cultural expressions representing humanity's rich and diverse living heritage. It is crucial for preserving cultural diversity and is deeply intertwined with human rights. The aim of the UNESCO Convention on the Safeguarding of Intangible Cultural Heritage in 2003 was to safeguard ICH, rooted in the self-identification of heritage creators and bearers, reflecting historical and social shifts. However, its practical implementation, mirroring the 1972 World Heritage Convention, appears insufficient to safeguard ICH effectively. This article proposes a solution by advocating for the parallel application of international human rights law. ICH is an integral facet of cultural human rights, playing a pivotal role in realising and enjoying individual and collective rights among its creators and bearers. This approach aligns with the intrinsic relationship between ICH and human rights, acknowledging the interdependence of cultural expressions and fundamental freedoms of individuals and communities. This literature review underscores the multifaceted significance of Intangible Cultural Heritage (ICH) as a dynamic representation of cultural diversity and identity. It highlights the limitations of the UNESCO Convention in fully safeguarding the intricacies of ICH and proposes a solution by advocating for the application of international human rights law. By recognising ICH as an essential element of cultural human rights, the review promotes a comprehensive approach to preserve the invaluable tapestry of intangible artistic expressions and to ensure the realisation of the rights of heritage creators and bearers.

Cominelli and Greffe (2012) first and foremost stressed the importance of ICH as a source of creativity and innovation in their study. They attempted to investigate various strategies for protecting heritage, mainly traditional craftsmanship, demonstrating the limitations of a system focused on individuals and creators and the difficulty of a larger one considering the unique nature of these abilities, knowledge, and cultural practices while constantly innovating and fostering creative processes. Intangible Cultural Heritage (ICH) is a dynamic tapestry of traditions, representations, expressions, knowledge, and skills intrinsic to communities and individuals. It plays a pivotal role in shaping contemporary global and competitive landscapes, and its preservation is essential for its transformation and innovation. ICH connects historical legacies to modern aspirations, facilitating a continuum of inspiration and inventive thought. This research aims to understand the myriad aspects of ICH, its role in fostering innovation, and the imperatives of its preservation. It analysed safeguarding mechanisms for ICH, drawing on diverse case studies and explored the efficacy of individual-centric protection systems. The study synthesises the interplay between ICH, creativity, and safeguarding strategies, highlighting the symbiotic relationship between heritage preservation and innovative dynamism. It aims to understand the holistic significance of ICH, bridging tradition and progress and fuelling ongoing social and economic creativity.

Yang et al. (2018) shed light on the hurdles faced by traditional craft heritage in developing countries and provide insights into tackling these challenges for the long-term preservation of this invaluable heritage. Through a comprehensive exploration of the available literature, the review identifies critical issues and seeks solutions to foster the sustainability of traditional crafts. The challenges identified include a lack of adequate data, problems with industrialisation and mass production, an absence of young people who are willing to stay in this profession of work, deficiency of basic infrastructure, a lack of innovation and technology, a shortage of facilities for education and training, and an absence of financial resources.

Tan et al. (2018) explored the concepts of place attachment and 'person-place bonding' and their relationship with the sustainability of intangible cultural heritage within the context of heritage tourism. The study focused on George Town and Melaka, Malaysia, where local communities play a pivotal role in safeguarding and maintaining the intangible heritage. Tourism has become a significant concern for heritage sites, particularly World Heritage Sites (WHSs), as it poses significant challenges to preserving intangible cultural heritage. This study underscored the critical role of local community voices in safeguarding intangible cultural heritage at WHSs. The study's findings were based on in-depth interviews and observations in the WHSs of George Town and Melaka, Malaysia, as the key players of the intangible heritage. Additionally, 32 relevant documents were analysed to provide a comprehensive overview of the cultural landscape. The study examined the elements of 'person-place-bonding' and how person, place, and contribution of bonding between person and business to the sustainability of intangible cultural heritage and heritage tourism. The study identified three themes: 'Sense of Loss' highlighted concerns about heritage erosion due to external influences, motivating active participation to prevent the loss; 'Sense of Justice' emphasised the importance of preserving heritage for future generations, reflecting an intergenerational commitment, and 'Sense of Mission', reflecting the community's responsibility as cultural stewards, shaping their dedication to safeguarding their heritage. These themes emphasise preserving heritage, intergenerational commitment, and cultural stewardship. The study was characterised by two key codes: 'awareness', fostering responsibility for intangible cultural heritage, and 'creativity', ensuring relevance and authenticity for the heritage ecosystem of WHSs. The study underscored the importance of local community voices in preserving intangible cultural heritage at World Heritage Sites. It explores place attachment and emotional, moral, and cultural dimensions that drive community engagement and participation. The emergent themes shed light on the motivations underlying community efforts to sustain their heritage and the importance of adaptive strategies within the heritage ecosystem.

Kamal (2019) delved into the intricate relationship between tourism planning and development, focusing on the case of Siwa Oasis in Egypt, examining the consequences of tourism development on the livelihoods of Siwan community members, with a particular emphasis on gender dynamics, providing insights into the potential pitfalls of development policies in rural minority contexts via discussing a case study in the context of the remote rural community, Siwa Oasis in the western desert of Egypt. Since Siwa Oasis is more than 800 kilometres from the capital city of Cairo, it was marginalised due to its remote location. Moreover, this study emphasised the empowerment of minority communities, conserving their intangible cultural heritage and promoting their internal and external values. A Date Palm Festival in Siwan aims to encourage the community's history and identity by hosting a trade exhibition for local small-scale growers. This event, lasting three days, aims to help them receive fair prices for their goods, allowing consumers to contact producers directly. The festival is significant for local growers who were previously unable to market their dates due to low prices from substantial producers. Tourism planning involves strategic strategies like infrastructure development, marketing, policy formulation, and community engagement to attract visitors to a destination, influenced by stakeholders like governments, local communities, businesses, and NGOs. Siwa Oasis' case study on tourism planning in rural minority communities highlighted the complex interplay of economic goals, cultural preservation, and gender dynamics. It emphasised the importance of considering potential consequences and engaging local communities in decision-making.

Kim et al. (2019) explored the priorities and perspectives of ICH practitioners in the context of developing ICH as a sustainable tourism resource, with a focus on the case study of South Korea with a single representative through twenty-five in-depth interviews of ICH master practitioners employing purposive sampling technique based on their expertise in the phenomenon. The interviewed data were transcribed into English. The data were analysed using QSR NVivo to generate content and themes to understand the quantitative expression of articles. The study reconnoitres the concept of ICH authenticity, highlighting the importance of training and mastering original performances to maintain authenticity. It emphasised the role of practitioners in safeguarding the authenticity of ICH resources. Most respondents showed a positive attitude towards commodification and transformation, emphasising the need for increased awareness and interest in ICH among the younger generation. Empowering ICH practitioners is crucial for sustainable tourism and developing strategies tailored to each ICH. However, a top-down development approach in Korea lacks empowerment in safeguarding and promoting ICH. It sheds light on the complex interplay between authenticity, transmission, and promotion of ICH in the tourism sector.

Eichler (2020) delved into the evolving landscape of indigenous and minority cultural rights within the context of ICH, examining the tension between recognition and the potential homogenisation of artistic expressions. It also highlighted the role of external factors such as tourism, commodification, and digitalisation in shaping this discourse. The Convention for the Safeguarding of the Intangible Cultural Heritage (CSICH) marked a significant milestone in recognising indigenous and minority communities' rights to their cultural heritage. In contrast, their engagement with their cultural identities has been complex. The study findings suggested that adopting CSICH recognises indigenous and minority artistic rights, emphasising preserving unique identities and traditions. However, homogenising language raises concerns about the erasure of distinct expressions, requiring careful consideration. Indigenous and minority communities faced other challenges in protecting their ICH due to tourism growth, commodification, and the digital age's potential to dilute authenticity and oversimplify complex narratives. The review examined the challenges faced by marginalised groups in Andean Carnival celebrations in Oruro and Barranquilla, focusing on their negotiation of cultural identities via respective case studies. These cases provided insights into how marginalised groups navigate the dynamics of participation in institutionalised ICH contexts. The review critically examined how these celebrations facilitate the negotiation of cultural identities for indigenous peoples, Afro-descendants, and other marginalised communities. In the face of external pressures and potential homogenisation, this study evaluated the spaces created by these celebrations as platforms for inclusion and identity negotiation. It assessed how these spaces empower marginalised groups to assert their unique cultural identities, challenge historical marginalisation, and reclaim agency over their representation. The study explored the evolving landscape of indigenous and minority artistic rights within the context of ICH, which is complex and multifaceted. However, ongoing vigilance is required to ensure that external pressures, including tourism, commodification, and digitalisation, do not undermine marginalised communities' genuine representation and agency in defining their cultural heritage.

Dragicevic (2021) sought to examine the significance of traditional crafts and their forms from the perspectives of Dubrovnik residents on the potential for creating new tourism product forms based on traditional arts. According to locals, traditional crafts could play an essential role in the long-term growth of the tourist attraction Dubrovnik. The researcher indicated that India is a significant exporter and supplier of goods for gifts and handicrafts to the global market. The Indian handicraft industry is highly labour-intensive and widely dispersed throughout the nation in rural and urban areas. The industry is regarded as the second-largest employer after agriculture due to the large number of artisans who work part-time in the craft industry. Over 6 million artisans, many of whom are women and members of underprivileged groups, are employed by the sector. India's current handicraft tradition primely illustrates how traditional designs and contemporary methods coexist harmoniously. The most popular crafts are metalwork, earthenware, pottery, sculpture, woodwork, embroidered and crocheted items, shawls, zari products, stone carving, and imitation jewellery. Other popular crafts include hand-printed textiles and scarves. Due to the rising demand for these products and the recognition of Indian arts both domestically and abroad, these industries are expanding.

Aktürk and Lerski (2021) delve into the multifaceted issue of climate displacement and its consequences for intangible heritage, emphasising the need for cultural preservation amidst environmental upheaval reflections based on observations. Researchers concluded from their discourses that climate-induced displacement is an intricate issue and global concern with far-reaching consequences, including the endangerment of intangible heritage and affecting communities worldwide through rising temperatures, extreme weather events, and sea-level rise due to greenhouse gas emissions. This affects uneven distribution among communities, with vulnerable regions bearing the brunt of disasters, exacerbating existing inequalities and social injustices. It also causes land loss, homes, and natural resources due to floods, sea-level rise, storms, droughts, and salinisation, leading to climate-induced displacement. Climate change relocation raises ethical and human rights concerns, including discrimination, conflict, and security, necessitating careful consideration of the well-being of individuals and communities.

Recognising the cultural significance of traditions, rituals, and practices is essential for building resilient communities in the face of climate change. Climate-induced displacement threatens intangible heritage in affected communities, threatening generations-deep connections to rituals, customs, ancestral ties, and cultural practices as people are uprooted from their traditional homelands. Climate displacement often overlooks intangible heritage, causing significant erosion of cultural traditions and practices despite the visible loss of land and infrastructure. This paper emphasised the role of intangible heritage in fostering resilient communities amidst climate change, emphasising the significance of incorporating indigenous cultural values in climate action planning and contributing to the emerging policies and practices aimed at preserving intangible heritage in the context of climate displacement. It underscored the importance of recognising the intrinsic value of cultural traditions in building resilience and fostering social cohesion.

Lonardi and Unterpertinger (2022) delve into the world of cultural tourism, explicitly focusing on the ICH of minority communities and the significance of minority language centres on the Ladin communities of South Tyrol, Italy, providing insights into the impact of cultural sustainability on tourism in the region using semi-structured interviews to German- and Italian-speaking tourists in Val Pusteria, with a video presenting a real-life situation in Ladin. The qualitative approach enables a nuanced exploration of tourists' perceptions and experiences of Ladin culture and language. The findings revealed that tourists in South Tyrol are captivated by the region's unique architecture, traditional lifestyle, events, clothing, and Ladin language, seeking authentic experiences that connect them with the region's essence. One of the study's most compelling findings pertains to the role of cultural sustainability in the region's tourism industry, highlighting the preservation of rich traditions and languages, particularly the Ladin language, which attracts tourists and fosters affection.

Examining the Process of Maintaining and Developing Skilled and Efficient Human Resources in the Sustenance of ICH

In their study, Weiler and Ham (2002) established the two-part premise that training local guides is a crucial strategy for sustainable development. This study delved into a pivotal paper that explores the unique role of interpretive tour guide training in the context of sustainable tourism development and management in developing countries, examining a comprehensive training program conducted in Panama, Patagonia (Argentina), and the Galapagos Islands (Ecuador), and evaluates the program's efficacy in enhancing interpretive guiding skills and fostering sustainable tourism practices. The function of training interpretive tour guides in the context of sustainable tourism growth has garnered increasing attention due to its potential to foster meaningful visitor experiences, promote environmental and cultural awareness, and contribute to the economic well-being of host communities. The training programmes were tailored to each destination's unique needs and contexts, promoting sustainable tourism practices through collaboration with local stakeholders and experts. The paper evaluated interpretive tour guide training programs using qualitative and quantitative methodologies. It assessed their impact on guide performance, visitor experiences, and tourism industry sustainability, promoting a "train-the-trainers" approach. The evaluation process considered long-term effects on guides' knowledge dissemination, promoting a "train-the-trainers" approach to ensure effective knowledge sharing. The study showcased a model for developing countries' interpretive guide training programs, emphasising context-specific curriculum development, experiential learning, community involvement, continuous evaluation, and sustainability principles.

Any educational experience—and the tourism experience falls into this category—suggested by Christie and Mason (2003). However, as the tourism industry has evolved, the importance of well-trained guides has become increasingly recognised. This study explored the evolution of tour guide training, its role as an adult education activity, and its potential to enhance visitor experiences through practical interpretation. The study revealed that historically, untrained guides often conveyed inconsistent information, but the tourism industry now recognises the value of knowledgeable guides, with guide training becoming common in developed countries. It was also emphasised that the training of tour guides should be considered an adult education activity involving practical skills development, interactive learning, and problem-solving, promoting self-directed learning and a deeper understanding of the material. (Tilden, 1957; Pond, 1993)

The study established that practical guide training enables a well-trained guide to educate and entertain tourists, engagingly presenting complex information and resulting in memorable and meaningful experiences. This article advocated for tour guides to be trained to critique their knowledge, attitudes, and behaviour, enabling them to identify biases, gaps in knowledge, and opportunities for improvement, thereby enhancing their interpretations. The article proposed a new model for guide training to enhance tour guiding quality, add value to the tourist industry, and ensure tourism site sustainability. This model included in-depth content knowledge, practical communication skills, critical thinking workshops, cultural sensitivity training, and continuous learning. Guides should comprehensively understand their subjects, engage visitors through storytelling, participate in critical thinking workshops, and be sensitive to diverse cultures.

Tsaur and Tu (2019) highlighted in their study the absence of adequate measurement scales and discussed developing and validating a novel Tour Leaders' Cultural Competence (TLCC) scale. They examined the relationship between tour leaders' cultural competence and job performance. Recognising the need for a comprehensive and reliable measurement tool for tour leaders' cultural competence, researchers embarked on a study to develop and validate the TLCC scale. The study utilised a multi-phase approach involving in-depth interviews and rigorous analysis to establish the initial set of scale items. With an extensive pool of initial articles, an exploratory factor analysis was conducted on 321 short-haul inclusive tour leaders. To validate the structure of the TLCC scale, a confirmatory factor analysis was performed on a separate sample of 284 long-haul inclusive tour leaders. Through this analysis, the research identified 36 items under the ten factors (Understanding local travel environment; Cross-cultural communication and interaction skills; Understanding local Culture; Language ability; Understanding local life habits; Understanding cultural backgrounds of tour members; Cultural Empathy; Cultural affinity; Cultural mediation; and Cultural adaptability) that collectively represented tour leaders' ability to navigate cultural diversity effectively to measure TLCC. Identifying the ten distinct factors of tour leaders' cultural competence enhances our comprehension of the nuanced skills required for effective cross-cultural interactions. Practically, the TLCC scale is a valuable tool for tour leaders, enabling them to self-assess their cultural competence and identify areas for improvement. Additionally, travel agencies and tour operators can utilise the TLCC scale to evaluate and train their tour leaders, ultimately enhancing the quality of tourists' experiences. The scale's development also underscores the need for continued research and emphasis on cultural competence within the tourism industry.

Lou (2020) explored the function of “edutainment”—a combination of educational and entertaining activities—in preserving intangible cultural heritage. The report focuses on a cultural park that transforms traditional ethnic practices into cultural goods and leisure experiences for public participation and display, drawing on fieldwork in southwest China’s Guizhou province. This article discusses the potentials and constraints of using interactive and experiential elements to facilitate the dynamic preservation of intangible heritage in creative urban environments. Edutainment is said to provide new learning opportunities for the general public and younger generations, but it also inevitably involves changes to traditional cultural forms and practices. In this setting, educational entertainment is a tactical tool to promote regional cultures while preserving traditional knowledge and skills. This article is based on the data gathered while conducting fieldwork in Guizhou in the summers of 2018 and 2019. Southwest China’s hinterland province of Guizhou has no external borders. According to official and general perceptions, a socioeconomically “remote” interior region has experienced long-term “disadvantage.” According to many of the event coordinators at the Colourful Guizhou Cultural and Creative Industry Park, the Park aims to provide an all-encompassing experience that encourages play, observation, and learning. The goal is to use first-hand knowledge and interactive design to create a close-knit environment where the public can learn about intangible cultural heritage, and social awareness will be raised. To accomplish its educational goals, the Park directs visitors to “look,” “play,” and ultimately “learn.” According to this model, the Park establishes a path for visitors to follow as they explore exhibition halls inside buildings and exterior spaces, such as public squares, and interact with the heritage content on display. Similar to the discussion of built-themed environments, this design instructs visitors to move through the space and time in order determined by the exhibits’ and performances’ placement. According to the exhibition hall, ICH was the first museum theme in Guizhou. The two Guizhou-related ICH items that have been added to the UNESCO List of Intangible Cultural Heritage, the “Grand Song of the Dong ethnic group” and the “Spring Storytelling in Shiqian” (as part of the “The Twenty-Four Solar Terms”), are introduced to tour groups and sporadic visitors at the start of their visits inside the exhibition hall, highlighting the symbolic significance of these heritage items on a global scale.

Chen (2021), in his study, uses the context of ICH in Macau. This city is listed as a World Heritage Site and looks into three typical ICH festivals to examine progressive thinking regarding the recently explored topic of tourism “experiencescape.” Over two years, interviews with 85 tourists and 30 important stakeholders were conducted to understand better the motivations behind the creation and utilisation of ICH experiencescape. Thanks to ICH’s distinctive features, a richer understanding of the local culture and people has been created, which inspires future tourist behaviour intentions. An experiencescape is regarded as a functional area encompassing a variety of influences (such as political and socio-cultural elements, relationships between supply and demand, etc.), all of which are constantly changing. When analysed, finding the helpful process by which tourism marketers, operators, and visitors work together to create such “experiencescapes” takes on new significance. Experiencescape is a comprehensive strategy involving numerous stakeholders and experience dimensions to analyse experiential goods’ production and consumption thoroughly. It must first consider the variety of stakeholders who impact experiencescape creation before examining each party's precise roles during the process. De-differentiation occurred in a variety of ways as a result of globalisation and the impact of postmodernism. Destinations must develop independent strategies for survival due to fierce competition in destination marketing. ICH offers centred on music, tradition, and history as promotional tools for raising destination awareness. Because of its commitment to maintaining ICH, Macau also best exemplifies ICH experiencescape. Since the Cultural Heritage Protection Law took effect and the Macau Government Tourism Office (MGTO) began to increase its efforts to promote ICH-related events, promoting and protecting Macau's ICH has taken precedence on the government's political agenda. Festival organisers, performers, public officials, and proprietors of nearby businesses are the primary stakeholders.

Bindhu and Panakaje (2023) mentioned in their study, "Adopt a Heritage: Apni Dharohar, Apni Pehchaan'' project, a collaboration between the Indian Ministry of Tourism, the Archaeological Survey of India, and state/UT governments, aims to promote tourism accommodations at heritage/ tourist sites throughout India in a planned and made tourist-friendly manner by incorporating India's cultural architectural heritage into mainstream development processes through corporate enlightenment. The purpose of this proposal was to inspire public and private sector businesses, organisational citizens, non-governmental organisations, private citizens, and other participants to become "Monument Mitras," attempting to assume responsibility for developing and improving primary and advanced visitor accommodations at these programs based on their involvement and survivability in terms of a CSR-compliant efficient portfolios. Additionally, participants would be in charge of maintaining and operating it. They might have been chosen as "Monument Mitras" by the ground-breaking concept of "Vision Bidding," in which the organisation with the tremendous potential for the historic site would have been allowed to connect pride to Socially Responsible activities. Additionally, their use of the Incredible India website and facilities would be restricted. The goal was to improve the visitor experience and promote incredible historic sites globally. The initiative prioritises basic amenities like cleanliness, community life quality, drinkable water, accessibility for seniors and people with developmental disabilities, highly standardized signage, illumination, and advanced amenities such as a surveillance system, night vision equipment, and a tourism enablement centre, all available and will increase domestic and international tourism. The idea has gained support from various organisations, including businesses and individuals in the public and private sectors, educational institutions and law firms. The project has so far permitted 31 organisations (referred to as Monument Mitras) to adopt a total of 95 monuments and tourist attractions spread out across India.

RESEARCH QUESTIONS

This study analysed scholarly articles on the importance of ICH in the tourism sector, the role of human resources, and, ultimately, the training and skills needed in maintaining the ICH to address research inquiries.

- What is the role of ICH in the tourism sector?
- How has the role of human capital evolved in the field of Cultural and Heritage Tourism?

- What are the critical essential skills and competencies required for individuals engaged in Cultural and Heritage (ICH) Tourism?
- How do they contribute to the sustenance and growth of ICH tourism?

METHODOLOGY

A systematic review of studies on ICH, the role of humans, and the skills and competencies required to sustain ICH gathered information from existing literature. The researchers conducted the review using appropriate criteria for inclusion and exclusion.

Research Strategy

The study searched for articles on ICH, the role of humans, and the skills and competencies required to sustain ICH. It used online databases and grey literature, focusing on Indian perspectives on skills and competencies needed for human capital development to preserve ICH.

Article Selection Criteria

The reliability and validity of systematic literature review-type studies depend on the articles' quality. A search approach identified 197 databases and 81 grey literature articles, eliminating redundant ones. Full-text articles were evaluated, and 17 research articles were selected using the PRISMA flow chart (Figure 1).

The following criteria were used to select the papers for this study to ensure the quality of the systematic review. The following are the inclusion and exclusion criteria:

- The analysis only considered papers that focused on ICH, the role of humans, and the skills and competencies required to sustain ICH.
- Articles that did not address ICH sustenance and the pivotal role of the human in the sustenance of the ICH perspective were excluded.
- The analysis excluded the thesis in favour of peer-reviewed publications from reputable journals.
- All studies, both research and review-based, were considered.
- The research articles chosen have been published from 2000 onwards, keeping in mind the paucity of research in the particular domain.

  For the study, only papers published in English were used.

FINDINGS

Role of Human Capital in Preserving ICH

Table 1.

Author(s)	Year	Type	Method	Findings
Francis-Lindsay, J.	(2010).	Exploratory - Qualitative	Literature review and Case Study Analysis	'Bottom-up' approach recommended instead of 'top-down' approach – increase in the involvement of the community in the sustainable development of tourism Building community resilience
Lenzerini, F.	(2011).	Exploratory - Qualitative	Literature review	ICH is: dynamic reinvents itself Archetypal, and "ephemeral" nature
Cominelli, F., & Greffe, X.	2012	Exploratory-Qualitative	Cross-sectional interviews (2008-2011) with policymakers, opinion leaders, and organisations working in the field of arts and crafts and craftsman	Knowledge Multitude of references Protocols for innovation Prospects for diversification Competitive pricing
Yang, Y., Shafi, M., Song, X., & Yang, R.	2018	Exploratory - Qualitative	Literature review	Poor approach towards industrialisation
Tan, S. K., Tan, S. H., Kok, Y. S., & Choon, S. W.	2018	Exploratory - Qualitative	Observations participants and non-participants), Interviews, and Analysis of textual documents	Sense of Place: 'Sense of Loss (SoL) and 'Sense of Justice (SoJ) are associated 'Sense of Justice (SoJ) and 'Sense of Mission' (SoM) are associated Awareness is the cause of: 'Revive' and 'Outcome' (associated with Sustainability) 'Revive' is a cause of 'Outcome' 'Creativity' is a source of 'Outcome and 'Sustainability' 'Challenge' is a barrier to 'Revive'
Kamal, A. M.	(2019).	Exploratory - Qualitative	A case study on Siwa Oasis (2016-2018)	Requires technical support Requires skill development Requires promotion
Kim, S., Whitford, M & Arcodia, C.	2019	Case Study - Interview	A case study on South Korea (June-July, 2016)	Majority respondents showed a positive attitude towards commodification and transformation of ICH.
Eichler, J.	(2020).	Exploratory - Qualitative	Review of Literatures Interviews	Fragmented international legal context. Requires legal understanding
Dragicevic Curkovic, M.	2021	Empirical	Structured Interviews Literature reviews	The respondents (consisting of residents of Dubrovnik) supposed traditional crafts factor of sustainable development of tourist destination
Aktürk, G & Lerski, M.	2021	Exploratory	Observations	Challenges: Relocation from homelands, coping with flood and climate change, living as immigrants, anti-immigrant policies by host communities
Lonardi, S & Unterpertinger, Y.	2022			Knowledge of language is fundamental from the pov. of a host as well as guest community

Examining the process of maintaining and developing skills and efficiencies of human capital in the sustenance of ICH:

Table 2.

Author(s)	Year	Type	Method	Findings
Weiler, B., & Ham, S. H.	2002	Experimental	A six-week training programme was dclivcrcd in Panama (with two shorter interpretive guiding courses) and two different places (Patagonia and Galapagos) and then evaluated in multi-level	Understanding of roles and responsibilities as a tour guide Understanding of how to maintain sustainability by minimising impact Understand visitors' profile and their expectations Communication and interpretation Effective customer service in terms of welcome address and orientation Leading and managing in guided tour Cross-cultural communication and understanding of special groups
Christie, M. F., & Mason, P. A.	2003	Review	Literature Review	Tour guide training must be a part of adult education There is a need for training the tour guide Training must lead to a change in the behaviour of the tour guide
Tsaur, S. H., & Tu, J. H.	2019	Empirical -Mixed (Qualitative and Quantitative)	Tour guides were interviewed to understand the factors, and then items were pooled, and a questionnaire was floated among them	The tour guide must be aware of the following: Local transportation Intercultural communication Recognizing local culture Language proficiency Local customs Cultural backgrounds Cultural awareness Cultural kinship Cultural harmony Adaptation to culture
Lou, Yu.	2020			Ethnic minorities are geared toward making their place Commodifying cultural resources
Chen, Z.	2021			Spaces where people meet and gain knowledge about the place
Bindhu, D., & Panakaje, N.	2023	Exploratory - Qualitative	Reviews of Literatures	Intellectual skill Interpersonal skill Individual Marketing skills

DISCUSSIONS

The study's findings also relate to the following case studies where the locals have been trained to showcase their traditions (handicrafts, culture, etc.) to sustain in the society.

Nurturing Human Capital for Sustainable ICH Tourism (Case Studies)

Case Let 1: Tribals of Thailand

The Local Aroi project aims to empower and train indigenous communities while promoting Thai cuisine. To give customers the authentic taste of Thailand, Local Aroi offers native cuisine from 5 to 10 native communities. Their objective is to assist the locality in producing income. Along with the social enterprise Local Alike, which focuses on ecotourism, and renowned Thai Chef Khun Book (Boonsmith Phukganasuk), who is well-known for his mouth-watering fusion dishes and Food TV shows, Local Aroi developed a fine-dining establishment that offers genuine indigenous menus to customers. Chef Khun Book also teaches the locals how to improve their culinary abilities. In their company, there are also Chef teams that create new dishes using ingredients from the community. They frequently take photos with members of the indigenous group to reach a wider audience and become more well-known. Because everyone was confined to their homes during the COVID-19 pandemic, Local Aroi launched their delivery service, significantly impacting their business. After receiving training, local community members deliver the food as well. Pha Mee Community, Prom Lok Community, Khok Muang Community, and Koh Klang Community are a few of the participating/partnering communities.

Case Let 2: Tribals of Odisha

The state of Odisha is known as the home of handicrafts, an essential component of our cultural heritage. Around 1.30 lakh artisans are dispersed throughout the state and engage in up to 50 different crafts to varying degrees. The State has the best craftsmanship in the nation and the most artisans who have received national awards, such as Padma Awards. Tribal tourism is a significant aspect of Odisha tourism. There are 62 tribal communities in Odisha. Storytelling is a unique aspect of the tribal culture; they have rich traditional food. They also have different varieties of rice. There are members of the community who guide the tourists on their own and tell the history of their tribes. In addition to paper mache, shola pith work, lacquer work, zari work, glass beads, cloth garlands, camphor garlands, jute carpets, rush mate, and, most notably, sea shell crafts manufactured in Puri and Gopalpur, Odisha produces a vast range of handcrafted goods.

The districts of Koraput, Kalahandi, Balasore, and Bolangir are where the paddy craft of Odisha is still practised. The Gond tribes have specialised in agriculture as their primary trade. Rice is a symbol of wealth, and because of this, the goddess Lakshmi is frequently created by artists. The bulk of created items are related to religion in some way. Representations of the Goddess Laxmi, elephants, chariots, tribal wedding crowns, representations of temples, hand fans, and baskets are among the intricately pieced-together items. Due to its religious significance, this craft was formerly prevalent. Today, only 15 to 20 households still practice it, and that too only during the poor agricultural months. Although some artists can access metropolitan craft bazaars, most products are sold in neighbourhood hats (markets). The state government assists the tribal groups in Odisha to develop their handicraft.

Case Let 3: Handicrafts of Gujarat (Kevadia)

The inauguration of the Statue of Unity took place in Kevadia. The government is distinctively developing Kevadia, with everything from a butterfly garden to a unity glow garden, a valley of flowers to a cactus garden and an ekta nursery, a jungle safari to a dino trail park. Khalwani and Zarwani offer eco-tourism tours to take in the area's natural beauty. Adventure activities like River Rafting, Nauka Vihar, and cycling tours between Zarwani and Khalwani exist. There is a valley of flowers and a unity glow garden there, but since it is a tribal area without nightlife, artificial attractions are promoted there to increase the nightlife.

A couple of families are producing this Rogan art, and the word "Rogan," which means "all faiths," is Persian. PM Modi presented A Rogan painting to former US President Obama during their meeting.

This art gained popularity following that act of kindness, and people began to travel to the artisans' villages. The word "Mushru," which in Arabic means ``prohibited," refers to a type of weaving. The Mashru weaving originated from the notion that wearing silk close to the body was wrong in Islam. Where two different threads are intertwined: the cotton threads are on the inside, and the silk threads are on the outside. When the fabric is worn, it helps to regulate body temperature by absorbing sweat due to the cotton layer inside and silk layer outside, which helps to keep you warm in the winter. A tribal cafeteria is located in the Ekta Nursery. There are tribal women employed there from the nearby villages.

Case Let 4: Sonajhuri Haat Kolkata (New Town)

For over a month now, artisans from all over West Bengal have gathered every weekend at the haat in New Town, more akin to an artisans' hub, to display their wares. Hours are 11 am to 9 pm from Friday through Sunday. The original haat of Shantiniketan of the same name inspires the Haat's name. This is a platform where the respective communities accumulate handicrafts from diverse communities. Through this mela ICH, different people are getting deserved attention and acknowledgement along with a way of earning. This mela may emerge as a tourism spot soon as it has shown much potential within such a short period.

At the New Town Community Zone, directly below the bridge over the Axis Mall, the mela has a permanent location. Khadi cotton sarees, embroidered kurtis, artsy cloth bags, hand-painted neckpieces, earrings, bangles, rings, dokra, and copper statement jewellery items are now housed in the decorated area. The Creative Bengal Foundation's fair, or haat, features a sizable selection of cane home decor items, including baskets, vases, sling bags, laptop sleeves, wallets, and more. Trays, plates, and spatulas are among the wooden products, while saucer and plate sets, wind chimes, glasses, candle holders, and other items are available in a wide range of ceramic products. The fashionable jute juttis and chappals are also very striking, along with wall decor pieces to wooden tables, designer clocks, wooden showpieces like owls, and lawn chairs. Additionally, some artists at the Sonajhuri Haat create pattachitra and paint T-shirts. The haat offers everything from 'bori' to 'banshi'. In a single location, one can locate "goyna bori," "narkel naru," and "ektara." The haat is where people can find many stories while listening to classic Bengali music. Many sides of the artists, some of whom are from Nadia, Dhakuria of Kolkata, Midnapore, and South 24 Parganas, could be seen, from open camaraderie to quiet practice.

Theme Identified from the Discussions

In the role of human capital in preserving ICH, themes have been generated such as: 'requires', 'community', 'legal', 'approach', 'development', 'sustainable', and 'sense'. The developmental approach of the community is required. Moreover, to get sustainable ICH, awareness or sensitisation is also required (Image:1).

The themes have been generated on the community or local people's need to develop their 'understanding ability and 'empathy' through training, which helps in the sustenance of ICH (Image:2). As much as they can empathise with their own as well as others, they can sustain their tradition by showcasing their talents to the globe.

To sum up, it can be said that human capital is crucial to the preservation and protection of ICH and that some action is required to maintain this role. For instance, the case studies presented in this paper demonstrate how locations develop into tourism hotspots and how to do so. The local communities that are the ICH bearers of that location are being encouraged by the government, NGOs, or other groups or individuals who are merely interested in the preservation of ICH and in supporting its bearers.

Challenges Faced by Cultural Bearers -A Critical Assessment

Cultural appropriation

The inappropriate or unacknowledged adoption of an element or elements of one culture or identity by members of another is known as cultural appropriation. According to Cambridge University, cultural appropriation is taking or utilising artefacts from a culture that is not your own, especially without demonstrating that you understand or appreciate that culture. The World Heritage process is shallow since it does not address underlying injustices meaningfully. Heritage commodification and the application of dissonance theory to heritage interpretation and management (Fraser, 1981; Tunbridge & Ashworth, 1996; Salemink, 2012). According to Hoffman and Wendland (2006), "Traditional knowledge and cultural manifestations are frequently the result of intergenerational and fluid social and communal creative processes that reflect and identify a community's history, cultural and social identity, and social values".

Eichler (2020) in her study focused on the cultural identity negotiations of marginalised groups during the Andean Carnival celebrations in Oruro and Barranquilla, examining the difficulties they faced. The review critically analysed how these celebrations help indigenous peoples, people of African descent, and other marginalised communities negotiate their cultural identities. The study assessed the spaces produced by these celebrations as venues for inclusion and identity negotiation in the face of external pressures and potential homogenisation. It evaluated how these spaces allow marginalised groups to assert their distinctive cultural identities, confront historical marginalisation, and reclaim control over their representation. The study investigated how intricate and multifaceted the changing landscape of indigenous and minority cultural rights is within the context of ICH. Continued vigilance is necessary to ensure that outside forces like tourism, commodification, and digitalisation do not undermine marginalised communities' genuine representation and agency in defining their cultural heritage.

Acts of appropriation occur around us in a wide range of creative disciplines when cultural influences combine, integrate, and synthesise. Intellectual property rights and ownership difficulties are numerous and complex, particularly in the tourism industry, where culture is firmly interwoven in the trip experience. Some will claim that tourism would not exist without culture. Local intangible cultural assets boost the tourism industry considerably. When a community evolves its collective and social value system - a social construct - intangible and tangible cultural components are acquired and reproduced. However, these cultural characteristics are turned into economic products for exchange through tourism. The tourism industry's refusal to recognise the monetary value of local intangible cultural heritage in tourism production and to appropriately reward local producers and community-based innovators is a severe moral and ethical challenge. Developing a more egalitarian and sustainable strategy that benefits all stakeholders is a logical next step. (Ziff & Rao, 1997; Jafari, 1996; Wanda et al., 2010)

Sustainability

The emphasis on sustainability in ICH illustrates the value of utilising natural and eco-friendly materials, encouraging the use of organic fabrics like cotton or hemp among consumers, and minimising the use of synthetic materials that cause pollution. Additionally, ICH employs sustainable practices that include reducing waste through recycling and upcycling methods, encouraging a circular economy mindset within the sector, and promoting the maintenance of conventional craftsmanship methods. The ICH industry can contribute to a more sustainable future for the environment and the communities involved by incorporating sustainable practices, such as waste reduction and using renewable energy sources, and encouraging a sense of responsibility for maintaining cultural heritage. Due to its potential to promote meaningful visitor experiences, advance environmental and cultural awareness, and boost the economic prosperity of host communities, the role of training interpretive tour guides in sustainable tourism growth has attracted growing attention. The training programmes promoted sustainable tourism practices by working with local stakeholders and experts. They were designed for each destination's needs and contexts (Weiler & Ham, 2002). The Tz'utujil people in Guatemala were relocated to Chuk Muk, chosen by local stakeholders. Cultural, ethnic, and social considerations were considered during the relocation's planning and execution, such as when designing homes to accommodate large families. Additionally, taken into account during the process were "indigenous cosmic worldview, ethics, and understanding of nature and disasters. (Aktürk & Lerski, 2021)

Community involvement

This paper aims to shed light on the community's involvement in safeguarding and preserving heritage passed on to them through generations. The literature shows that communities actively participate in various activities to preserve the cultural heritage, and governments are also providing opportunities to those stakeholders, aiding their development. On the one hand, man was an object, like every other object in the natural world, committing to the arbitrary rules of physical laws. On the other hand, man was a subject, an agent with the singular ability to comprehend and change his worldly condition to become more fully, more essentially, himself Foucault.

Globalization

To comprehensively analyse experiential goods' production and consumption, "experiencescape" is a comprehensive strategy that must involve numerous stakeholders and experience dimensions. Before analysing each party's precise roles during the process, it must consider the various stakeholders who impact experiencescape creation. The effects of postmodernism and globalisation led to various forms of de-differentiation. Due to the intense competition in destination marketing, destinations must develop independent strategies to survive. (Chen, 2021)

Tan et al. (2018) observed the components of "person-place-bonding" and how people, places, and the bonds they form with them contribute to the sustainability of intangible cultural heritage and heritage tourism. Three themes were found in the study: "Sense of Loss," "Sense of Justice," and "Sense of Mission." Two key codes served as the study's defining characteristics: "awareness," which fosters responsibility for intangible cultural heritage, and "creativity," which guarantees the study's relevance and authenticity for the heritage ecosystem of WHSs.

Policy & Governance

The safeguarding practice Safeguarding Strategy of Traditional Crafts for Peace Building in Colombia was chosen for this list because it aims to teach future generations about building crafts and skills through a workshop and by putting traditional crafts policy into place.

The promotion and protection of Macau's ICH have taken precedence on the government's political agenda ever since the Cultural Heritage Protection Law was enacted and the Macau Government Tourism Office (MGTO) started to step up its efforts to promote ICH-related events. The main participants are the festival's organisers, performers, government representatives, and owners of nearby businesses. (Chen, 2021)

According to the paper, the "Adopt a Heritage: Apni Dharohar, Apni Pehchaan" project, a collaboration between the Indian Ministry of Tourism, the Archaeological Survey of India, and state/UT governments, aims to promote tourism accommodations at heritage/tourist sites throughout India in a planned and made tourist-friendly manner by incorporating India's cultural architectural heritage into mainstream development processes through corporate enlightenment.

Education & Awareness

According to ICH practitioners, transmitting genuine ICH and promoting ICH as a tourism resource can potentially have a beneficial symbiotic relationship. It is crucial to spread awareness of ICH among locals, especially the younger generation, as its survival is at risk if locals do not recognise and use it. It is crucial to spread ICH and increase its economic, social, and cultural value. Some ways to do this include hosting events, festivals, and performances, creating ICH hubs or centres, and running required educational programmes. (Kim et al., 2019)

BREAKTHROUGHS FROM RESEARCH QUESTIONS

RQ 1: What is the role of ICH in the tourism sector?

Humans (community/local people) play a critical role in ICH in the tourism sector, which is very evident (Francis-Lindsay, 2010; Lenzerini, 2011; Cominelli, & Greffe, 2011; Cominelli, & Greffe, 2012; Yang, et al., 2018; Tan, et al., 2018; Kamal, 2019; Eichler, 2020; Dragicevic, 2021; Weiler, & Ham, 2002; Christie, & Mason, 2003; Tsaur, & Tu, 2019; Bindhu, & Panakaje, 2023)

RQ 2: How has the role of human capital evolved in the field of Cultural and Heritage Tourism?

Since ICH is a nascent concept added to the tourism sector, earlier, it was believed in the concept of a tour guide. A tour guide can be from that community or can be from external. However, nowadays, it is evident that the community must be trained with the requisite skills to sustain ICH. Since ICH is itself a tangible and intangible resource. The community's cultural heritage plays a significant role in preserving intangible cultural heritage. The community people need to showcase their cultural heritage and their talents, and hence, their active involvement is crucial. (Francis-Lindsay, 2010; Dragicevic, 2021; Weiler & Ham, 2002; Christie & Mason, 2003; Tsaur & Tu, 2019; Bindhu, & Panakaje, 2023)

RQ 3: What are the critical essential skills and competencies required for individuals engaged in Cultural and Heritage (ICH) Tourism?

The critical skills and competencies required for individuals engaged in ICH Tourism are understanding of their own culture, understanding cross-culture, communication skills, service, understanding the importance of minimising the use of resources, and their marketing and leadership skills. (Weiler & Ham, 2002; Tsaur, & Tu, 2019; Bindhu, & Panakaje, 2023)

RQ 4: How do they contribute to the sustenance and growth of ICH tourism?

Creativity Innovations attract people everywhere, as they offer unique and refreshing experiences. These innovations can range from groundbreaking technological advancements to artistic expressions that challenge traditional norms. People are naturally drawn to creativity as it sparks curiosity, ignites inspiration, and opens up new possibilities for personal growth and societal progress.

From a marketing perspective, customers are attracted to crafts and handwork because these aspects showcase the unique cultural heritage of a community. Crafts and handwork demonstrate the skills and traditions passed down through generations, preserving cultural identity. Additionally, the focus on sustainability in ICH highlights the importance of using natural and eco-friendly materials, promoting environmental consciousness among customers by encouraging organic fabrics, such as cotton or hemp, and minimising synthetic materials that contribute to pollution. Furthermore, sustainable practices in ICH also reduce waste through recycling and upcycling techniques, foster a circular economy mindset within the industry, and encourage the preservation of traditional craftsmanship techniques. Incorporating sustainable practices into ICH ensures the longevity of cultural heritage and contributes to our planet's overall well-being for future generations. Moreover, they are fostering a sense of responsibility towards preserving cultural heritage. By incorporating sustainable practices, such as reducing waste and implementing renewable energy sources, the ICH industry can contribute to a more sustainable future for the environment and the communities involved.

When community people actively participate and are involved with the environment, the importance of using sustainable and innovative products increases (Wongmasa, 2020, 1:06:21); (Rao, 2020, May, 59:11), (Brar, 2020, August, 57:58), and (Mitra, 2023, July) as well as the sustenance and growth of the community and ICH increase.

CONCLUSION

This study seeks to fill this knowledge gap by examining both the critical skills and competencies required for the growth of a thriving and long-lasting ICH tourist business, as well as the process of retaining and growing talented and productive human resources with a holistic purview on the numerous characteristics of human capital in ICH tourism by compiling insights from a wide range of scholarly studies. The conclusion demonstrates that the human aspect is the primary source of protecting ICH. Humans create intangible cultural heritage, passed down from generation to generation, and protected from extinction by humans. Several questions arose from the research. What part does ICH play in the travel and tourism industry? What role has human capital had in the growth of cultural and heritage tourism? What core skills and knowledge are required of those working in the cultural and heritage (ICH) tourism industry? How do they support and advance ICH travel and tourism? In the knowledge economy, intangible variables play a crucial role in inventing and manufacturing new items. On the other hand, they inspire creativity and give various references regarding signs, forms, colours, and symbols as a source of continuously evolving legacy for all economic sectors. The arts and heritage can offer norms or protocols for innovation that can be employed in other industries because they are creative endeavours. The global economy, on the other hand, improves chances for diversification by expanding the markets for some goods. Fundamentally speaking, ICH is far more active than material heritage. On the one hand, it enables ICH to transform itself continuously to reflect the cultural character of its creators and owners. Due to its intrinsic ability to transform and mould qualities parallel with the communities' continuing cultural development, such heritage can reflect the communities at any time. Because of its intrinsic flexibility, or, to put it another way, its "ephemeral" nature, ICH, on the other hand, is particularly prone to being absorbed by the archetypal cultural forms that are trendy at any given time. The results also demonstrate that discourses around cultural heritage have primarily developed within a disjointed international legal setting. ICH regimes nevertheless fall short when addressing victim-oriented legal forums, despite criteria for (tangible) cultural heritage being able to shed their environmentalist, preservationist position.

Theoretical Implications

Combining knowledge from various scholarly works, this paper aims to provide a comprehensive perspective on the multifaceted dimensions of human capital in ICH tourism. It does this by investigating both the crucial skills and competencies necessary for the cultivation of a vibrant and enduring ICH tourism industry, as well as the process of preserving and developing skilled and efficient human resources. The Intangible Cultural Heritage (ICH) was defined by Lenzerini (2011) as a collection of intangible cultural expressions that stand in for humanity's rich and varied living heritage. In addition to being closely related to human rights, it is essential for maintaining cultural diversity. Intangible cultural heritage (ICH), which is rooted in the self-identification of heritage creators and bearers and reflects historical and social shifts, was the focus of the 2003 UNESCO Convention on the Safeguarding of Intangible Cultural Heritage. However, its practical application, which follows the 1972 World Heritage Convention, appears insufficient to protect ICH adequately.

According to Aktürk and Lerski (2021), as people are uprooted from their traditional lands, climate-induced displacement threatens intangible heritage in affected communities, endangering generations-deep ties to rituals, customs, ancestral connections, and cultural practices. Despite the apparent loss of land and infrastructure, climate change frequently ignores intangible heritage, significantly eroding cultural traditions and practices.

Practical Implications

This article highlights the various skills and competencies of the stakeholders that are protecting their traditional heritage or Intangible Cultural Heritage. The stakeholders are artisans, craftsmen, tour guides, and so on. They are provided different training programs to enhance their skills and educate the tourists about their responsibilities to their peers. Christie and Mason (2003) argued that tour guides should be taught to critically evaluate their knowledge, attitudes, and behaviour to improve their interpretations. This would allow them to spot biases, knowledge gaps, and areas for growth. The article proposed a new model for guide training to improve tour guide quality, benefit the tourism industry, and ensure the sustainability of tourism sites. This model included workshops on critical thinking, training in cultural sensitivity, in-depth subject knowledge, practical communication skills, and ongoing education. In addition to engaging guests through storytelling, guides should participate in critical thinking workshops and respect various cultural traditions. There are currently 138 national ICHs in South Korea, each designated by the Cultural Heritage Association, a public organisation. Korea also has a well-organized ICH apprenticeship programme. There are four levels of the practitioner training system: Beginner, Advanced, Assistant Instructors, and Living Human Treasures (Kim. et al., 2019). A combination of educational and entertaining activities, "edutainment" plays a role in preserving intangible cultural heritage. Drawing on fieldwork conducted in the Guizhou province of southwest China, the article focuses on a cultural park that converts traditional ethnic practices into cultural goods and leisure activities for public participation and displays. (Lou, 2020)

Limitations and Future Research Direction

This study is qualitative because it was carried out by examining prior works. The local community members involved in ICH were not directly engaged, so the paper could not provide any independent information.

Since few articles were accessible for the study and its scope was narrow, the research is not absolute. Future research might be done in this area to learn more about ICH from the viewpoint of the residents and how new ways can be used to help the community grow

REFERENCES

Aktürk, G., & Lerski, M. (2021, May 8). Intangible cultural heritage: A benefit to climate-displaced and host communities. *Journal of Environmental Studies and Sciences*, *11*(3), 305–315. doi:10.100713412-021-00697-y PMID:33996378

Bindhu, D., & Panakaje, N. (2023). Adopt a Heritage Scheme: A CSR Initiative for Preserving our Past Glories to Future. [IJCSBE]. *International Journal of Case Studies in Business, IT, and Education*, *7*(2), 250–262. doi:10.47992/IJCSBE.2581.6942.0272

Brar, R. (2020, August 3). *Heritage Tourism In Gujarat* [Video]. Digital India Learning. YouTube. https://www.youtube.com/watch?v=A4i8q_6pRUQ

Chen, Z. (2021, February 26). Visualizing experiencescape – from the art of intangible cultural heritage. *Current Issues in Tourism*, *25*(4), 559–578. doi:10.1080/13683500.2021.1892040

Christie, M. F., & Mason, P. A. (2003). Transformative Tour Guiding: Training Tour Guides to be Critically Reflective Practitioners. *Journal of Ecotourism*, *2*(1), 1–16. doi:10.1080/14724040308668130

Cominelli, F., & Greffe, X. (2012). Intangible cultural heritage: Safeguarding for creativity. City. *Cultura e Scuola*, *3*(4), 245–250. doi:10.1016/j.ccs.2012.10.003

Dragicevic Curkovic, M. (2021). The Role of the Traditional Crafts as Intangible Heritage on the Global Tourist Market. *SHS Web of Conferences*. Ecosciences. 10.1051hsconf/20219206005

Eichler, J. (2020). Intangible cultural heritage, inequalities, and participation: Who decides on heritage? *International Journal of Human Rights*, *25*(5), 793–814. doi:10.1080/13642987.2020.1822821

Folklife at the International Level: Intangible Cultural Heritage Defined (2020, January 31). *Folklife Today*. https:////blogs.loc.gov/folklife/2020/01/folklife-at-the-international-level-intangible-cultural-heritage-defined

Francis-Lindsay, J. (2010, June 28). *From fashion to 'tangible-intangible' action: local communities 'culturizing' new tourism development. WIT Transactions on Ecology and the Environment*. WIT Press. doi:. doi:10.2495/st100421

Fraser, N. (1981). Foucault on modern power: Empirical insights and normative confusions. *Praxis International*, *3*, 272–287.

Jafari, J. (1996). Tourism and culture: an inquiry into paradoxes. In *Culture, Tourism Development: Crucial Issues for the XXIst Century*. UNESCO.

Kamal, A. M. (2019, February 28). Intangible Cultural Heritage as a Tool for Community Empowerment: A Case Study of the Date Palm Festival in Siwa Oasis, Egypt. *Ottoman Journal of Tourism and Management Research*, *4*(1), 361–377. doi:10.26465/ojtmr.2018339516

Kim, S., Whitford, M., & Arcodia, C. (2019, February 19). Development of intangible cultural heritage as a sustainable tourism resource: The intangible cultural heritage practitioners' perspectives. *Journal of Heritage Tourism*, *14*(5–6), 422–435. doi:10.1080/1743873X.2018.1561703

Lenzerini, F. (2011). Intangible Cultural Heritage: The Living Culture of Peoples. *European Journal of International Law*, *22*(1), 101–120. doi:10.1093/ejil/chr006

Lonardi, S., & Unterpertinger, Y. (2022). The Relevance of Intangible Cultural Heritage and Traditional Languages for the Tourism Experience: The Case of Ladin in South Tyrol. *Sustainability*, *14*(5), 2729. doi:10.3390u14052729

Luo, Y. (2020, June 15). Safeguarding intangible heritage through edutainment in China's creative urban environments. *International Journal of Heritage Studies*, *27*(2), 170–185. doi:10.1080/13527258.2020.1780463

Mitra, P. (2023, July, 2). In pictures: Sonajhuri Haat — a slice of Santiniketan in the heart of New Town, Saris, jewellery, trinkets, paintings – this Friday-to-Sunday (af)fair with a 'mela' vibe is Kolkata's newest weekend shopping destination. *My Kolkata/ Telegraph India/ The Telegraph Online*. https://www.telegraphindia.com/my-kolkata/places/kolkatas-sonajhuri-haat-a-slice-of-santiniketan-in-the-heart-of-new-town-photogallery/cid/1949245?slide=1

Rao, S. (2020, May 12). *Odisha – India's Best Kept Secret!* [Video]. Digital India Learning. YouTube. https://www.youtube.com/watch?v=iKsvlx_Vr5k

Rosengren, K. E. (1993, September 1). From Field to Frog Ponds. *Journal of Communication*, *43*(3), 6–17. doi:10.1111/j.1460-2466.1993.tb01271.x

Salemink, O. (2012). Appropriating culture: The politics of intangible cultural heritage in Vietnam. In *State, society and the market in contemporary Vietnam* (pp. 158–180). Routledge.

Tan, S. K., Tan, S. H., Kok, Y. S., & Choon, S. W. (2018). Sense of place and sustainability of intangible cultural heritage – The case of George Town and Melaka. *Tourism Management*, *67*, 376–387. doi:10.1016/j.tourman.2018.02.012

Tilden, F. (1957). *Interpreting our heritage: Principles and practices for visitor services in parks, museums, and historic places*. University of North Carolina Press.

Tsaur, S. H., & Tu, J. H. (2019). Cultural competence for tour leaders: Scale development and validation. *Tourism Management*, *71*, 9–17. doi:10.1016/j.tourman.2018.09.017

Tunbridge, J. E., & Ashworth, G. J. (n.d.). *(1996). Dissonant Heritage: The Management of the Past as a UNESCO - Intangible Heritage Home*. Press., https://ich.unesco.org/en/home.University

Wanda George, E. (2010). Intangible cultural heritage, ownership, copyrights, and tourism. *International Journal of Culture, Tourism and Hospitality Research*, *4*(4), 376–388. doi:10.1108/17506181011081541

Weiler, B., & Ham, S. H. (2002). Tour Guide Training: A Model for Sustainable Capacity Building in Developing Countries. *Journal of Sustainable Tourism*, *10*(1), 52–69. doi:10.1080/09669580208667152

Wongmasa, N. (2020, May 9). *AEN E-learning 05 on Covid-19: Adaptation in Thai Ecotourism sharing cases from TEATA* [Video]. Asian Ecotourism Network AEN. YouTube. https://www.youtube.com/watch?v=oGrWWDCq0gQ

Yang, Y., Shafi, M., Song, X., & Yang, R. (2018). Preservation of Cultural Heritage Embodied in Traditional Crafts in the Developing Countries. A Case Study of Pakistani Handicraft Industry. *Sustainability*, *10*(5), 1336. doi:10.3390u10051336

Ziff, B., & Rao, P. V. (Eds.). (1997). Borrowed Power: Essays on Cultural Appropriation. Rutgers.

Chapter 17
Virtual Museums in the Conservation of Cultural Heritage and Satisfaction Level With Visits to Virtual Museums

Funda Varnaci Uzun
https://orcid.org/0000-0002-4915-9894
Aksaray University, Turkey

ABSTRACT

In this part of the book, the importance of virtual museums and their place in the conservation of cultural heritage and cultural heritage tourism will be emphasized. In this part, first, the concept of museum, the emergence and development of virtual museums, the importance of museums and virtual museums in the conservation and cultural heritage tourism and transfer of cultural heritage to future generations, the importance of virtual museums in terms of education will be conceptually explained. The increasing importance of virtual museums during the Covid-19 pandemic will also be mentioned by giving examples around the world. Within the context of the current study, students attending Aksaray University (Turkey) will be made to visit virtual museums of various museums around Turkey and the world. After the visit, the Virtual Museum Satisfaction Scale will be administered to the students. The data to be collected through the scale will be evaluated by using descriptive analysis and difference analyses applied in relation to different variables.

INTRODUCTION

Museums are one of the most important places in a society that preserve cultural heritage. Museums are not only places where culture is protected and displayed and is conveyed to future generations but also institutions that establish communication with people, generate ideas for society, preserve knowledge for generations and contribute to the development of society. Museums have an important mission in transmitting both tangible and intangible cultural heritage to future generations (Batuhan, 2023).

DOI: 10.4018/978-1-6684-9957-3.ch017

The rapid development of technology has also had an impact on the understanding of the concept of museum. With the technological advantages brought by the information age, the works and objects displayed in museums have gained a digital characteristic. As a result of the emergence of virtual museums, museums have been able to exhibit their collections in electronic environments, enabling them to reach a wider audience (Çınar et al., 2021). Internet technologies that have improved the use, sharing and access of information have also affected museums. With these developments, the "cultural heritage artefacts" in museums, which are among the world's cultural assets, have been presented in a visually enhanced format and made available for the common use and sharing of humanity. In this way, museums have transformed into universal exhibition halls in a virtual environment (Peker, 2014). We are living in the age of Industry 4.0. Its impact in the field of Cultural Heritage is under our eyes and is still to be assessed. The potentialities of Virtual Reality (VR) and Mixed Reality (MR) technologies are undoubtedly opening new perspectives for experiencing cultural heritage in terms of visualization, representation, interaction, exploitation, enhancement and dissemination. With particular reference to museums and cultural and historical sites, the benefits to the Creative Industry sector can be envisaged in the possibility of addressing new audiences, introducing new means of interaction, creating new engaging stories to tell, changing and innovating our perception and understanding of cultural assets (Aiello et al., 2019). The concept of a virtual museum refers to an information system having the characteristics of a museum and including electronically consolidated collection of objects. Virtual museums can include unique and rare objects of cultural heritage that are important for preservation, documentation and representation for public use (Povroznik, 2018). Virtual tours offered by virtual museums to their visitors through online applications are created by combining panoramic images (Alav, Altıngövde, Kaplan, 2006; Kalıncı, 2015). These images are obtained by merging several camera shots into a single 360-degree image, using wide-angle (fisheye) lenses, or by capturing images with the help of curved reflective surfaces (Sungur & Bülbül, 2019). Virtual museums offer a "parallel reality" to their users. The meaning of the word "virtual", which comes from the Latin root "virtus", means "decision-making ability", "potential" and "freedom", but today it means "information and communication". If we look at it from a different perspective, the cybernetic spaces created in virtual museums are the mutation of the real space in a representational but not actual universe. That is, the real space undergoes a metamorphosis and transforms. It becomes visual by turning into an algebraic expression and creates a virtual reality. The user has uninterrupted access to virtual museums from anywhere in the world (Vargün, 2020). The idea of a virtual museum was first introduced by André Malraux in 1947 as a museum without walls, location and spatial boundaries. A virtual museum is a collection composed of digitally recorded images, audio files, text documents and other data of historical, scientific or cultural significance, accessed through electronic media. Virtual museums, developed as an extension of physical museums, are examined in 3 categories: Brochure museum, Content museum and Educational museum. The brochure museum aims to inform future visitors about the museum and is used as a marketing tool with basic information such as location, opening hours and events calendar etc. in order to create motivation to visit. The content museum is a website created to provide information about museum collections. It is a database that contains detailed information about museum collections by presenting the content in an object-oriented manner. The educational museum is a website that offers different access points to its virtual visitors depending on their age, background and knowledge. Information is presented in a context-oriented rather than object-oriented manner. The goal of the educational museum is to make the virtual visitor come back and establish a personal relationship with the online collection (Styliani et al., 2009).

Virtual museology is developing today as a new museology approach that serves the function and content of contemporary museology in the most beautiful way. These museums can make their collections and their exhibitions more easily accessible to wider audiences (Kahraman, 2021). A key component of virtual museum tours is a controllable viewing window. Users can easily and interactively navigate the entire location by moving as close to or far away from the viewing window as they want. Users can look in any direction in an app, see a full circle of space, and navigate by clicking on specific points. Web-based virtual tours provide a different experience to visitors like museums and other cultural heritage institutions using virtual reality and augmented reality technologies. Through virtual tours, visitors have the opportunity to visit and explore the museum in any environment they wish, while it is also possible to influence their decision to visit the museum on-site after the virtual visit. In this sense, virtual museum tours are also considered as an important marketing communication tool (Aytekın & Aktaş, 2023). Visitors to virtual museum exhibitions are neither passive nor deprived of opportunities to develop their critical skills. A virtual museum provides visitors with the freedom to explore and become active participants by creating their own virtual tours and paths. In the virtual museum environment, the visitor is not only an observer, but also constructs the knowledge himself/herself by interacting with the objects. Museum visitors use the virtual museum environment and engage in interactive dialogue through which they gain access to thematic information and explanations about museum objects (Styliani et al., 2009). In a virtual exhibition, communication with the virtual guide, museum experts, or other visitors is an essential communication factor. Furthermore, virtual visitors and museum experts may easily cooperate and share information by creating a group or platform in the virtual exhibition environment. As several users can at the same time join the virtual exhibition through a chat system or avatars, users can communicate and deliberates on exhibitions. For example, the Virtual Egyptian Museum provided a rich museum experience to people living in areas remote from cultural centers (with a particular attention to schools) and uses the collection of antiquities as the basis for an educational journey in egyptology and history (Kamariotou et al., 2021).

The first efforts to establish a virtual museum in Turkey began in 1990 with the attempt to move some collections in the Topkapı Palace to the virtual environment. The aim of this project, known as the Topkapı Palace project, is to provide easy access to photographs, reliefs, drawings or animations. In 1996, the "Sabancı Paintings and Sculpture Collection" owned by Sakıp Sabancı was transferred to virtual environment. This was followed by the transfer of "Rahmi Koç Museum", which is a private museum, to the virtual environment in 1997 (Peker, 2014). Today, many private museums in Turkey and museums affiliated to the Ministry of Culture can be visited virtually.

BACKGROUND

Cultural heritage is a bridge that connects the past and the present, has significant social, economic, scientific and educational value that can activate national pride, social belonging and belief in life, and plays an important role in regional stability and social development (Yang et al., 2021). Cultural heritage encompasses objects, places or practices that have a connection to the society's past traditions and history, and they are not only important for the present but they also are desired to be transferred to future generations. Cultural heritage is evidence of the tangible and intangible qualities of creative effort passed down from past generations. It is highly valuable as the source of a society's memory and the formation of its local and national identity (Batuhan, 2023).

Museums collect cultural heritage items, catalogue them into collections and serve the society by presenting them. In this way, the elements that constitute the cultural heritage of a society on a small scale and the cultural heritage of all humanity on a large scale can be seen in museums. Museums, centres where cultural heritage items, whether ancient or contemporary, are collected, preserved and exhibited, have evolved from the practice of individuals collecting valuable items and have transformed over time to become what we see today. The number and diversity of museums, which serve the purpose of acquiring and preserving artworks while also educating society and individuals, have increased over time due to the proliferation of the groups of artworks to be exhibited and the collection of artworks for various purposes (Bulut & Uzun, 2021).

The preservation, perpetuation and transmission of cultural heritage as a legacy to future generations have become a global policy since the past century, marked by national and international agreements, laws, regulations and directives. Success in this endeavour is considered one of the indicators of modernization. Sustainability in preservation depends on the awareness, consciousness and education of the individual who is the key factor in the preservation process. The conditions of the information age bring technology to the forth as a tool for fostering awareness and consciousness in this regard. Virtual reality, which is predicted to significantly expand its boundaries in the near future, comes to the fore as one of the prominent technologies in this regard. In addition to various disciplines, virtual reality technologies are also widely used in the field of cultural heritage (Barlas Bozkuş, 2014).

Museums play an important role in the preservation and transmission of cultural assets to society. In today's world, where computer technologies have advanced significantly, utilizing these technologies is considered an important opportunity for promoting and transmitting cultural heritage elements, especially to younger generations (M. Yıldırım, 2021). The preservation, repurposing, presentation and transmission of historical structures considered as cultural heritage is a lengthy and intricate process that involves a variety of methods and techniques. In the efforts made for documentation, restoration, reconstruction, restitution and presentation, virtual reality methods and systems are primarily employed (Erbaş & Tuncel, 2022). The presentation and preservation of both tangible and intangible cultural heritage have significantly advanced in modern times through the use of information and communication technologies (ICT), especially with the utilization of computer graphics and multimedia (Rizvic, 2014). The opportunities provided by the information age have enabled museums to exist in virtual environments. An important and widely accepted form of use in the field of cultural heritage is virtual museums. Virtual museums are effective in promoting cultural assets among wider audiences (Sürücü & Başar, 2016). In today's world, one common way of utilizing culturally valued structures is through the establishment of virtual museums. The opportunity for visitors to actively engage in and become part of the presentation by experiencing it has a significant impact on their better understanding of cultural values and enhanced sense of ownership. With the developing technology in museums, people have become more active rather than passive. Interactive methods are seen as successful in transferring information because memorizing information is seen to be difficult for the new human model. Addressing the new generation has become more possible with interactivity, a modern approach. The main purpose of this technology is to move people to an active position. Through these applications, interest in museums is also increasing. It is thought that a more permanent and healthy cultural heritage awareness will develop when interactive applications are made for children and adults in most museums (Erbaş & Tuncel, 2022). In the context of museology, virtual museums provide a new way to understand and promote cultural heritage. By means of virtual museum technology, the user has the opportunity to experience the museum without coming into contact with real objects. While viewing the artwork, the virtual visitor can refer to content useful for the learning process. Another revolutionary aspect brought about by new technologies is the ability to exhibit inaccessible or invisible artworks through virtual exhibitions. Its impact on cultural heritage lies in offering new perspectives for experiencing cultural heritage through visualization, representation, interaction, utilization, development and dissemination (Aiello et al., 2019). Virtual museums are a means to enrich the museum experience. Users have a holistic vision of the heritage of a historical period with a virtual museum visit. Moreover, virtual museums not only provide information about cultural assets but also stimulate users' curiosity and participation (Giglio et al., 2015).

Virtual museums are not intended to replace walled museums. These are not standalone entities but rather an extension of physical museum exhibition spaces, making the ideas, concepts and messages of the real museum accessible. In other words, they can be described as "digital reflections" of physical museums (Styliani et al., 2009). Virtual museums often represent a wide variety of resources in terms of organizational and functional content. It therefore offers many different possibilities for the representation of cultural heritage. Virtual museums have great potential in terms of scientific use of these resources, but they often cannot be used effectively due to limitations in the organization of basic information sources. Therefore, opportunities for the effective use of virtual museums and digital cultural heritage depend on a number of factors, such as the quality of digital content and the documentation of data on virtual museums and their resources. When creating new virtual museums, paying attention to important principles such as bringing together valuable resources that serve different purposes and expanding their usability can make virtual museums more widely accessible (Povroznik, 2018). The use of augmented reality technology in virtual and spatial museums will also increase the function of museums in preserving cultural heritage. Providing the opportunity to travel to past periods and experience the moments of those periods is considered to be one of the important features that make augmented reality technology significant. In the context of preserving and perpetuating cultural heritage, places such as museums, historical sites, ruins and protected areas hold significant importance. The use of augmented reality technologies to authentically enhance these places makes them interesting places to visit. Visitors have the opportunity to make their visit in a more interesting environment by means of virtual animations such as being able to use the exhibited objects, wearing historical garments and re-enactment of historical events. Through realistic re-enactments like these, the perception and interest in museums can be enhanced and thus more people can be attracted to museums. It is thought that the number of visitors to museums will increase significantly with the use of this technology. Given that the interest of the young generation in museums will increase with the integration of augmented reality technology into museums, it seems likely that it will emerge as an important element in the promotion and transfer of cultural heritage. Beyond museums, the accurate virtual restoration and enhancement of historical ruins and archaeological sites using this technology are crucial for the promotion of cultural heritage (M. Yıldırım, 2021).

The COVID-19 epidemic, which started in December 2019 and turned into a pandemic in March 2020, had a huge impact on museology and virtual museums. Many museums around the world implemented full closures and stopped accepting visitors between March 2020 and November 2020 (Kahraman, 2021). However, this has led to increased interest in virtual museums. The COVID-19 pandemic has elevated the importance of technological content such as virtual reality, augmented reality or mobile applications, as well as web-based museums that enable remote visits (Aytekin & Aktaş, 2023). While the number of visitors of the world's 100 most important museums in 2019 was 230 million, it decreased by 77% to 54 million visitors in 2020. On the other hand, the number of virtual museum visitors has increased significantly compared to the pre-pandemic period (Kahraman, 2021). Museums increased their virtual museum opportunities by 80% during quarantine, resulting in a 40% increase in virtual museum visits (Vayanou et al., 2020). Museums with a rich history like the Louvre Museum have decided to continue their activities online through virtual museums, not only catering to their own country's residents but also welcoming visitors from all over the world with their open virtual museums. In addition to the Louvre Museum, museums with high visitor numbers around the world, such as the British Museum, Van Gogh Museum, Vatican Museum, Dali Museum and Metropolitan Museum, have served as virtual museums. In Turkey, the Turkish Ministry of Culture and Tourism, General Directorate of Cultural Heritage and Museums has opened virtual museums for visits, including the Independence War Museum-Ankara; Republic Museum-Ankara; Ephesus Museum-İzmir; Troy Museum-Çanakkale; Museum of Anatolian Civilizations-Ankara; Gazi Museum-Samsun; Göbeklitepe Archaeological Site-Şanlıurfa; Ethnography Museum-Ankara; Antalya Museum-Antalya; Boğazköy Museum-Çorum; Archaeological Museum-Gaziantep; Zeugma Museum-Gaziantep; Çorum Museum-Çorum; Şanlıurfa Museum-Şanlıurfa; Adana Museum-Adana; Archaeological Museum-Hatay (Kasapoğlu Akyol, 2020). Before the pandemic, museums had made a small portion of their collections accessible online, but during this period, they digitized larger portions of their collections and made them available for online access. The educational functions of virtual museums have become prominent during this period. The opportunities offered by virtual museums during the pandemic have brought museums a new and diverse audience. People who, for various reasons such as distance, living in another country, physical disabilities or health conditions, cannot physically visit museums have been able to virtually visit museums and, at times, digitally access many objects beyond the collections displayed in museum spaces. It is believed that some of these visitors may become the permanent visitors of museums in the future (Kahraman, 2021). During the pandemic, there were numerous online initiatives at both local and international levels after visiting virtual museums, emphasizing the role of cultural heritage, history, collective memory, shared experiences, culture and heritage as a source of knowledge. These efforts demonstrate the continued sensitivity towards the preservation of cultural heritage (Ginzarly & Jordan Srour, 2022).

During the global COVID-19 pandemic, which led to a shift in societal paradigms worldwide since March 2020, the educational function of virtual museums has become notably prominent (Kahraman, 2021). The use of information and communication technologies in the education and promotion of cultural heritage, along with virtual applications, has popularized the topic by supporting individual learning and offering a deeper perspective through these new learning approaches. Virtual museum applications facilitate easier access to cultural heritage artefacts, promote tourism and learning and lead to the emergence of new tools, methods and educational models related to technology (Vargün & Nuhoğlu, 2019). Cultural heritage education also encompasses students' understanding of the economic and social conditions of people from different time periods and tracing technological developments. Hence, cultural heritage education nurtures the concepts of change and continuity, enabling students to understand the continuity in our historical and cultural experiences and fostering a sense of connection to culture (Avcı & Memişoğlu, 2016). An individual's participation in cultural heritage education is directly related to his/her adoption and preservation of that value. Therefore, cultural heritage education has become a joint effort of all institutions and administrations, with schools being at the forefront. With the advancement of information and communication technologies, innovative methods and virtual applications have been integrated into cultural heritage education, making learning more enjoyable and lasting (Vargün, 2020). When the history of museology is examined, it is seen that museums have, from the past to the present, simultaneously fulfilled their functions of preservation, research and knowledge generation while also taking on the role of widespread educational institutions. Today, museums do not leave this role to chance; they aim to fulfil their educational function in an organized and structured manner (T. Yıldırım & Tahiroğlu, 2012). In today's world, virtual museums, which are the digital counterparts of real museums, play an active role in both in-museum and out-of-museum educational activities, just like physical museums. Virtual museums serve as a valuable learning environment in education today, particularly due to their absence of resource and time constraints, the ability to transcend the limits of a classroom and providing access to rich and diverse learning materials. Museums are institutions that are seen as functional educational environments today. They provide students with an experiential learning environment, making it easier for students to internalize all the values accumulated in various disciplines, from science to art, from culture to history. However, various reasons such as financial constraints, transportation issues, and the difficulty of obtaining permissions make it challenging to organize educational museum trips. Virtual museums, which are independent of time and space, are seen as an option in education when physical museum trips cannot be conducted (Sungur & Bülbül, 2019). The use of technologically supported virtual museums is believed to help students learn about historical and cultural heritage topics in a more meaningful way.

Literature Review

In this part of the study, the importance of virtual museums, the relationship of virtual museums with cultural heritage, and studies in which virtual museums are used as an educational environment are addressed.

In the study conducted by Çınar et al. (2021), the opinions of 7th grade middle school students about the use of virtual museums in Social Studies classes were examined. One hundred students who were in the 7th grade of middle school were provided with information about virtual museums and were taken on a virtual museum tour during online classes conducted through the Zoom application. Then, virtual tour links of the War of Independence Museum and the Republic Museum were sent to the students and they were allowed to take a virtual museum tour. After the trip, the students answered the open-ended questions in the virtual museum opinion form. The collected data were evaluated using descriptive analysis (Çınar et al., 2021).

In another study, the opinions of pre-service primary teachers about virtual museum applications were determined. An open-ended questionnaire was administered to 41 fourth-year students attending the Department of Primary Teaching at Giresun University and the data were analyzed using the content analysis technique (Sungur & Bülbül, 2019). In the study conducted by Yıldırım & Tahiroğlu (2012), the effect of virtual museum visits made within the scope of the 5th Grade Social Studies course on students' attitudes towards the course was determined. The study employed a pre-test post-test control group experimental design. In the study, the "Scale of Attitude towards Social Studies Course" was administered to both groups at the beginning and end of the experimental research process, and the data were evaluated using quantitative analysis techniques (T. Yıldırım & Tahiroğlu, 2012).

In the study conducted by Şen (2023), the aim was to gather teacher opinions on the use of virtual museums in the context of Social Studies classes, particularly in the context of teaching the subjects related to the 5th-grade Cultural Heritage learning area and the 6th-grade Cultural Heritage learning area. The study group of the study was composed of Social Studies teachers working in public schools located in the central districts of Karatay, Meram and Selçuklu in the city of Konya. The data were collected using a semi-structured interview form, and content analysis was employed in the analysis of the collected data.

In the study conducted by Saatci İzci & Halaç (2021), it was aimed to determine the city virtual tours prepared for promotional purposes and the rate of inclusion of cultural heritage in these virtual tours. In another study, the contribution of information and communication technologies (ICT), augmented reality (AR) and mobile applications to cultural heritage education was examined by evaluating the creative opportunities offered by them and the shortcomings of these applications (Vargün & Nuhoğlu, 2019).

Styliani et al. (2009) examined various types of virtual museums and discussed the advantages and limitations of the tools used in their creation. In another study, the role of virtual museums as a source of information in the preservation of cultural heritage was examined (Povroznik, 2018). Antonaci et al. (2013) focused on the use of virtual museums by teachers in schools for educational purposes. Barlas Bozkuş (2014) examined the increasing role of virtual museums, which have become educational institutions with the reforms and innovations implemented, in international cultural collaborations. Kahraman (2021), on the other hand, examined the intellectual foundations of virtual museums that emerged as a result of the pressure put on museums to change, the concepts related to virtual museums, their classification and examples of virtual museums in the world and in Turkey. Vargün & Nuhoğlu (2019) classified the technological possibilities of current applications for the promotion of cultural heritage and evaluated their negative aspects as well as their educational positive aspects from a critical perspective.

Rizviç (2014) defines the concepts of interactive digital storytelling in virtual museums and presentations of cultural heritage, discussing the advantages and disadvantages of these concepts as perceived through user evaluations. Giglio (2015) aimed to demonstrate how Information and Communication Technologies (ICT) can be used to draw on and disseminate cultural heritage, providing learning experiences for different target users, especially young people.

Vayanou et al. (2020) evaluated the current virtual museum practices during the pandemic period and the challenges that may arise in the post-pandemic period. Another study aimed to understand different emotional and cognitive dynamics associated with cultural heritage contents shared virtually during the COVID-19 epidemic (Ginzarly & Jordan Srour, 2022).

MAIN FOCUS OF THE CHAPTER

Method

In this part of the book, it is aimed to reveal the importance of virtual museums and their place in the preservation of cultural heritage. The second aim of this part is to determine students' level of satisfaction with a virtual museum visit and to evaluate their opinions about the virtual museum visit. Within the scope of the study, 24 students enrolled in the Social Studies Teaching Department at Aksaray University (Turkey) were given the opportunity to participate in a sample virtual museum tour. Then, each student was given a museum to take their own virtual museum tour. When the students completed their tour, the Virtual Museum Satisfaction Scale developed by D'Alba (2012) and adapted into Turkish by Teker and Özer (2016) was administered to the students. After the administration of the scale, the students were interviewed to learn about the contribution of the museum visit to them and their feelings during the visit. The data obtained from the scale were evaluated by using descriptive analysis and comparative analysis. For the comparative analysis, the normality assumption value was examined and Shapiro-Wilk Sig. value was found to be .124. Although this showed that the data showed a normal distribution, Mann

Whitney U test was used for comparison due to the small number of students in the groups. The interview data were interpreted with the help of direct quotations

Findings

In this part, findings regarding the application conducted on 24 students within the scope of the study will be presented. After the students had a virtual museum tour, the Virtual Museum Satisfaction Scale was administered. As a result of the evaluation of the data obtained from the scale, the following findings were reached:

The minimum score to be taken from the 19-question, 5-point Likert-type scale is 19 and the maximum score is 95. When the descriptive statistics analyzed on the basis of the total score are examined, it is seen that the highest score among 24 students is 91.0 and the lowest score is 47.0. The mean virtual museum satisfaction score of the 24 students is 75.58. For this scale, 19.0-44.33 represents the low score range, 44.34-69.67 represents the medium score range and 69.68-95.00 represents the high score range. Therefore, it is possible to say that these students' virtual museum satisfaction level is high (Table 1).

Table 1. Descriptive statistics regarding the total scores of the students from the virtual museum satisfaction scale

Descriptive Statistics		
	Statistic	**Std. Error**
Mean	75.5833	2.37355
Variance	135.210	
Std. Deviation	11.62799	
Minimum	47.00	
Maximum	91.00	

Table 2 shows the mean scores from the students' responses to the virtual museum satisfaction scale items. When the mean scores of the students are examined on a question-by-question basis, the items with the highest mean scores are; *"I think I learned something from my experience using the virtual museum"* (M:4,37), *"I learned how to use the virtual museum quickly"* (M:4,37), *"It is easy to remember how to use the virtual museum* (M:4,37)", *"I did not need special instructions to use the virtual museum* (M:4,29)" and "*Using the virtual museum is effortless* (M:4,25)". The fact that these items have high means suggests that students found the virtual museum application instructive and perceived it as an easy-to-learn and easy-to-use tool (Table 2). The items with the lowest means on the other hand are; "*The virtual museum offers enough information about the artist/pieces of the collection* (M:3,04)", "*Using the virtual museum is frustrating* (M:3,25)", "*I feel I would have learned more by exploring the virtual museum without the need to follow a narrative* (M:3,25)", "*Using the virtual museum is confusing* (M:3,41)" and "*The narrative in the virtual museum helped me to learn more about the artist and the pieces* (M:3,54)" (Table 2). When the items with low means are examined, it is seen that students did not find virtual museum applications frustrating and confusing. However, students also believe that virtual museum applications are inadequate in terms of accessing more information.

Table 2. Descriptive statistics for the responses given by students to the virtual museum satisfaction scale items

	Descriptive Statistics			
No	Items	N	Mean	Std. Dev.
1	Using the virtual museum is captivating	24	4.00	1.179
2	Using the virtual museum is enjoyable	24	4.16	.963
3	Using the virtual museum is confusing	24	3.41	1.212
4	Using the virtual museum is frustrating	24	3.25	1.224
5	Using the virtual museum is fun	24	4.12	.946
6	I think I learned something from my experience using the virtual museum	24	4.37	.923
7	The narrative in the virtual museum helped me to learn more about the artist and the pieces.	24	3.54	1.382
8	I feel I would have learned more by exploring the virtual museum without the need to follow a narrative	24	3.25	1.224
9	My experience with the virtual museum encouraged me to seek more information about the artists/pieces in the tour	24	3.83	.916
10	I want to learn more about the artists/pieces of the tour as a result of my experience using the virtual museum	24	4.12	.899
11	The virtual museum offers enough information about the artist/pieces of the collection	24	3.04	1.082
12	I would like to explore other collections using the virtual museum	24	3.91	1.212
13	The virtual museum is easy to use	24	4.20	.721
14	I did not need special instructions to use the virtual museum	24	4.29	.750
15	I learned how to use the virtual museum quickly	24	4.37	.710
16	It was easy to learn how to use the virtual museum	24	4.33	.816
17	It is easy to remember how to use the virtual museum	24	4.37	.710
18	Using the virtual museum is effortless	24	4.25	.794
19	The virtual museum is a complement for my visit to the museum	24	3.70	.907

As a result of the analysis conducted to determine whether the scores received by the students from the virtual museum satisfaction scale varied significantly depending on gender, it was determined that the mean score of the female students is M: 75.70 and the mean score of the male students is M: 75.28 (Table 3).

Table 3. Mean scores taken from the virtual museum satisfaction scale by the male and female students

Gender	N	Mean	Std. Dev.	Std. Er. Mean
Female	17	75.7059	10.93026	2.65098
Male	7	75.2857	14.12698	5.33949

As a result of the Mann Whitney U test, the mean difference between the female students and male students was not found statistically significant (Table 4).

Table 4. Mann Whitney U test results regarding the significance of the difference between virtual museum satisfaction means of the students by gender

Ranks					
Gender	**N**	**Mean Rank**	**Sum of Ranks**	**Mann-Whitney U**	**Asymp. Sig.**
Female	17	12,41	211,00	58,000	924
Male	7	12,71	89,00		
Total	24				

After the administration of the scale, the students' opinions about the virtual museum visit were taken. The responses given by the students to the questions "What were the contributions of the virtual museum visit to you?" and "How did you feel during the virtual museum visit?" confirm their high level of satisfaction with the virtual museum visit. The students' opinions about the contributions of the virtual museum to them are as follows:

"Participating in this tour has been beneficial to me both in terms of museum culture and education, and I have also gained knowledge about the exhibited artworks (S1)".

"The virtual museum application provided me with convenience since I didn't have the opportunity to visit Çorum and explore this museum in person. It allowed me to establish a connection between the past and the present. While touring the museum, I felt the development of social life and science. I also discerned the cultural differences between the past and the present (S2)".

"As I didn't have the opportunity to visit and see the museum, I was very curious about it, the three-dimensional representation allowed me to see in detail what was inside, what cultural heritage was present, how it was designed, and where everything was located. I was able to explore the surroundings as I wished and also gained knowledge about the museum's content (S6)".

"Taking this tour provided me with a preliminary knowledge of the Ankara Ethnography Museum. If I have the opportunity to visit the museum in person, it will help me about the exhibition rooms and the displayed artworks (S8)".

"Through the virtual museum, I had the opportunity to explore a museum that I might never have been able to visit. I acquired knowledge about artworks from various periods. I grasped what was emphasized in different eras and how artworks diversified in various fields. I also gained information about how to use virtual museums. In my future career, I will be able to actively engage my students in lessons through virtual museums and provide them with an easy way to explore museums. This will not only help students better understand their lessons but also enhance their observational skills (S10)".

"During this virtual museum tour, I had the opportunity to see places I couldn't physically visit or see. While I couldn't feel the environment, the displayed objects and settings made the subject more memorable. Unlike traditional museum visits where we have to physically go to the museums, with virtual museums, we can do it from anywhere we want, which demonstrates the convenience of accessibility it offers (S14)".

"Thanks to this tour, I was able to see a historical place I hadn't visited from the comfort of my home. Applications like these contribute to increasing people's curiosity about their own history and help us rediscover the richness of our culture (S23)".

"Taking this tour provided me with a preliminary knowledge of the Istanbul Archaeology Museums. In the future, when I have the opportunity to visit the museum in person, it will help me about the exhibition rooms and the displayed artworks. Despite some gaps in information, it gave me a feeling of having taken a museum tour and piqued my curiosity due to its exhibitions of historical artefacts. If I ever find myself in Istanbul, I would like to visit the museum in person as well (S20)".

"Based on the exhibitions within the fortress and the impressive structure of the castle, it made me realize the importance the Ottoman Empire placed on castle construction. Additionally, it encouraged me to conduct further research on the castle online (S18)."

"Having learned a lot and saw many things about the cultural structures of the people of the period, their city life, the materials they used in their daily lives, the works of their traditions, the historical events that took place during the period, etc., I experienced a transition between periods and it was a trip in which I was able to interpret historical developments (S9)."

Students emphasized that they gained more information about the museum in terms of the contributions of the virtual museum tour to them. They stated that they learned in detail about the values we have, especially in the field of history and cultural heritage. They also stated that it was beneficial for them to see and experience the museum in detail from their homes when they did not have the opportunity to visit it in person. The desire to conduct more comprehensive research about the artworks and historical periods they encountered in the museums they visited is also one of the positive aspects of the visit to virtual museums. In addition, they became willing to learn how to use virtual museums, understand the importance of using virtual museums and convey this knowledge to their own students.

The students' opinions about what they felt during the virtual museum visit are as follows:

"Thanks to the three-dimensional presentation, I felt like I was actually there, and being able to move around freely inside was enjoyable and fun (S8)".

"I had never seen the Göbeklitepe Archaeological Site before, but by exploring it through the virtual museum, I felt as if I were actually there, even though I couldn't read the texts. During the virtual museum tour, I was transported back thousands of years, trying to think like one of these people, pondering how they built these structures, why they carved these symbols, and what purposes or needs drove them. It was a fascinating experience (S10)".

"Despite the lack of complete information, it gave me the feeling of having taken a museum tour, sparking a sense of curiosity and admiration within me. It made me feel like I had physically visited the museum and instilled a desire to explore the artefacts more extensively and gain knowledge about them in person (S6)".

"While touring the Izmir Atatürk Museum through the virtual museum, I felt as if I were actually there, just like in a real visit. I found the 360-degree viewing capability to be quite successful. The application's ability to showcase museum sections made navigation easy, and it created an impression that I was really present during the tour (S24)".

"I was happy to have the opportunity to visit museums and examine works on subjects I was curious about through the virtual museum (S18)".

"When I went full screen on the museum's website and focused on the screen, I truly felt like I was wandering within the walls of that museum. Exploring a museum located in Ankara from where I was and seeing the artefacts clearly was a wonderful experience (S13)".

"I had never been to Çanakkale before. However, after this virtual museum tour, it increased my desire to visit this fortress and the province of Çanakkale in real life. I had done very few virtual museum tours before, but this virtual journey made me think that I should be more active in such applications (S5)".

"Exploring this museum transported me back to the early years of our Republic. With this experience, it allowed me to feel and understand more closely the events and circumstances of that time (S3)".

"Seeing the significant differences during transitions between eras and being able to observe these changes through the museum's artefacts was both enjoyable and fulfilling. It's not always easy to visit museums in person, but thanks to this application, we can say that museums have come to our doorstep. This demonstrates how technology has provided convenience in human life, and contemplating the potential future of technology is promising in many ways (S11)".

Students often emphasized the feeling of being as if it were real and they were actually there during virtual museum visits. They also mentioned feeling entertained, curious, amazed, happy, delighted and hopeful during virtual museum visits. Students' opinions about the virtual museum visit are positive and their satisfaction with the visit is high.

SOLUTIONS AND RECOMMENDATIONS

In this part of the book, it is aimed to reveal the importance of virtual museums and their place in the preservation of cultural heritage, and to evaluate students' level of satisfaction with and their opinions about a virtual museum visit. In this part, firstly, the concepts of museum, virtual museum and the development of virtual museums are mentioned. A detailed conceptual framework regarding the importance of virtual museums in preserving and teaching cultural heritage and their increasing importance during the pandemic period is also discussed. According to the results of the "Virtual Museum Satisfaction

Scale" administered after a virtual museum tour conducted with 24 students enrolled in the Department of Social Studies Education at Aksaray University (Turkey), the satisfaction level of the students with the virtual museum tour was found to be high. According to the scale results, students found virtual museum visits instructive, but they found the scope of some museums insufficient. According to the students' opinions, virtual museums are educational in terms of history and cultural heritage, and being able to visit museums remotely, especially those that cannot be visited physically, is highly beneficial. Students mentioned in their feedback that they felt like they were actually touring the museum, and they expressed their willingness to implement virtual museum applications for their own students. The results of this study are consistent with the findings of previous studies in the literature. In the study of Çınar et al., (2021), it was concluded that students think that virtual museum tours in the Social Studies course contribute to their learning, that virtual museum tours create an alternative to museum trips, eliminating time and space restrictions, and make the learning-teaching process fun by enriching the course environment. In another study, it was determined that activities supported by virtual museum visits had a positive effect on students' attitudes towards the Social Studies course (T. Yıldırım & Tahiroğlu, 2012). It has been suggested that virtual museum applications are beneficial for students in primary school education and that various courses taught in the primary teaching undergraduate program should be associated with virtual museum applications (Sungur & Bülbül, 2019). In a study in which teachers' opinions about virtual museums were taken, it was stated that teachers were aware of the concept of virtual museums and that virtual museums could be used in the teaching of the subjects in the Culture and Heritage learning area in the 5th and 6th grade curriculums of the Social Studies course (Şen, 2023).

Based on the results of this study and studies in the literature, the importance of virtual museums in preserving cultural heritage and transferring it to future generations is indisputable. Participating in cultural heritage education is directly related to the individual's adoption and protection of that value. For this reason, cultural heritage education has become a joint effort of all institutions and administrations, especially schools. With the development of information communication techniques, innovative methods and mobile applications used can be integrated into cultural heritage education, making learning more enjoyable and permanent. Thus, more space should be given to applications with virtual museums, especially in the curriculum. Content should be developed with virtual museum applications at all levels, from pre-school to university. Efforts should be made to increase awareness of virtual museums at the public level, as they provide access to distant museums and provide preliminary information before the visit. The number and quality of virtual museums in the world and in Turkey should be increased. In cases where virtual museum applications cannot be implemented, interactive museum web pages with enriched content should be developed.

FUTURE RESEARCH DIRECTIONS

This study is limited to 24 students attending Aksaray University. Studies that enable more students to visit virtual museums in order to raise awareness about both cultural heritage and virtual museums are recommended for the future. In future studies, studies can be conducted on the use of other technologies in museums other than virtual museums. Studies should be conducted discussing the effects of the use of augmented reality and mobile applications in education on teachers and students.

CONCLUSION

Museums, which are one of the places where cultural heritage is best preserved, have changed within the framework of the development of technology and the changing understanding of museology. In this context, virtual museums have been created. The importance of virtual museums has increased even more during the pandemic. With the pandemic, the contribution of virtual museums to visiting cultural heritage all over the world and increasing awareness of cultural heritage has become visible. In addition, the use of virtual museums for educational purposes facilitates the achievement of objectives related to cultural heritage in the curriculum.

REFERENCES

Aiello, D., Fai, S., & Santagati, C. (2019). Virtual Museums as a Means for Promotion and Enhancement of Cultural Heritage. *The International Archives of the Photogrammetry, Remote Sensing and Spatial Information Sciences, XLII-2*(W15), 33–40. doi:10.5194/isprs-archives-XLII-2-W15-33-2019

Antonaci, A., Ott, N., & Pozzi, F. (2013). Virtual Museums, Cultural Heritage Education and 21st Century skills. Learning & Teaching with Media & Technology, Genoa, Italy.

Avcı, M., & Memişoğlu, H. (2016). Kültürel Miras Eğitimine İlişkin Sosyal Bilgiler Öğretmenlerinin Görüşleri. *İlköğretim Online, 15*(1), 104–124. doi:10.17051/io.2016.42123

Aytekin, H., & Aktaş, G. (2023). Türkiye'de Sanal Müze Araştırmaları: Bibliyometrik Analiz. *Journal of Business in the Digital Age, 6*(1), 1–10.

Barlas Bozkuş, Ş. (2014). Kültür ve Sanat İletişimi Çerçevesinde Türkiye'de Sanal Müzelerin Gelişimi. *The Journal of Academic Social Science Studies, 26*(26), 329–344. doi:10.9761/JASSS2408

Batuhan, T. (2023). Kültürel Miras ve Müze İlişkisi: Mardin Müzesi. *Journal of Literature and Humanities, 70*, 85–91. doi:10.5152/AUJFL.2023.22014

Bulut, P., & Uzun, E. B. (2021). Sanal Müzeler ve Yaratıcı Drama için Bir Atölye Önerisi: Müze Sınıfa Geliyor! *Yaratıcı Drama Dergisi, 16*(1), 131–156.

Çınar, C., Utkugün, C., & Gazel, A. A. (2021). Sosyal Bilgiler Dersinde Sanal Müze Kullanımı Hakkında Öğrenci Görüşleri. *Uluslararası Sosyal ve Eğitim Bilimleri Dergisi, 16*(16), 150–170. doi:10.20860/ijoses.1017419

D'Alba, A. (2012). *Analyzing Visitors' Discourse, Attitudes, Perceptions, and Knowledge Acquisition In An Art Museum Tour After Using A 3d Virtual Environment.* [Unpublished PhD Thesis, Unıversity of North Texas].

Erbaş, H. E., & Tuncel, D. (2022). Kültür Varlıklarının Sanal Ortamda Yeniden İşlevlendirilmesi: Yerebatan Sarnıcı Örneği. *European Journal of Science and Technology, 41*, 362–372. doi:10.31590/ejosat.1179021

Giglio, S., Gabriele, L., Tavernise, A., Pantano, P., Bilotta, E., & Bertacchini, F. (2015). *Virtual museums and Calabrian cultural heritage: Projects and challenges.* Digital Heritage. doi:10.1109/DigitalHeritage.2015.7419603

Ginzarly, M., & Jordan Srour, F. (2022). Cultural heritage through the lens of COVID-19. *Poetics*, *92*, 101622. doi:10.1016/j.poetic.2021.101622 PMID:34725533

Kahraman, Z. (2021). Sanal Müzecilikte Yeni Yaklaşımlar. *Güzel Sanatlar Fakültesi Dergisi*, *3*(2), 145–160.

Kamariotou, V., Kamariotou, M., & Kitsios, F. (2021). Strategic planning for virtual exhibitions and visitors' experience: A multidisciplinary approach for museums in the digital age. *Digital Applications in Archaeology and Cultural Heritage*, *21*, 1–11. doi:10.1016/j.daach.2021.e00183

Kasapoğlu Akyol, P. (2020). COVID-19 Küresel Salgın Dönemi Ve Sonrası Müze Etkinlikleri. *Milli Folklor*, *16*(127), 72–86.

Peker, N. (2014). *Sosyal Bilgiler Dersinde Sanal Müze Kullanımı Ve Sosyal Bilgiler Öğretmeni Adaylarının Sanal Müze Kullanımına Yönelik Tutumları.* [Unpublished Master Thesis, Aksaray Üniversitesi].

Povroznik, N. (2018). *Virtual Museums and Cultural Heritage: Challenges and Solutions.* The Digital Humanities in the Nordic Countries 3rd Conference Helsinki, Helsinki, Finland.

Rizvic, S. (2014). Story Guided Virtual Cultural Heritage Applications. *Journal of Interactive Humanities*, *2*(1), 1–15. doi:10.14448/jih.02.0002

Saatci İzci, B., & Halaç, H. H. (2021). Kent Sanal Turlarında Kültürel Mirasın Değerlendirilmesi. *Turk Turizm Arastirmalari Dergisi*, *5*(2), 985–1004. doi:10.26677/TR1010.2021.749

Şen, H. (2023). *Sosyal Bilgiler Öğretiminde 5 Ve 6. Sınıf Kültür Ve Miras Öğrenme Alanlarında Sanal Müze Kullanımına Yönelik Öğretmen Görüşleri.* [Unpublished Master Thesis, Necmettin Erbakan Üniversitesi].

Styliani, S., Fotis, L., Kostas, K., & Petros, P. (2009). Virtual museums, a survey and some issues for consideration. *Journal of Cultural Heritage*, *10*(4), 520–528. doi:10.1016/j.culher.2009.03.003

Sungur, T., & Bülbül, H. (2019). Sınıf Öğretmeni Adaylarının Sanal Müze Uygulamalarına Yönelik Görüşleri. *Abant İzzet Baysal Üniversitesi Eğitim Fakültesi Dergisi*, *19*(2), 652–666. doi:10.17240/aibuefd.2019.19.46660-492112

Sürücü, O., & Başar, M. E. (2016). Kültürel Mirası Korumada Bir Farkındalık Aracı Olarak Sanal Gerçeklik. *Artium*, *4*(1), 13–26.

Teker, N., & Özer, A. (2016). Sanal Müze Sanal Tur Memnuniyet Ölçeğinin Türkçe'ye Uyarlanması: Geçerlik Ve Güvenirlik Çalışması. *Milli Eğitim Dergisi*, *209*, 314–335.

Vargün, Ö. (2020). *Kültürel Miras Eğitiminde BİT'in Rolü ve AR Mobil Uygulama Projesi* [Unpublished PhD Thesis, Yıldız Teknik Üniversitesi].

Vargün, Ö., & Nuhoğlu, M. (2019). Kültürel Miras Eğitiminde Bilgi İletişim Teknolojilerinin Rolü ve Mobil Uygulamalar. *UNIMUSEUM*, *2*(2), 45–53.

Vayanou, M., Katifori, A., Chrysanthi, A., & Antoniou, A. (2020). *Cultural Heritage and Social Experiences in the Times of COVID 19.* CHI '05 Extended Abstracts on Human Factors in Computing Systems, Island of Ischia, Italy. https://dl.acm.org/doi/10.1145/1056808.1056881

Yang, Y., Zhang, H., Chen, M., Jiang, Y., & Chai, H. (2021). An Inheritance Mode of Rural Cultural Heritage Based on Virtual Museum in China. *International Journal of Computer Games Technology, 2021*, 1–14. doi:10.1155/2021/4787991

Yıldırım, M. (2021). Sanal Mekânda Kültür Coğrafyası Çalışmaları ve Artırılmış Gerçeklik Teknolojisinin Kullanılması. *Afyon Kocatepe Üniversitesi Sosyal Bilimler Dergisi, 23*(3), 861–874. doi:10.32709/akusosbil.887804

Yıldırım, T., & Tahiroğlu, M. (2012). Sanal Ortamda Gerçekleştirilen Müze Gezilerinin İlköğretim Öğrencilerinin Sosyal Bilgiler Dersine Yönelik Tutumlarına Etkisi. *Elektronik Sosyal Bilimler Dergisi, 11*(39), 104–114.

KEY TERMS AND DEFINITIONS

Cultural heritage: Heritage is a general term referring to works created by previous generations and believed to possess universal value.

Museum: It is a non-profit, permanent institution that researches, collects, preserves, interprets and exhibits tangible and intangible heritage in the service of society. Public, accessible and inclusive museums promote diversity and sustainability. They offer a variety of experiences for education, entertainment and knowledge sharing; operate ethically and professionally with community participation; and engage in communication.

Virtual Museum: Virtual museums are collections consisting of digitally recorded images, audio files, text documents, and other historical, scientific, and cultural data that can be accessed through electronic media.

Chapter 18
Sustaining Livelihoods and Culture Through Tourism Development:
The Case of Sriniketan in West Bengal, India

Priyakrushna Mohanty
Christ University, Bengaluru, India

Rupa Sinha
https://orcid.org/0000-0001-7174-7708
Amity Institute of Travel and Tourism, Amity University, Kolkatta, India

Jyothi Kumar
Christ University, Bengaluru, India

ABSTRACT

The rural area of Sriniketan in West Bengal, India is full of cultural embodiments that can not only serve as a base to develop tourism but also generate sustainable livelihoods. However, the Sriniketan region suffers from chronic poverty and its unique culture is getting depleted thanks to the lack of awareness and interest among locals. With the help of the DFID Sustainable Livelihood Framework (DFID-SLF), this study tries to analyse the contributions that culture-based tourism can make towards generating sustainable livelihoods at Sriniketan. A modified SLF has been prepared with an added element of cultural capital as a contribution to the existing livelihood literature and guiding sheet for future practitioners. Based on the primary (in-depth interviews of fifteen households and five key respondents) and secondary data collected, this paper concludes that tourism development in Sriniketan can not only aid its cultural preservation but also generate an alternate source of livelihood and thereby, making both (culture and livelihoods) sustainable

DOI: 10.4018/978-1-6684-9957-3.ch018

1. INTRODUCTION

The tourism industry has been known for generating sustainable sources of livelihoods (Tao & Wall, 2009; Swain, S. K., & Batabyal, D., 2016)) while supplementing indigenous cultures of the rural and isolated areas which are currently facing deterioration (Butler & Hinch, 2007). Many forms of tourism like indigenous tourism, ecotourism and cultural tourism can play a major role in providing additional or alternative livelihood opportunities, elevating poverty, preserving culture, improving community participation and promoting environmental sustainability (Masud, Aldakhil, Nassani & Azam, 2017; Batabyal et al., 2020).

Meanwhile, the majority of rural India depends on agriculture for generating its livelihood which often results in over-dependence on the only source, making the whole livelihood system unsustainable due to lack of diversification opportunities (Roy, 2005). However, the same rural areas possess numerous man-made and natural resources which require preservation and can act as rudimentary elements of tourism development (Ma, Wang, Dai, & Ou, 2020). Keeping this in mind, the authors of this paper have tried to exhibit the Tourism-Sustainable Livelihoods-Culture Linkage in India based on the insights from the in-depth interviews of key respondents and review of pertinent secondary literature collected from Sriniketan. The DFID sustainable livelihoods framework has been adopted to provide a conceptual base for the paper. It has been argued that tourism can be a source for generating additional and/ or alternative livelihood opportunities in the rural areas which in return can galvanize the process of preservation of culture, ensuring the sustainability of both livelihoods and culture.

Attempts have been made to find out how the implementation of cultural tourism projects can provide opportunities for diversification of livelihoods in the rural area of Sriniketan situated in the state of West Bengal in Eastern India. The place, Sriniketan has been taken as the case study mainly because of two reasons; the region possesses several cultural resources which are unexplored and there is an existence of persistent poverty in the region. The assessment of livelihood structure at Sriniketan has been done by the sustainable livelihood framework that tries to understand the factors which aid and hinder the sustainability of livelihood.

A modified SLF has been proposed as a contribution to the existing livelihoods literature and to guide the upcoming practitioners in the paths of generating sustainable livelihoods at destination Sriniketan work falls into the category of empirical research and the majority of this paper is qualitative in nature. Appreciative inquiry and in-depth interviews have been conducted to collect the primary data for the study. The sources for secondary data for the study are the published materials collected from Sriniketan, journal articles, books, Govt. dossiers, reports, and press releases. Based on the empirical shreds of evidence collected, it was concluded that tourism development in Sriniketan can act towards the preservation of its indigenous culture and simultaneously provide an additional source of livelihood which will make the whole livelihood system sustainable. However, contradictors like lack of awareness, over dependency on agriculture, the unwillingness of stakeholders and stigmatized thinking possess many challenges in the path of tourism development.

2. THE SUSTAINABLE LIVELIHOOD APPROACH

The concept of Sustainable Development (SD) made a grand entry into the development discourse with the Brundtland Commission Report framed in the year 1987 (Redclift, 2005) which defined SD as

"Development that meets the needs of the present without compromising the ability of the future generations to meet their own needs. It contains within it two key concepts: the concept of "needs," in particular the essential needs of the world's poor, to which overriding priority should be given; and the idea of limitations imposed by the state of technology and social organisation on the environment's ability to meet present and future needs. (World Commission on Environment and Development [WCED], 1987, p.43).

However, over time, the concept of SD had to face much criticism for various reasons. Scholars like Rennie and Singh (1996) argue that it is very difficult to recognize sustainability as a criterion for development both in academics and practice. The U.S. National Science Foundation (2000), states that SD is "is laden with so many definitions that it risks plunging into meaninglessness, at best, and becoming a catchphrase for demagogy at worst. It is used to justify and legitimate a myriad of policies and practices ranging from communal agrarian utopianism to large scale capital-intensive market development". It is also believed that the concept of SD has largely ignored the idea of defining based on culture (Norgaard, 1998). Lastly, SD has been criticized for providing scope for capitalism and inducing trickle down.

Subsequently, to counter the limitations of SD with regards to its broader nature, many countries started focusing more on individual and micro-based economic-developments during the 1980s. These forms of development were coupled with the emergence of 'participatory methods' in tourism academics (Krantz, 2001). Against this background, new improved mechanisms to evaluate the development process of a particular place was called for and thus, the idea of Sustainable livelihoods came to the forefront. The fundamental advantage of the SL thinking was its focus on the household- based appreciative inquiry approach (Gutierrez-Montes, Emery, & Fernandez-Baca, 2009). The idea of SL was first mentioned in the Brundtland Commission Report on Environment and Development (Brundtland, 1987). However, the formal definition of SL was first proposed. In a working paper jointly authored by Robert Chambers and Gordon in the paper, both argued that, unlike the urban livelihoods, rural livelihoods are complex which demands special attention (Chambers and Conway,1992). The definition of SL in that working paper is still now one of the most celebrated definitions of SL. It states

"A livelihood comprises the capabilities, assets (stores, resources, claims, and access) and activities required for a means of living: a livelihood is sustainable when it can cope with and recover from stresses and shocks, maintain or enhance its capabilities and assets, and provide sustainable livelihood opportunities for the next generation; and contribute net benefits to other livelihoods at the local and global levels in the long and short term."(Chambers and Conway 1992, p.7)

In this definition, the focus has been provided to the assets, capabilities and activities which are means for generating a livelihood. The amalgamation of livelihood strategies is referred to as a livelihood portfolio while livelihood strategies describe the adaptation of particular activities over a while (Scoones 1998). The definition also talks about stresses (the impact of which is felt over some time like low monsoons) and shocks (sudden impact like a flood) which as barriers to achieving SL.

Based on the definition by Chambers and Conway (1992), the Department for International Development (DfID) developed the Sustainable Livelihood Framework (SLF) [fig. 1] which is "an analytical structure to facilitate a broad and systematic understanding of the various factors that constrain or enhance livelihood opportunities and to show how they relate to each other" (Krantz 2001). The SLF provides a multilevel, holistic and people-centered understanding of the whole livelihood portfolio of any household or community or geographical area depending on the scale of the study.

Figure 1. SLF model

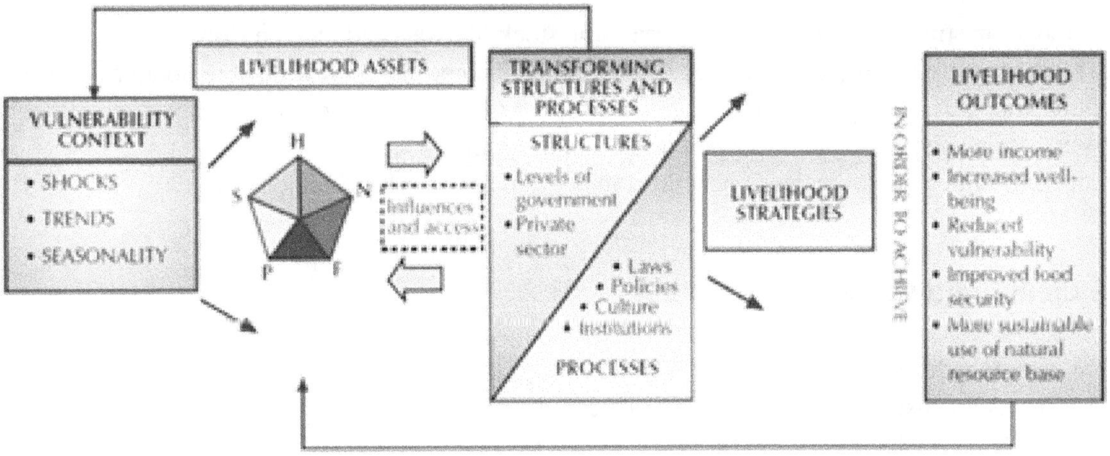

At the basic level, the SLF is divided into five major parts i.e. vulnerability context, livelihood assets, transforming structure and processes, livelihood strategies and livelihood outcomes. Before, moving into the next section of the paper, these five elements must be discussed thoroughly.

Vulnerability context: In-depth analysis of trends, conditions, assessment and shocks of policy setting that affect livelihoods
Trends: population (growth), governance, technological developments, national/international economic
Shocks: economic, human health, livestock/crop health, conflict, natural disaster
Seasonality: production, prices, employment, health
Livelihood Assets: Resource or Asset Analysis
Human capital: knowledge, skills, good health, ability to work
Social capital: relationships of trust, general relationships (horizontal and vertical), membership (e.g., groups, networks), exchange and reciprocity, kinship
Natural capital: land, water, wild resources, forests, biodiversity
Physical capital: producer goods, infrastructure and means
Financial capital: available stocks, cash at bank, cash in hand
Cultural Capital: Indigenous culture, monuments, heritage, cuisine, dress etc.

2.1 Transforming Structure and Processes

Structures: includes the public sector, civil society e.g., political (legislative) bodies, commercial enterprises, private /commercial companies, executive agencies (ministries, departments), NGOs
Processes: power relations, policy, institutions, legislation, culture

2.2 Livelihood strategies

Accumulation strategies; strategies to increase the stock of assets and income flows
Adaptive strategies; strategies to distribute risk through income diversification or livelihood adjustments
Coping strategies; strategies to minimise the influence of livelihood shocks
Survival strategies; strategies to prevent death and destitution.
Competing strategies: strategies to beat the competition e.g., markets and jobs

2.3 Livelihood Outcome

More income reduced vulnerability
More sustainable use of renewable resources etc.

3. THE TOURISM-SUSTAINABLE LIVELIHOOD-CULTURE LINKAGE IN INDIA

The tourism industry has often been referred to as the 'sunrise industry' due to its mammoth share of contribution to the world economy and employment (Hall, 2005). In the year 2019, the Travel and Tourism (T&T) sector accounted for 10.4% of the global GDP and 10.6% of the total employment of the world while growing steadily with time (WTTC, 2020). It also accounts for 10% of the world's economic output and one out of eleven jobs (WTO, 2016). The sector is also celebrated for being labour intensive and providing various sources of livelihood especially in marginalized and rural places of developing countries (Goodman, 2002). It is also the third-largest export industry after fuels and chemicals (Goffi, Cucculelli & Masiero, 2019). Other than its direct impact on the economy, the T&T industry additionally yields other indirect advantages through government investments in infrastructure, investment flows and domestic purchases of goods and services. Additionally, there are reverse linkages with farming, fisheries, utilities and media communications (WTO, 2016).

In India, tourism is one the biggest sources of the economic driver with an approximate 9.4% of total contribution to Indian GDP and 8.0% of all the employment generated in India (WTTC, 2018). Even though India receives a fairly low share of 1.2% of foreign tourist arrivals (FTAs), its international tourism receipts (ITRs) stands at a 2.5% mark highlighting the huge potential of the sector in India (Ministry of Tourism, Govt. of India, 2019). Starting from Kashmir in the north to Kanyakumari in the south, India exhibits a true sense of diversity that almost no other destination in the world can offer. Among all its tourism potentials, culture is one of the biggest assets of India when it comes to tourism resources. This country is the birthplace of four major religions of the world. It is a country that was ruled by the great kings of India, the Mughals and the British all of whom have contributed marvelous monuments and cultural impediments to this mystic land. Gantait, Mohanty & Swamy, (2018) have argued that are being vital tools for job creation, economic growth and induced infrastructure growth, the conservation of these cultures and monuments is emerging out as a challenging task for the future stakeholders because of inefficient management skills, negligence, lack of co-ordination and disinterest among stakeholders. Further, it is a fact that some forms of tourism like agro-tourism, rural tourism and cultural tourism can play a major role in generating economic benefit and providing livelihood to the local community (Salazar, 2012). A part of these benefits can be directed towards the conservation of the culture and heritage (Gantait, Mohanty & Swamy, 2018). Similar views have been expressed by UNESCO which states "...unless economic development has a cultural basis it can never lead to truly lasting development. Culture is 'not' something 'to be taken into consideration. It is fundamental. " Another interesting of traditions and culture is their connection to is there is their connection to 'pastness' (Appadurai, 1981) which carries treasured 'renewable resources' and 'portents of future' rather than 'scars of past' (Daskon & Binns, 2010).

However, the aspect of culture has been mostly viewed as a hindrance in the implementation of development initiatives and inhibition to entrepreneurship while ignoring to recognize the complex relationship between culture and economic performance (Jenkins, 2000). Not only there has been a scarcity of practical methods to integrate culture into the political and social measures, but also a lack of willingness among planners and governments to develop them (Huffer, 2006). Therefore, there cannot be a better time to recognizing the power of culture in generating economic benefit and creating jobs.

Further, many regions in India suffer from chronic poverty and the situation is more tragic in rural areas where 77% of India's poor and three out of four Indians reside (World Bank, n.d.). The majority of the rural areas in India depends on agriculture and allied services for their income and livelihood. However, agriculture and allied sectors contribute only 17-18% to the Indian GDP while more than half of India's population are employed under it (India Economic Survey, 2018).

This uneven input-output equation points towards the inefficiency of and over-dependency on agriculture and allied sectors in India. It also indicates the unsustainability in livelihood structure in rural areas and warrants for techniques to measure and provide guidelines for making the livelihoods sustainable. One of such techniques is the Sustainable livelihood (SL) approach that is considered a highly efficient tool to study the livelihoods of a particular region by arranging the factors that enhance or constrain livelihood (Serrat, 2017). The SL approach emphasizes the point that the more diverse the livelihood profile, the more sustainable it is. Therefore, there is a clear need to diversify livelihood strategies that can reduce the seemingly high burden on the agriculture sector in India and the authors of this paper believe that tourism based on culture can fill in this gap. Culture as such has been recognized as a livelihood resource in many studies and there is increasing attention being given to the aspect of culture in the doctrine of sustainable livelihood (Addinsall, Weiler, Scherrer & Glencross, 2016). In their recent study about the addition of culture into the SLF Ma et al. (2020, p. 13) mention that '' Understanding of the value of their culture in attracting tourists helps villagers utilize their traditions in tourism businesses, generate economic benefits, and improve their living standards''.

4. RESEARCH SETTING

Sriniketan is a community development block of the Bolpur subdivision of Birbhum District. Birbhum District is a part of the ancient Rarh Region. Geographically, the western portion of the district is an expansion of the Chota Nagpur plateau. The district comprises several rivers namely, Ajay, Mayurakshi, Kopai & Bakreswar a tributary of Mayurakshi River, Brahmani, which have squashy alluvial soil. The economic condition of Birbhum is largely controlled by the agriculture sector.

Figure 2. Maps

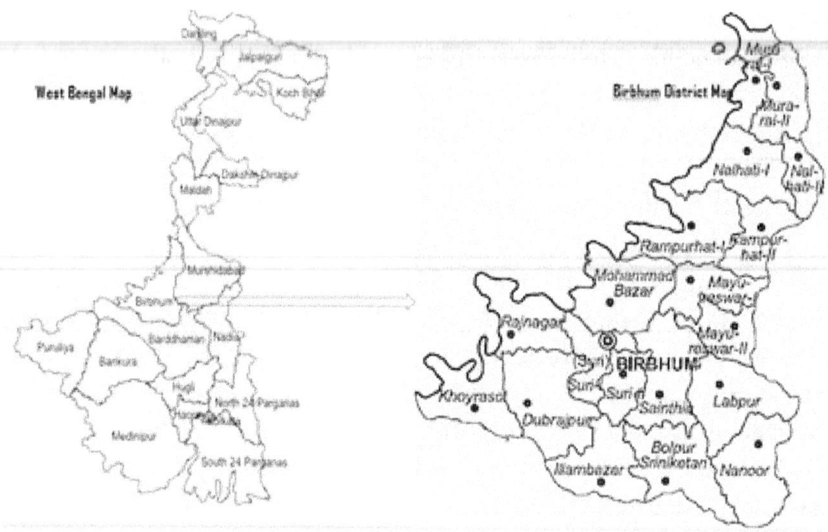

Culturally rich Birbhum is home to many cultural landmarks like Joydeb Kenduli, the birthplace of Sanskrit Poet Jayadeva (emerged as a religious spot) and comprises of temples and ashrams, a hosting place for Baul fair that is also organized on the occasion of Makar Sankranti (Birbhum District Official Website, n.d.).It is also the place where Noble laureate Sri Rabindranath Tagore founded Visva- Bharati as a reaction to the colonial education system and to promote the idea of an open and integral education system. Visva-Bharati has two main parts namely Sriniketan and Santiniketan. While Santiniketan is widely recognized as Tagore's concept of 'village Reconstruction', Sriniketan too stands as the epitome of Tagore's ideology (Das, 2017). Various such cultural establishments are situated in Sriniketan and have been depicted in the figure below.

Figure 3. Google map

4.1 Demographic Profile of Sriniketan

As per the census of 2011, Sriniketan encompasses 156 inhabited villages and 1 census towns namely Surul Census towns. Some of the villages like Surul, Ballavpur Danga, Raktokarobi Karugram (Under Roopore Village) are culturally enriched in terms of handicrafts like pottery, Kantha Steach, Batik print, Woodwork, leatherwork etc. and other cultural attractions, tremendous scopes to develop cultural tourism.

Table 1. Census 2011

	Total	Rural	Urban
Population	2,02,553	1,90,393	12,160
Children (0-6 years)	23,160	21,994	1,166
Literacy	70.67%	70.10%	79.45%
Sex Ratio	972	968	1,027
Scheduled Caste	30.60%	30.70%	28.50%
Scheduled Tribe	19.10%	19.40%	14.10%

[Source: https://www.censusindia.co.in/villagestowns/bolpur-sriniketa n-block-birbhum-west-bengal-2264]

5. RESEARCH METHODOLOGY

This study can be classified into the category of empirical research as it based on measured or observed phenomena and is derived from the knowledge from direct observation of the researcher (Goodwin, 2005). A qualitative approach has been adopted for the study given the rationality that the majority of the works on SL suggest that the five elements of the SLF vary differently for different places and the main focus of the researcher is to trace out the underlining factors rather than verifying or generalizing the fact. Given the exploratory nature of the context, the qualitative inquiry was preferred over quantitative evaluation (see Mason, Augustyn, & Seakhoa-King, 2010). Also, one of the biggest proponents of SL research Scoones (1998,p. 7) notes that '' clearly these five indicators of sustainable livelihoods are quite different in scope, with a range from very precise measures, amenable to quantitative assessment, to very broad and diffuse indicators requiring more qualitative techniques for assessment''.

Data for the study were collected during the period from September 2020 to November 2021. The appreciative inquiry which is a form of participatory action research has been adopted to derive the empirics of the study (see Reed, 2006). The use of appreciative inquiry in understanding the realities of tourism in rural areas is on the rise with notable works like Aziz, Hashim, and Awang (2018), Espiner, Stewart, and Lama (2017), Koster and Lemelin (2009), and Nyaupane and Poudel (2012) coming to the limelight. Further, since the SLF acknowledges the assets and capabilities of any unit of study rather thappreciative inquiry will better fit the study needs an its problems & challenges, (Serrat, 2017).

Both primary and secondary data has been collected for the study. The primary data have been collected from 12 households and 5 key respondents through in-depth interview and unstructured questionnaire. The secondary data for the study been collected from journal articles, path-breaking books, Govt. dossiers, reports and press releases etc. The data were collected whenever the respondents were free.

All of the respondents reported were of the age group of 18 or above. The familiarity with the place and command over Bengali (local language), provided the researchers with an advantage while collecting the data. While the first author had a short stay at the place, the third author of this paper had booked room in Sriniketan and stayed there till the time data was collected.

The form of participation while conducting the data collection process was kept as moderate(semi-participation) to bring out as many facts as possible while not getting biased in the study. Before conducting any interview, the respondents were informed about the aims & objectives of the study.

Purposive sampling, which is a form of non-probability sampling is mostly preferred for qualitative studies (Etikan, Musa, & Alkassim, 2016) and was taken as the sample selection strategy for this research. Suen, Huang, and Lee (2014, p. 105) in connection with purposive sampling note that '' Researchers who use this technique carefully select subjects based on study purpose with the expectations that each participant will provide unique and rich information of value to the study. As a result, members of the accessible population are not interchangeable and sample size is determined by data saturation not by statistical power analysis''. Thus, the researchers kept on approaching the participants till data saturation was achieved. Out of the 19 households and 7 key respondents approached, 12 households and 5 key respondents agreed to be interviewed. The approached households and key respondents were selected based on the recommendation of one of the oldest member (age – approximately 85, living in the village since birth) in the community such that they represented the whole village.

All the respondents chose to be anonymous. This may be because the study involved data regarding their and the community's economic condition. Throughout the whole data collection process, no conflicts were reported. Following the collection of the data during the day and mostly evening (respondents were comparatively free then), the researcher would translate and transcribe the data collected in a diary. At the end of the data collection process, these entries were verified with the respective respondents and any modifications called for were incorporated. Finally, thematic content analysis was adopted to bring out the major themes which were then put under the headings of various elements of the SLF. Before the preparation of the final draft, the authors met online to discuss any other modifications to be done. The authors did not receive any funding for this study, which reiterates the unbiased approach carried out in the study.

6. ACHIEVING SL THROUGH CULTURE-BASED TOURISM: A CASE OF SRINIKETAN

Table 1 depicts the overview of the SL elements at Sriniketan. Then, the findings about each element have been discussed separately and the summary of findings has been listed out.

Table 2. Summery of SLF elements at Sriniketan

Elements of SLF	Description of Elements	Summarized Discussions and Findings
Vulnerability	Trends, Shocks and Seasonality	Moderate Vulnerability High Seasonality
Livelihood Assets	Human, Natural, Social, Physical, Financial, Cultural	Lack of diversification of resources Sustainable usage of resources Underutilization of Cultural Capital
Transforming Structure and Processes	Structures, Policies, Institutions, Processes, Governance	Lack of willingness to promote Assistance from institutions No proper policies in place
Livelihood strategies	Accumulation, Adaptive, Coping Survival, Competing	High survival strategies Lower adaptive, accumulation, coping and competing for strategies
Livelihood Outcome	Diversified non and off-farm income, food security, Sustainability, equity, reduced vulnerability	Unsustainable livelihood culture High poverty

[Source: Data collected and compiled by the author]

6.1 Vulnerability

In terms of the element of vulnerability at Sriniketan, most of the respondents revealed that due to the drying of the river Kopai and Khoai, local people are worried. Though there are other sources of fresh-water harvesting, the drying up of the river has created a lot of speculations in the minds of the local people. Barring a few respondents majority of the respondents viewed that there are almost no shocks that Sriniketan has faced yet. However, when it comes to the aspect of seasonality 12 out of 17 respondents stated that Sriniketan experiences a great deal of seasonality in which the farmers get better livelihood during harvesting season and the handicraft makers and tour operators earn a massive amount of money during the Pausa Mela and Sit Mela time. It was reported that a room was rented to a visitor at the price of INR 35,000 for 3 nights and during lean season hotels find it almost impossible to fill the rooms.

6.2 Livelihood Assets

In terms of livelihood assets, it was found out that there is an over-dependence on the natural capital (land) as most of the residents in Sriniketan depends on agriculture for their livelihood. All of the 12 households interviewed during the data collection were fully or partially dependent on agriculture for deriving their livelihoods. Migrating to the state capital Kolkata for better income was also reported by 8 respondents. Out of the 17 respondents interviewed only 6 had concrete or semi-concrete houses and most of the houses (n=13) were inherited ones. These facts is show the presence of persistent poverty and lack of diversification in SL assets.

The social thread of the village is still found to strong with 15 respondents reporting a good relationship among the community members. This attribute was further reinforced by the fact that people who are combatively better off in society helped the poor members at the time of crisis. When it comes to human capital, the respondents reported a lack of interest in education due to the lack of proper educational infrastructure.

6.2.1 The Missing Link of Cultural Capital

All the respondents interviewed, acknowledged the presence of a large number of cultural embodiments at Sriniketan. During the interview, the respondents spoke in lengths about the handicrafts, places of cultural interest, indigenous people, fairs and festivals, arts & crafts, traditional knowledge and cuisines of Sriniketan. Barring just one, other respondents agreed to the fact that these rural cultural elements are getting depleted thanks to the lack of funding opportunities, the inclination of the new generation towards the modern (urban) culture and disinterest among locals quite honestly because of no financial benefits from these cultural elements. Another important element noted by the respondents is the use of traditional knowledge at work. 14 respondents viewed that traditional knowledge is still practised in the day to day life. While responding so, the majority of the respondents noted the transferring and exchanging of traditional knowledge across generations and within the community. However, some of the respondents showed concern that tourism may deplete the cultural values of the place. All in all, Tthe aspect of cultural resource is certainly underutilized as Sriniketan is a village full of cultural embodiments, but and the economic benefits them it haves not been reaped yet.

6.3 Transforming Structure and Processes

In terms of institutional support, most of the respondents referred to the initiatives taken up by the National Bank for Agriculture and Rural Development (NABARD), when it comes to selling handicrafts and souvenirs. Few of the NGOs are also working towards the preservation of culture in Sriniketan. One of the main institutions that were recognized by almost all respondent was the Viswbharati University in Shantiniketan, West Bengal. Concerning policies and governance, it was observed that residents of Sriniketan are very conscious about the local environment and hygiene and there are a lot of community rules which are being practiced for the betterment of the village and its residents as a whole. However, the support from Govt. institutions were reported to be unsatisfactory in the view of upliftment of agriculture.

6.4 Livelihood Strategies

A livelihood strategy is the outcome of the livelihood assets, resistance to vulnerability and transforming policies. In terms of accumulation strategy, the respondents reported that due to the lower income they couldn't save a lot. When it comes to adaptation strategy, most of the respondents were of the view that they are not ready to change their livelihood whatever happens. This suggested a lower adaptation strategy and also a coping strategy because if something goes wrong with agriculture then they would not be able to cope with the loss suddenly. However, survival strategy was found out to be the only satisfactory factor. The competing strategy was also found out to be higher as the resources for agriculture are limited and hence the competition is high.

6.5 Livelihood Outcome

Overall, when it comes to livelihood outcome, the current scenario of Sriniketan looks fragile because of resistance to change, traditional mindset and lack of opportunities to diversify the livelihood. Though the village is a bright model in terms of the 'rural reconstruction' principle upon which it was created, other livelihood resources are underutilized because many respondents reported that they are afraid that if a tourist comes, then they may lose their authentic culture and the whole ideology of the village may collapse.

However, the tourism industry by its very nature of inception believes in the principle of preservation of culture if promoted sustainably. Upon stating the above statement almost all respondents replied that if that happens then there is no harm in welcoming the tourist. The young generation of the village also reported about their high aspirations to develop further in career and if that has to happen tourism is truly the most go-to option. If sustainable tourism is promoted then the livelihood outcome will be the preservation of culture, economic upliftment and higher social cohesion. Based on the above remarks, the authors of the paper have prescribed a modified Sustainable Livelihood Framework which is depicted below;

Figure 4. Culture

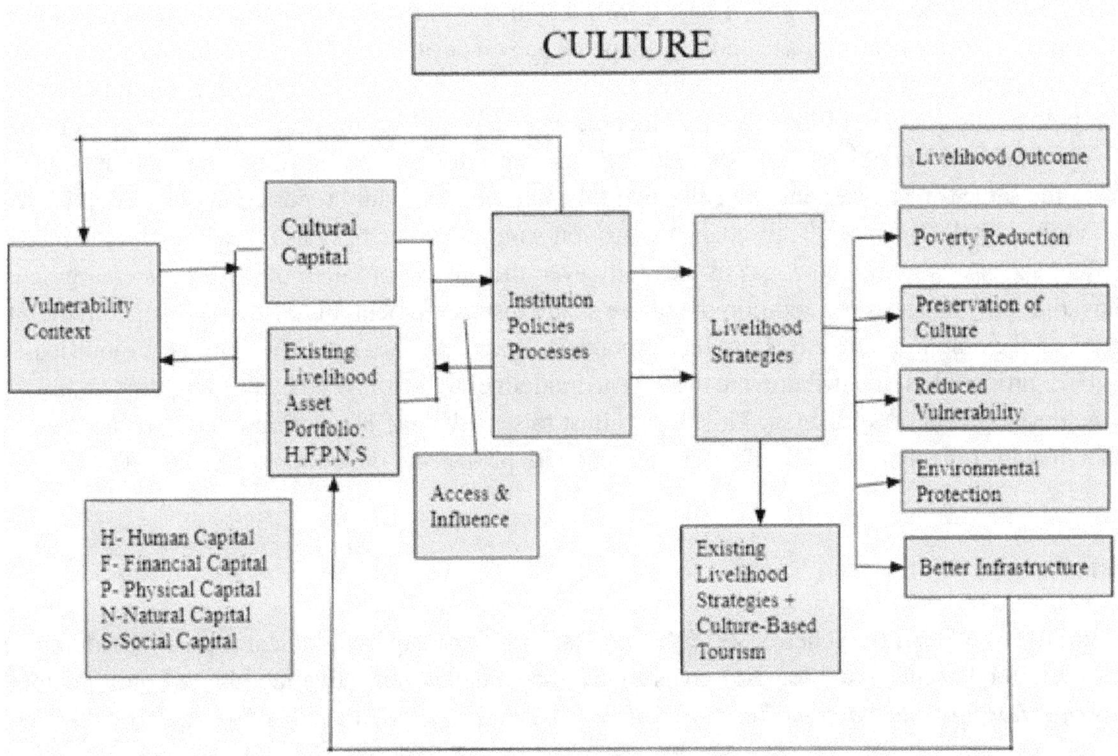

The framework suggests the addition of culture as not just an element, but also as a paradigm within which the livelihood systems of a place exist. Culture is more than just capital and studies like Addinsall (2017); Cahn (2006) attest to the same argument. Despite being such an important element, the majority of the livelihood studies have not adequately recognized the cultural element in the SL literature. This modified framework bridges this gap. However, much like every framework, this modified SLF does acknowledge the versatility of livelihood elements in terms of their demography and thus may not be generalized to all the geographical contexts. However, it does make a substantial contribution to the understanding of SL through this modified framework. Future researchers can test the applicability of this framework in different contexts and add or nullify any of the elements based on the demography or any other context of the research

7. CONCLUSION

Shriniketan was created on the principle of village reconstruction and it continues to grow with the same ideology. However, change is the only constant and over time, many new mechanizations and other livelihood opportunities have been introduced into the village system which Shriniketan has to adopt if it has to sustain its economy and culture. By adopting the DFID SLF this paper makes an investigation into the livelihoods system at Sriniketan. It was found that with rural poverty is still persistent in Sriniketan, many of the cultural elements are getting depleted thanks to the lack of funding and interest among residents. Migration to Kolkatta or more profitable occupations are also rampant which is adding oil to the fire. Further, the village of Sriniketan also suffers from higher vulnerabilities due to the drying up of the nearby rivers and over-dependence on agriculture for the livelihood generation. In this context, many of the respondents agree that culture-based tourism can in fact act as an alternative or supplementary source of income. This income can very well act towards the conservation of these depleting cultural embodiments. Additionally, the power of tourism in uplifting the economy of a place and at the same time conserving the culture of this placehas been duly reported in the academic works (Goodwin, 2007). However, Tthis change transition wouldn't come easy as the willingness to adapt to these changes was found to be less in the study even though the problem of poverty is emerging out. Thus, the study suggests the addition of culture as not just an element, but also as a context in the SLF. Based on the empirics collected from the respondents and observations of the authors, a modified SLF has been proposed as a contribution to theory and guidelines to be followed by the practitioners involved in the development of Sriniketan. This paper aimst to sensitize the local community, practitioners and academicians to reap the benefit of tourism along with preserving the culture of Shriniketan.

References

Addinsall, C., Weiler, B., Scherrer, P., & Glencross, K. (2016). Agroecological tourism: Bridging conservation, food security and tourism goals to enhance smallholders' livelihoods on South Pentecost, Vanuatu. *Journal of Sustainable Tourism*, *25*(8), 1100–1116. doi:10.1080/09669582.2016.1254221

Appadurai, A. (1981). The Past as a Scarce Resource. *Man*, *16*(2), 201. doi:10.2307/2801395

Aziz, R. C., Hashim, N., & Awang, Z. (2018). Tourism development in rural areas: Potentials of appreciative inquiry approach. *Journal of Tourism. Hospitality & Culinary Arts*, *10*(1), 59–75.

Batabyal, D., Mitra, A., & Goswami, S. (2020). Ecotourism Product Formulation for New Normal Tourism Practices: A Substitutability Approach for Selected Areas of Tamil Nadu, India. *Dogo Rangsang Res. J*, *10*(7), 200–218.

Birbhum D.O.W. (n.d.). *Brief History of Birbhum*. Birbhum DOW. http://birbhum.gov.in/geog1.htm

Brundtland, G. H. (1987). Our common future—Call for action. *Environmental Conservation, 14*(4), 291–294. doi:10.1017/S0376892900016805

Brundtland, G. H. (1987). World commission on environment and development. *Environmental Policy and Law, 14*(1), 26–30.

Butler, R., & Hinch, T. (2007). *Tourism and indigenous peoples: Issues and implications*. Routledge. doi:10.4324/9780080553962

Cahn, M. (2006). *Sustainable rural livelihoods, micro-enterprise and culture in the Pacific Islands: case studies from Samoa*. [Thesis, Massey University].

Chambers, R., & Conway, G. (1992). *Sustainable rural livelihoods: practical concepts for the 21st century*. Institute of Development Studies.

Das, M. (2017). Sriniketan: A Model Centre for Rural Reconstruction. *International Journal of Humanities And Social Science Invention, 6*(11), 57–61.

Daskon, C., & Binns, T. (2009). Culture, tradition and sustainable rural livelihoods: Exploring the culture-development interface in Kandy, Sri Lanka. *Community Development Journal: An International Forum, 45*(4), 494–517. doi:10.1093/cdj/bsp019

Espiner, S., Stewart, E. J., & Lama, L. T. (2017). Assessing the effectiveness of 'Appreciative Inquiry' (AI) in Nepali pro-poor tourism (PPT) development processes. *Tourism Planning & Development, 14*(3), 369–388. doi:10.1080/21568316.2016.1243148

Etikan, I., Musa, S. A., & Alkassim, R. S. (2016). Comparison of convenience sampling and purposive sampling. *American Journal of Theoretical and Applied Statistics, 5*(1), 1–4. doi:10.11648/j.ajtas.20160501.11

Gantait, A., Mohanty, P., & Swamy, A. (2018). Conservation and Management of Indian Built-Heritages: Exploring the Issues and Challenges. *South Asian Journal of Tourism and Heritage, 11*(1), 5–21.

Goffi, G., Cucculelli, M., & Masiero, L. (2019). Fostering tourism destination competitiveness in developing countries: The role of sustainability. *Journal of Cleaner Production, 209*, 101–115. doi:10.1016/j.jclepro.2018.10.208

Goodman, R. (2002). Pastoral Livelihoods in Tanzania: Can the Maasai Benefit from Conservation? *Current Issues in Tourism, 5*(3-4), 280–286. doi:10.1080/13683500208667924

Goodwin, C. J. (2005). *Research in Psychology: Methods and Design*. John Wiley & Sons, Inc.

Goodwin, H. (2007). *Indigenous tourism and poverty reduction* (Vol. 104). Elsevier.

Gutierrez-Montes, I., Emery, M., & Fernandez-Baca, E. (2009). The sustainable livelihoods approach and the community capitals framework: The importance of system-level approaches to community change efforts. *Community Development (Columbus, Ohio), 40*(2), 106–113. doi:10.1080/15575330903011785

Hall, C. M. (2005). Rural wine and food tourism cluster and network development. *Rural tourism and sustainable business, 26*, 149-164.

Huffer, E. (2006). Regionalism and cultural identity: Putting the Pacific back into the plan. *Globalisation, governance and the Pacific islands: State, society and governance in Melanesia*, 43-55.

Jenkins, T. (2000). Putting postmodernity into practice: Endogenous development and the role of traditional cultures in the rural development of marginal regions. *Ecological Economics*, *34*(3), 301–313. doi:10.1016/S0921-8009(00)00191-9

Koster, R. L., & Lemelin, R. H. (2009). Appreciative inquiry and rural tourism: A case study from Canada. *Tourism Geographies*, *11*(2), 256–269. doi:10.1080/14616680902827209

Krantz, L. (2001). The sustainable livelihood approach to poverty reduction. SIDA. *Division for Policy and Socio-Economic Analysis, 44*.

Ma, X., Wang, R., Dai, M., & Ou, Y. (2020). The influence of culture on the sustainable livelihoods of households in rural tourism destinations. *Journal of Sustainable Tourism*, *29*(8), 1–18.

Mason, P., Augustyn, M., & Seakhoa-King, A. (2010). Exploratory study in tourism: Designing an initial, qualitative phase of sequenced, mixed methods research. *International Journal of Tourism Research*, *12*(5), 432–448. doi:10.1002/jtr.763

Masud, M., Aldakhil, A., Nassani, A., & Azam, M. (2017). Community-based ecotourism management for sustainable development of marine protected areas in Malaysia. *Ocean and Coastal Management*, *136*, 104–112. doi:10.1016/j.ocecoaman.2016.11.023

Ministry of Finance. Govt. of India. (2019). *India Economic Survey, 2018*. Ministry of France. http://mofapp.nic.in:8080/economicsurvey/

Ministry of Tourism. Govt. of India. (2019). *India Tourism Statistics, 2019*. (pp. 7-10). Ministry of Tourism. https://tourism.gov.in/sites/default/files/Other/ITS_Glance_2018_Eng_Version_for_Mail.pdf

Norgaard, R. B. (1988). Sustainable development: A co-evolutionary view. *Futures*, *20*(6), 606–620. doi:10.1016/0016-3287(88)90003-1

Nyaupane, G., & Poudel, S. (2012). Application of appreciative inquiry in tourism research in rural communities. *Tourism Management*, *33*(4), 978–987. doi:10.1016/j.tourman.2011.10.009

Nyaupane, G. P., & Poudel, S. (2012). Application of appreciative inquiry in tourism research in rural communities. *Tourism Management*, *33*(4), 978–987. doi:10.1016/j.tourman.2011.10.009

Redclift, M. (2005). Sustainable development (1987-2005): An oxymoron comes of age. *Sustainable Development*, *13*(4), 212–227. doi:10.1002d.281

Reed, J. (2006). *Appreciative inquiry: Research for change*. Sage publications.

Rennie, J. K., & Singh, N. C. (1996). *Participatory research for sustainable livelihoods: a guidebook for field projects*. Iisd.

Roy, T. (2005). *Rethinking economic change in India: labour and livelihood*. Routledge. doi:10.4324/9780203023105

Salazar, N. (2012). Community-based cultural tourism: Issues, threats and opportunities. *Journal of Sustainable Tourism*, *20*(1), 9–22. doi:10.1080/09669582.2011.596279

Scoones, I. (1998). Sustainable rural livelihoods, A framework for analysis []). Brighton, IDS.]. *IDS Working Paper*, 72.

Serrat, O. (2017). The sustainable livelihoods approach. In *Knowledge solutions* (pp. 21–26). Springer. doi:10.1007/978-981-10-0983-9_5

Suen, L., Huang, H., & Lee, H. (2014). A comparison of convenience sampling and purposive sampling. *Hu li za zhi The journal of nursing, 61*(3), 105.

Swain, S. K., & Batabyal, D. (2016). Sustainable livelihood and its diversification through community-based ecotourism at Pichavaram and surroundings: An investigative study. *South Asian Journal of Tourism and Heritage*, *9*(2), 82–97.

Tao, T., & Wall, G. (2009). Tourism as a sustainable livelihood strategy. *Tourism Management*, *30*(1), 90–98. doi:10.1016/j.tourman.2008.03.009

U.S. National Science Foundation. (2000). *Towards a comprehensive geographical perspective on urban sustainability.* New Brunswick, NJ: Rutgers—The State University of New Jersey: http://web.worldbank.org/WBSITE/EXTERNAL/TOPICS/EXTPOVERTY/EXTPA/0,contentMDK:20208959~menuPK:443290~pagePK:148956~piPK:216618~theSitePK:430367~isCURL:Y,00.html

WTO. (2016). *Tourism sector highlighted as important contributor to trade and development* (pp. 1-2). UNWTO. Retrieved from https://www.wto.org/english/news_e/news16_e/bus_13jul16_e.htm

WTTC. (2018). *Travel and Tourism Economic Impact 2018: India* (pp. 1-3). United Kingdom: WTTC. https://www.wttc.org/-/media/files/reports/economic-impact-research/countries-2018/india2018.pdf

WTTC. (2020). *Economic Impact Reports*. WTTC. https://wttc.org/Research/Economic-Impact

Chapter 19
The Impact of Human Resource Challenges at Cultural Destinations and the Way Forward for Proper Destination Management

Udaidip Singh Chauhan
Parul University, India

ABSTRACT

The travel industry globally involves the movement of people from one place to another due to various reasons. Every human travelling has different needs and aspirations that they expect to get fulfilled at the places they travel. The involvement of humans as the key factor of travel acts as the indicators for the most complex and dynamic factors. It is not only the attraction that fulfills the note of satisfaction in the traveler, but every interaction at various places during the travel that creates an experience. Human Resource Management involves people at the centre of its study and helps develop the quality of the individual at every step. A destination with exemplary people friendliness and quality interaction induce repeat visits from travelers. Human Resource plays a vital role in the growth, development, quality and sustainability of any destination. The chapter shall work on the nuances of how Human Resource Management can be a differentiator in tourist experience at any destination and leave a positive footprint for the destination continuously.

INTRODUCTION

Destinations attract the unforeseen through its visual impact and leave the footprints in the mind of the traveler to follow. The visual imprint when translated into an experience for the tourist reflects the Moment of Truth created by the destination. There is a lot of cohesive and collaborative effort involved to make the experience happen for the traveler in reality. The strategic approach undertaken by the promot-

DOI: 10.4018/978-1-6684-9957-3.ch019

ers of any destination to carry forward the WOW effect is known as Destination Management. Tourism industry contributes to10% of global GDP (Neto, 2011). Tourism industry has created opportunities of jobs and entrepreneurship for various local communities at different destinations. More than 917 million tourists travelled to international destinations in the year 2022. Tourism has turned out to be a catalyst for socio-economic growth and development after the Covid-19 pandemic (Li et al., 2022). The involvement of many organizations and interests focused towards a single objective through the array of activities leads to efficient management of any destination.

People across the globe define the detail of coping with the world through a bond (Nash, 2023) or a bridge called culture. Travel acts as the facilitator for the human to find and explore the realms of ignorance and blend it within, what the world calls experience. Cultural and Heritage tourism is all about the tangible and un-tangible experiences that tourists embark on during a visit to a particular destination (Craik, 2002). The rich analogy of tangible and the un-tangible exchange between the tourist and the host population form a linkage between the known and the unknown. An immersive familiarity towards enrichment of knowledge, ideologies and rational scaffold breaks the premonitions and inhibitions of a tourist and transcends the tourist to a new plane. Experiences are the key to immersion and the right approach in forming the experiences depend on various attributes displayed by the host population during every single exchange of transaction. The cohesive nature of human service delivery leads to the basic crux of human centricity, the core essence of tourism (Green, 1997). Human Resource is a strategic and coherent approach of managing people and this realm of management of people lays the fundamentals of a cohesive activity in the arena of tourism too. A destination that is well trained is well gained. Future orientation is the main characteristic of the Human Resource Management (Cerovic, 2019). The better understanding and relationship between the tourist and the host population makes a destination enriching and sustainable. Human Resource is an inseparable constituent in tourism and each sector catering to the needs of the traveler require specialized skill sets as a resource for achieving the deliverables as required by the destination.

OBJECTIVES OF THE CHAPTER

- Understand the concept of culture and role of human resource at any tourist destination (cultural and heritage) and its positive impact on tourists and their experience.
- List the issues that act as constraints to human resource at various destinations of cultural value.
- Understand how a destination can be a growth engine culturally, socially and economically through the suggested remedies on the influence of better human resource aspects.

The word culture is derived from the Latin term *cultura*, from which the English word culture derives. This phrase was frequently used in 18th-century England to describe to the improvement of one's character through the refinement of judgment, taste, and intelligence; and, by extension, to those activities believed to express and maintain this sophistication (Williams, 1983). In 1871, British anthropologist Edward Taylor blended parts of both notions to define culture as "that complex whole that includes knowledge, belief, art, law, morals, custom, and any other skills and practices acquired by man as a member of society" (as cited in Kroeber & Kluckhohn, 1952, p. 81). This is largely regarded as the first formal anthropological definition of the term, as it introduced the concept of culture as a taught, shared, and inclusive framework that encompasses practically every aspect of human social life.

CHARACTERISTICS OF CULTURE

1. Passed by ancestors: The culture one follows is passed by the previous generation or is acquired through heredity. People inherit certain characteristics and qualities from their forefathers (Boyd & Richerson, 1982). People learn from their family members or from their society. Social and physical environments also have a great impact on the culture of a particular community.
2. Dynamic in nature: Culture differs from place to place, people to people, different people follow different religions, have different eating habits, people dress up differently, follow different customs & rituals and speak a different language (Hodder, 2010). Shared by community: Different communities around the world share a common culture whether it's their language, food, dress or religion.
3. Culture is learned from other humans: it is neither inherited nor biological. People learn about culture from their family, friends, society, peers, media and institutions (Arnett, 2005). The process of learning continues throughout the life of a human.
4. Culture is associated with symbols: Culture is often linked to a particular symbol for instance language, dialect or sign associated with a religion, institution (Agha, 2004).
5. Culture is interlinked: The components of culture are linked with each other, in order to understand a particular culture of a community one must explore all its components.

SKILLS AND COMPETENCIES IN RELIGIOUS, CULTURAL, AND HERITAGE TOURISM

Cultural tourism is one of the most popular forms of tourism throughout the world and research indicates that the same has lead to regional developments in certain parts of the world. The essential motivation of the traveler involved in cultural tourism includes tangible and intangible cultural attractions and products. Human interaction is very important in this form of tourism as it encompasses understanding of art, architecture, historical heritage, living cultures, lifestyles, value systems, beliefs, traditions and other aspects. The chapter shall incorporate the importance of human interfaces in tourism with reference to cultural tourism. Skills and competencies in religious, cultural and heritage tourism are inseparable entities. Culture and tourism are driving force in many countries of the world for growth and development of tourism. Many countries have gained a strong position through tourist inflow due to their architectural marvels and famous world heritage sites. Religion is a facet of cultural tourism and the faith or belief of different communities lead to exchange of humans from one place to another making it a prominent aspect for development of tourism in different regions. According to Richards (2018), Cultural tourism is an activity in which the main motivation of the visitor are to learn, discover, experience and consume both material and intangible cultural products of the place being visited. It is a confluence of a group of products in a synergy that blends culture, society, territory which in economic terms translates into high revenues per day (Moere 2017). Religious tourism on the other hand is quite a new tourism product globally where people travel to religious places or visit sacred sites not only for religious purpose but also to engage in cultural tourism based around religious themes.

Religious monuments play a major role in tourism; the religious architectures are enduring and impressive. The rituals followed by different religions carry a wealth of tradition and culture with them. The need for human interface at cultural and religious destinations manifolds because of multiple and varied interactions that travelers have with various elements present in the dynamics of the destination. In 1989, UNESCO General Conference gave a definition of traditional and popular culture that transcended to the term "Culture". According to the definition, "the totality of tradition based creations of a cultural community expressed by a group or individuals and recognized as reflecting the expectations of a community in so far as they reflect its social and cultural identity; its standards and values are transmitted orally, by imitation or by other means. Human resource the coherent approach of effectively managing people is defined according to Megginson as "the knowledge, skills, creative abilities, talents and aptitudes obtained in the population; whereas from the viewpoint of individual enterprise, they represent the total of the inherent abilities, acquired knowledge and skills as exemplified in the talents and aptitudes of its employees". The role of human resource at any cultural destination is manifold and it is layered under every interaction the traveler undertakes at the destination and the needs fulfilled. If exemplified concisely the role of human resources is important in any destination because of some of the under mentioned reasons. They are

- **The Connect Competence Factor**

Connecting with travelers at a destination with the right competency to understand and fulfill the needs requires human resource that is trained. Trained manpower at any destination understands the needs faster and delivers faster and hence connects faster leading to satisfaction and exemplary experience levels for the traveler.

- **The Add Value Factor**

Trained human resources make travelers better aware about the importance of any cultural destination leading to value addition of the destination. The same propagates in a positive direction leading to higher popularity index of the destination.

- **The Creative Factor**

There is a huge predominance of new concepts, products and services when human resources are trained. These concepts often make the destination unique adding up to the mark up value. Travelers pick up the creative tinge and it acts as a major pull factor. The uniqueness and creative bent can be in any aspect and may be the dominant pull factor or a subset of the core tourism product.

- **The Economic Factor**

Trained Human workforce at any destination adds up to better money flow and utilization at the destination. The economy of the region strengthens leading to higher income and better standard of living for the host population. Training of workforce happens in multifarious directions leading to multiple ancillary subsets strengthening under the umbrella of the core product. A typical hub and spoke model makes the destination strong and self reliant. The economic development affects various other needs too. One of the main issues often prevalent in many cultural destinations especially at remote locations is the development of infrastructure. Economic strengthening leads to development of a strong infrastructural set up.

- **The Diversity Factor**

Economic growth coupled with innovation and a strong social structure in any cultural destination leads to diversification of opportunities. Diversification requires competency intervention and human resource training. The innovation spree leads to more improved products and services and better marketing of the products to travelers.

- **The Social Factor**

When culture interacts with other cultures and brings economic development, there arise the needs to sustain the existing culture for future economic sustenance. The social structure strengthens further paving the path for an improved society. A better society strengthens the development of future generations.

- **The Growth Factor**

The development of human resources at any cultural destination happens through training and making the host population understand the positive outcomes. With continuous and diverse training and development a destination prepares itself with a team of professionals. The skill set availability for different roles makes the destination more investment lucrative and leads to growth.

- **The Adaptability Factor**

Human Resource development in cultural sites leads to better adaptability of the host population to the needs and the culture of the traveler. It leads to better acceptance and the host population is ready to change according to the needs. With change in the circumstances there is a change amongst the host population leading to better acceptance and understanding.

HUMAN RESOURCE ISSUES AT CULTURAL HERITAGE AND PILGRIMAGE SITES

The interest in describing contemporary travel to cultural and sacred sites (Nolan and Nolan, 2009) has led to proliferation of composite terms such as Religious tourism, Pilgrimage tourism etc. The backward and forward linkages of tourism leading to the growth of other sectors increases the income of the local people as well as effective demand (Gokovali, Bahar, 2014). The high level of human involvement for its survival, growth, development and delivery of various services characterizes its distinct phenomena from another. With the demand of skilled manpower required at any destination especially cultural sites, the problems associated with the same are also manifold. In modern times, human capital represents overall quantitative and qualitative value of its human resources. This capital is crucial for modern organization due to direct connection with performances. The index of global competitiveness measured by World Economy Forum indicates that intellectual capital impact level to country competitiveness is 58 - 70%. This level indicates that influence of human capital (education, skills, and abilities), organization and technologies is dominant, so it represents the most important parameters that determine competitive position of countries. Countries that invest in education have higher GDP, which also implies that education development and expertise are crucial in economy. (Albijanić, M., 2011.30). The Quantum of knowledge and skills determines the essence of competence for the tourism industry as a whole. The issues related to human resource at cultural heritage and pilgrimage sites can be elucidated as

- The huge migration of population at cultural destinations especially skilled labour from particular pockets leads to formation and concentration of power within the migrated labour. The same leads to resistance of manpower from other regional pockets and there is always a tussle for dominance and power.
- Housing and Basic necessities are often a big challenge to the migrated labour population. There is either concentration of labour in particular areas that lead to challenges of hygiene and health. The poor condition of living leads to multifarious challenges amongst the labour population and often a big issue for the destination.
- There is often a mass exodus of the labour workforce during peak seasons leading to severe shortage of manpower to work and look after the needs of the tourists at the destinations. The same often leads to poor service and fluctuations in the delivery factors to travelers. Employers who get cheap labour are majorly dependent on the same and are unable to find alternatives in times of crisis. The crisis affects the allied industries too that act as feeder to the tourism industry often hampering the growth and development of a complete linkage between the destination and the traveler. The cost of hiring is also very expensive at many destinations leading to higher price index of goods and services at destinations. The gap makes the destination incomplete and often do not qualify for the yardsticks of sustainability.
- The availability of cheap migratory labour population often leads to mass exploitation by the owners of various businesses. The exploitation leads to poor delivery of services and continuous rotation of manpower in different establishments. Exploitation leads to exposure of untrained personnel for any particular job and the same is unable to fulfill the guests needs and expectations.
- With a huge population of people working at any cultural destination from outside and within and each one fighting for sustenance, there is always a chance of earning quick money using wrong practices. Many destinations have earned a bad repute in terms of overcharging travelers for even basic services. There is also a gross chance of increase in crime rate at destination and prevalence of antisocial activities making the destination difficult for tourists. Illegal activities are also prevalent in many destinations leading to disrepute of the destination leading to a lot of travelers avoiding the same. The skirmishes between locals and visitors are also a major issue in many destinations. The dominance of the local population and their non-acceptance of travelers or certain selected group of travelers often induce skirmish.
- With continuous exposure to different cultures and fulfilling the needs of the travelers from different parts, the host population is majorly affected leading to copying of practices and customs followed by travelers. Often the host population is also inspired to follow. The same often leads to change of food habits, modification of culture and beliefs affecting generations.
- The challenges of the external environment of tourism organizations (such as increased competition, technology innovations, fast-evolving markets, and economic constraints) lead to a continuing change in the kind and quality of the Knowledge, Skills and Attitude (KSAs) needed in the human resource, besides a growing demand for multi- skilled employees (Foster, McCabe, & Dewhurst, 2010). Such needed variation in workforce KSAs would enable tourism organizations to cope with the external environment challenges (Zehrer, 2011).

- The lack of education related to tourism industry within the host population creates aversion and only a few works and follow the unprofessional norms as prevalent. There is no initiative for a dynamic change.

THE WAY AHEAD FOR HUMAN RESOURCE ISSUES AT CULTURAL HERITAGE AND PILGRIMAGE SITES (WORKABLE SOLUTIONS)

Cultural destinations are the backbone of tourism for many countries and especially a country like India. With human resource challenges thriving majorly within the tourism industry at cultural destinations there have been many research and implementable solutions as worked out by many professionals and researchers. Some of the workable solutions include

- Standardized human resource practices and strategies aimed at personnel working in these destinations, host community and tourists, underpinned by the basic concept of sustainability of the cultural destination.
- Addressing of issues that are rapidly pertinent in the industry like inequitable salaries in comparison to other industries, lack of right training and development of workforce, strategic selection of workforce, quality of life, less salaries, long working hours, work life balance and many others may reduce attrition and lead to better and quality manpower at destinations.
- Increasing tourism education and its quality at destinations leading to inclusive insight into the needs of the destination. The same shall provide a better planning for resource strategy (Armstrong, 2012) to meet the destination needs and fulfillment of the gaps in terms of human resource pitfalls.
- Forecasting future demand/supply for employees in terms of numbers and skills, determining the needed actions when forecasts indicate the deficit or surplus of workforce (Jackson, 2012).
- Analyzing labour turnover statistics in order to use these data as an input for supply forecasting, tracking the reason for workforce exodus, resignation, and proposing actions to improve retention rates.
- Analyzing the operational effectiveness in order to make better use of the employees and adopting a flexibility approach that ensures satisfaction of the work force needs and effectiveness at work.
- Making human resource planning and development a focal activity at cultural destinations that creates a link between business and human resource and support the effective integration. (Lam and Schroebock, 2008).
- Create a pool of dynamic component of different variables through engagement bringing in a pool of skilled manpower within the destination delivering higher productivity effectiveness (Baum, 2018).
- Incorporation of philosophy that investment in quality human resource at cultural destinations is not a cost but an investment towards better outcome of the results as need to be achieved by the destination and also for its sustainability.
- The transition of organizations from traditional to knowledge-based ones, especially in tourism services (Shaw & Williams, 2009), shall transform human resource from assets- to value-creating entities and a key competitive advantage. Thus, value-adding measures should be included in the total human resource assessment at any destination (Mayo, 2012).

IMPLICATIONS AND FURTHER SCOPE OF FUTURE RESEARCH

Practical Implications

A key "take-away" of this study is that organizations working at cultural tourism destinations can use this model to assess their key aspects and variables related to human resource in the organization. Whether dominated by strategic context (i.e., outlining goals and estimating future cost and profit without much intervening in the practices of the line management), or whether more oriented towards human resource management context (i.e., investing in the procurement and development of human capital to achieve a competitive advantage), this study can aid different organizations and stake holders at cultural destinations to rethink their human resource management context and reconsider the practices that eliminate the balance between strategic goals and employees' needs. This study, also, offers the stakeholders main human resource practices considered to lie beneath ineffective HR planning. To achieve credibility in human resource planning at tourism destinations, stakeholders can target the reorganization of such practices, especially that they are all interrelated, and a change in one practice that would strongly affect the rest. This means, for example, when stakeholders focus on the antecedents of the core challenges, the effect of the consequential challenges on human resource planning and managing process would be, accordingly, limited. The researcher suggests that stakeholders start considering the factors that have a single influence relation with the bigger factor which in turn provide an opportunity for stakeholders to enhance the accuracy of implementation of the plans by paying much attention to these factors. On the contrary, other factors have double way influence relations and are dramatically affected by other factors. In terms of "Ignoring daily administrative," stakeholders should change their attitude towards human resource management by focusing on the day-to-day management of employees' wellbeing, problems, performance and behaviour. The stakeholders should put much effort in enhancing human resource information system in at all levels within the destination in parallel with awareness of the probable factorial antecedents thus developing procedures that eliminate their negative impact on human resource constraints. Finally, this study suggests that, if a destination can create a context that emphasizes managing human capital as an asset through good and hygienic practices and styles, it can move towards reliable human workforce planning that provides human capital with significant value to the destination and a competitive advantage.

SUMMARY

This chapter explains the concept and role and impact of Human Resource Management in Destination Management of Cultural and Heritage Sites. Cultural heritage of the destination plays an important role in creating tourism products which are offered as unique selling tourism products to the tourists (Nagy, 2012). Cultural heritage of a destination attracts tourists to visit a destination. The study also explains the concept of the importance of Human Resource Management and challenges that cultural and religious destinations encounter in terms of human resource. The growth in Cultural heritage tourism has helped many destinations to protect and preserve their cultural heritage. Positive aspects and negative aspects of cultural heritage tourism have also been highlighted. The positive side of cultural heritage tourism helps in protection & promotion of local cultural heritage, strengthen social values, revive local art forms, create opportunities for business as well as employment (Zhuang,2019). On the other hand, cultural heritage tourism brings the negative side which includes convergence of culture, overcrowding, commodification of culture, creation of hybrid culture, overpricing and arise in difference of opinions amongst the local residents and tourists (Mele, 2000). The chapter opens a number of avenues for future research. Having in mind the lack of empirical data on human resource planning at cultural tourism destinations that are more dominated by the unorganized workforce sector, the model can guide future researchers to provide necessary quantitative evidence on the relationships among human resource planning challenges, determine the range of impact such challenges have on human resource planning process at these destinations; define the challenges of high priorities, and, thereby, contribute to the knowledge of possible solutions and steps of solutions for implementation.

REFERENCES

Agha, A. (2004). *Registers of language. A companion to linguistic anthropology, Armstrong, M. (2012). Armstrong's handbook of human resource management prac-tice* (12th ed.). Kogan Page Publishers.

Archer, B., Cooper, C., & Ruhanen, L. (1998). The positive and negative impacts of tourism. *Global tourism, 2*, 63-81.

Bascom, W. (1965). The forms of folklore: Prose narratives. *Journal of American Folklore, 78*(307), 3–20. doi:10.2307/538099

Bhatia, A. K. (2002). *Tourism development: Principles and practices*. Sterling Publishers Pvt. Ltd.

Buhalis, D. (2000). Marketing the competitive destination of the future. *Tourism Management, 21*(1), 97–116. doi:10.1016/S0261-5177(99)00095-3

Chalana, M., & Sprague, T. S. (2013). Beyond Le Corbusier and the modernist city: Reframing Chandigarh's 'World Heritage'legacy. *Planning Perspectives, 28*(2), 199–222. doi:10.1080/02665433.2013.737709

Cole, S. (2006). Cultural tourism, community participation and empowerment. *Cultural tourism in a changing world: Politics, participation and (re) presentation*, 89-103.

Craik, J. (2002). The culture of tourism. In *Touring cultures* (pp. 123–146). Routledge.

Dahles, H. (2002). Tourism, small enterprises and community development. In *Tourism and sustainable community development* (pp. 172–187). Routledge.

Edgell, D. L. Sr. (2019). *Managing sustainable tourism: A legacy for the future*. Routledge.

Eliot, T. S. (2010). *Notes towards the Definition of Culture*. Faber & Faber.

Green, T. A. (Ed.). (1997). *Folklore: an encyclopedia of beliefs, customs, tales, music, and art* (Vol. 1). Abc-clio.

Kurtz, L. R. (2015). *Gods in the global village: the world's religions in sociological perspective*. Sage Publications.

Li, Y., Lau, C., & Su, P. (2020). Heritage tourism stakeholder conflict: A case of a World Heritage Site in China. *Journal of Tourism and Cultural Change, 18*(3), 267–287. doi:10.1080/14766825.2020.1722141

Li, Z., Wang, D., Abbas, J., Hassan, S., & Mubeen, R. (2022). Tourists' health risk threats amid COVID-19 era: Role of technology innovation, Transformation, and recovery implications for sustainable tourism. *Frontiers in Psychology, 12*, 769175. doi:10.3389/fpsyg.2021.769175 PMID:35465147

Mele, C. (2000). *Selling the lower east side: Culture, real estate, and resistance in New York City* (Vol. 5). U of Minnesota Press.

Munjal, P. G. (2021). Charting the role of digital platforms for cultural heritage tourism in India. *Worldwide Hospitality and Tourism Themes, 13*(2), 186–198. doi:10.1108/WHATT-10-2020-0126

Neto, F. (2003, August). A new approach to sustainable tourism development: Moving beyond environmental protection. []). Oxford, UK: Blackwell Publishing Ltd.]. *Natural Resources Forum, 27*(3), 212–222. doi:10.1111/1477-8947.00056

Oklevik, O., Gössling, S., Hall, C. M., Steen Jacobsen, J. K., Grøtte, I. P., & McCabe, S. (2019, December 02). Overtourism, optimisation, and destination performance indicators: A case study of activities in Fjord Norway. *Journal of Sustainable Tourism, 27*(12), 1804–1824. doi:10.1080/09669582.2018.1533020

Compilation of References

Yuen, B. (2005). Strengthening Urban Heritage In Singapore: Building Economic Competitiveness And Civic Identity. *Global Urban Development, 1*(1). https://docplayer.net/41368762-Strengthening-urban-heritage-in-singapore-building-economic-competitiveness-and-civic-identity.html

Aayog, N. (2019). *Improving Heritage Management in India.* Government of India.

Abbate, C. S., & Di Nuovo, S. (2013). Motivation and personality traits for choosing religious tourism. A research on the case of Medjugorje. *Current Issues in Tourism, 16*(5), 501–506. doi:10.1080/13683500.2012.749844

Abdollahzadeh, G., & Sharifzadeh, A. (2014). Rural residents' perceptions toward tourism development: A study from Iran. *International Journal of Tourism Research, 16*(2), 126–136. doi:10.1002/jtr.1906

Abdullah, M. R. (2018).. . *Culture, Religion, and Freedom of Religion or Belief., 16*(4), 102–115. doi:10.1080/15570274.2018.1535033

Abeysinghe, A. R. G. M. I., & Ariyawansa, R. G. (2017). Competitiveness of Kandy City as a Heritage Tourist Destination in Sri Lanka. *Sri Lankan Journal of Real Estate.* http://dr.lib.sjp.ac.lk/handle/123456789/8421

Abouelmagd, D. (2023). Sustainable urbanism and cultural tourism, the case of the Sphinx Avenue, Luxor. *Alexandria Engineering Journal, 71*, 239–261. doi:10.1016/j.aej.2023.03.041

Abou-Shouk, M., & Soliman, M. (2021). The impact of gamification adoption intention on brand awareness and loyalty in tourism: The mediating effect of customer engagement. *Journal of Destination Marketing & Management, 20*, 100559. doi:10.1016/j.jdmm.2021.100559

About West Zone Cultural Center. (n.d.). WZCC. https://wzccindia.com/about-west-zone-cultural-center-wzcc/

Addinsall, C., Weiler, B., Scherrer, P., & Glencross, K. (2016). Agroecological tourism: Bridging conservation, food security and tourism goals to enhance smallholders' livelihoods on South Pentecost, Vanuatu. *Journal of Sustainable Tourism, 25*(8), 1100–1116. doi:10.1080/09669582.2016.1254221

Adie, B. A. (2019). *World Heritage and tourism: Marketing and management.* Routledge. doi:10.4324/9780429506659

Agapito, D., & Chan, C. S. (2019). A multisensory approach to responsible management in community-based tourism: A case study in Hong Kong. *Journal of Tourism Quarterly, 1*(1/2).

Agha, A. (2004). *Registers of language. A companion to linguistic anthropology, Armstrong, M. (2012). Armstrong's handbook of human resource management prac-tice* (12th ed.). Kogan Page Publishers.

Aiello, D., Fai, S., & Santagati, C. (2019). Virtual Museums as a Means for Promotion and Enhancement of Cultural Heritage. *The International Archives of the Photogrammetry, Remote Sensing and Spatial Information Sciences, XLII-2*(W15), 33–40. doi:10.5194/isprs-archives-XLII-2-W15-33-2019

Ainsworth, M. D. S., & Bowlby, J. (1991). An ethological approach to personality development. *The American Psychologist, 46*(4), 331–341. doi:10.1037/0003-066X.46.4.333

Aisyah, S. (2023). Implications of Planning and Management for the Development of Sustainable Cultural Heritage Tourism in Malaysia. *Advances in Tourism Studies, 1*(2).

Akal, M. (2010). Economic Implications of International Tourism on the Turkish Economy. *Tourismos: An International Multidisciplinary Journal of Tourism, 5*(1), 131–152.

Akhtar, N., Khan, N., Mahroof Khan, M., Ashraf, S., Hashmi, M. S., Khan, M. M., & Hishan, S. S. (2021). Post-covid 19 tourism: Will digital tourism replace mass tourism? *Sustainability (Basel), 13*(10), 5352. doi:10.3390u13105352

Akpojaro, J., & Bello, R.-W. (2019). Smart Healthcare System: A Primer. In *Peer-Reviewed, Refereed, Indexed Journal with IC,* (5). https://www.kernelsphere.com/four-internet-things-communications-models/

Aktürk, G., & Lerski, M. (2021, May 8). Intangible cultural heritage: A benefit to climate-displaced and host communities. *Journal of Environmental Studies and Sciences, 11*(3), 305–315. doi:10.100713412-021-00697-y PMID:33996378

Al Amri, I. T., Sbai, A., & Ben Rabia, K. (2019). Tourism constituents and tourism marketing in Oman. *Dhofar Governorate as a Model, 216*(2), 1-17. https://mrk.journals.ekb.eg/article_100405_165853468a3d5f06b9156f88895c8596.pdf

Al Riyami, H., Almuhrzi, H., & Scott, N. (2022). Heritage hotel experiences in Nizwa, Oman. *International Journal of Tourism Research, 24*(1), 169–180. doi:10.1002/jtr.2492

al-Busaidi, Y. S. (2004). The protection and management of historic monuments in the Sultanate of Oman: the historic buildings of Oman. Proceedings of the Seminar for Arabian Studies, Al-Ghafri, A. S. (2008). Traditional water distribution in Aflaj irrigation systems: Case study of Oman. *What Makes Traditional Technologies Tick? A Review of Traditional Approaches for Water Management in Drylands*, 74.

Alexis, T. (2022). Heritagization of religious sites: In search of visitor agency and the dialectics underlying heritage planning assemblages. *International Journal of Heritage Studies, 28*(9), 1036–1065. doi:10.1080/13527258.2022.2122535

Al-Hashim, A. (2015). Embracing sustainable tourism in Oman: case study of Mirbat settlement. *International Journal of Cultural and Digital Tourism, 2*(2). 8-16.

Alkathiri, N. A., & Soliman, M. (2022). Examining foreign direct investment determinants of tourism industry in Oman and Egypt: The moderating role of investment environment. International Journal of Finance & Economics, 27(4), 4722–4740.

Allen, L. R., Long, P. T., Perdue, R. R., & Dieselbach, S. (1988). The impact of tourism development on resident's perception of community life. *Journal of Travel Research, 27*(1), 16–21. doi:10.1177/004728758802700104

Al-Marshoudi, A. S., Bin Sibly, S., & Khelghat-Doost, H. (2021). Successfully mutual cooperative collective action: Principle of institutional arrangement of the aflaj irrigation system in Sultanate of Oman. Water resources in arid lands: Management and sustainability.

Almeida, F., Buzady, Z., & Ferro, A. (2021). Exploring the role of a serious game in developing competencies in higher tourism education. *Journal of Hospitality, Leisure, Sport and Tourism Education, 29,* 100347. doi:10.1016/j.jhlste.2021.100347

Almuhrzi, H., Hughes, K., & Ballantyne, R. (2020). Exploring Arab and Western visitors' interpretive experiences at an Omani heritage site: Does one size fit all? *Journal of Heritage Tourism, 15*(2), 180–199. doi:10.1080/1743873X.2019.1632319

Alnafeesi, A. M. (2013). The Role & Impact of Preservation Architectural Heritage on Domestic Tourism. *Proceedings of International Conference on Tourism Development*, 93-100. https://www.semanticscholar.org/paper/The-role-%26-impact-of-preservation-architectural-on-Alnafeesi-Mohamed/ed9c1eb33a237f7b34e2654936a383055eac1520

Al-okaily, N. S., Alzboun, N., Alrawadieh, Z., & Slehat, M. (2023). The impact of eudaimonic well-being on experience and loyalty: A tourism context. *Journal of Services Marketing*, *37*(2), 216–231. doi:10.1108/JSM-08-2021-0282

Alvaro Fierro, I. A. (2018). Airbnb Branding: Heritage as a Branding Element in the Sharing Economy. *Sustainability*.

Amaro, S., Antunes, A., & Henriques, C. (2018). A closer look at Santiago de Compostela's pilgrims through the lens of motivations. *Tourism Management*, *64*, 271–280. doi:10.1016/j.tourman.2017.09.007

Amaro, S., Antunes, A., & Henriques, C. (2018). A closer look at Santiago de Compostela's pilgrims through the lens of motivations. *Tourism Management, 64*, 271-280..

Amaya-MolinarC. M.Sosa-FerreiraA. P.Ochoa-LlamasI.Moncada-JiménezP. (2017). The perception of destination competitiveness by tourists. *Revista Investigaciones Turísticas,* 14, 1-20. https://dx.doi.org/ doi:10.14461/INTURI2017.14.01

Amin, S. (2021). *An Overview of South Asian Tourism Industry and Significance of the Regional Tourism Integration.* Springer Books. In *The Economy of Tourism in Bangladesh* (pp. 157–178). Springer. doi:10.1007/978-3-030-72900-4_7

Andares, L., & Dimanche, F. (2014). Co-Creation of experience value: A tourist behavior approach. In N. Prebensen, J. Chen, & M. Uysal (Eds.), *Creating experience value in tourism Bloomington* (pp. 95–112). CABI.

Andriotis, K. (2009). Sacred Site Experience: A Phenomenological Study. *Annals of Tourism Research, 36*(1), 64–84. doi:10.1016/j.annals.2008.10.003

Antonaci, A., Ott, N., & Pozzi, F. (2013). Virtual Museums, Cultural Heritage Education and 21st Century skills. Learning & Teaching with Media & Technology, Genoa, Italy.

Apostolakis, A. (2003). The convergence process in heritage tourism. *Annals of Tourism Research, 30*(4), 795–812. doi:10.1016/S0160-7383(03)00057-4

Appadurai, A. (1981). The Past as a Scarce Resource. *Man, 16*(2), 201. doi:10.2307/2801395

Archer, B., Cooper, C., & Ruhanen, L. (1998). The positive and negative impacts of tourism. *Global tourism, 2*, 63-81.

Aref, F., & Redzuan, M. (2009). Religious Tourism and Sustainable Tourism Principles: A Case Study of Iran. Tourism Management, 30(3), 465–472.

Arora, R., & Singer, J. (2006). Cognitive and affective service marketing strategies for fine dining restaurant managers. *Journal of Small Business Strategy, 17*, 51–61.

Arumugam, A., Nakkeeran, S., & Subramaniam, R. (2023). Exploring the factors influencing heritage tourism development: A model development. *Sustainability (Basel), 15*(15), 11986. doi:10.3390u151511986

Ashworth, G. J., & Larkham, P. J. (2013). *Building a new heritage: Tourism, culture, and identity in the new Europe*. Routledge. https://books.google.co.in/books?hl=en&lr=&id=qWS6oMt5x7MC&oi=fnd&pg=PR1&dq=64.%09Ashworth,+G.+J.,+%26+Larkham,+P.+J.+(2013).+Building+a+new+heritage:+Tourism,+culture,+and+identity+in+the+new+Europe.+Routledge.&ots=nWmKE1Fb0h&sig=OIFvGq6QwxhcQACUzvYTNbQlqHA&redir_esc=y#v=onepage&q&f=false

Ashworth, G., & Tunbridge, J. (2000). The tourist city. *The Tourist-Historic City*, 53–81. doi:10.1016/B978-0-08-043675-3.50007-7

AŞKIN, R. A. D. (2017). New Consumption Trends Among Muslims In Turkey: Understanding Muslims In Terms Of Modern Consumptions. *PROCEEDINGS BOOKS*, 299.

Asokan, G. V., Srihari, K. R., & Kumar, S. S. (2000). Bull racing injuries in southern India. *Veterinary*, *147*(19), 556. PMID:11095398

Asral, I., Woosnam, K. M., Baldwin, E. D., & Backman, S. J. (2010). Residents as travel destination information providers: An online community perspective. *Journal of Travel Research*, *49*(4), 400–413. doi:10.1177/0047287509346856

Assari, A., & Assari, E. (2012). Urban spirit and heritage conservation problems: Case study Isfahan city in Iran. *The Journal of American Science*, *8*(1), 202–209. https://www.academia.edu/1314591/Urban_spirit_and_heritage_conservation_problems_case_study_Isfahan_city_in_Iran

Avcı, M., & Memişoğlu, H. (2016). Kültürel Miras Eğitimine İlişkin Sosyal Bilgiler Öğretmenlerinin Görüşleri. *İlköğretim Online*, *15*(1), 104–124. doi:10.17051/io.2016.42123

Aytekin, H., & Aktaş, G. (2023). Türkiye'de Sanal Müze Araştırmaları: Bibliyometrik Analiz. *Journal of Business in the Digital Age*, *6*(1), 1–10.

Ayuningtyas, D., & Ariwibowo, D. A. (2020). The strategic role of information communication technology in succeeding medical tourism. *Enfermeria Clinica*, *30*, 170–173. doi:10.1016/j.enfcli.2020.06.039

Aziz, R. C., Hashim, N., & Awang, Z. (2018). Tourism development in rural areas: Potentials of appreciative inquiry approach. *Journal of Tourism. Hospitality & Culinary Arts*, *10*(1), 59–75.

Babic, D. (2015). *Socially responsible heritage management - empowering citizens to act as heritage managers*. Social and Behavioral Sciences.

Babin, B. J., Darden, W. R., & Babin, L. A. (1998). Negative emotions in marketing research: Affect or artifact? *Journal of Business Research*, *42*(3), 271–286. doi:10.1016/S0148-2963(97)00124-0

Babu, V., & Munjal, S. (2015). Oachira Panthrandu Vilakku: A study of culturally embedded festival aligned with economic benefits. *Worldwide Hospitality and Tourism Themes*, *7*(4), 403–416. doi:10.1108/WHATT-03-2015-0009

Bachleitner, R., & Zins, A. H. (1999). Cultural tourism in rural communities: The Residents perspective. *Journal of Business Research*, *44*(3), 199–209. doi:10.1016/S0148-2963(97)00201-4

Backman, S., Uysal, M., & Backman, K. (1991). Regional Analysis of Tourism Resources. *Annals of Tourism Research*, *8*(1), 323–327. doi:10.1016/0160-7383(91)90016-5

Bagri, S. C. (1999). Status of Travel & Tourism Industry in India, A Futuristic View of Growth. *Journal of Tourism*, *41*.

Baharak Ashrafi, M. K. (2020). Heritage Impact Assessment, beyond an Assessment Tool: A comparative analysis of urban development impact on visual integrity in Four UNESCO World Heritage Properties. *Journal of Cultural Heritage*.

Bakhshipour-Roudasri, A. (2008). Relationship between the life satisfaction, problem solving style and religious attitude with mental health. Journal of Psychology. *University of Tabriz*, *9*, 8–26.

Baloglu, S., & Brinberg, D. (1997). Affective images of tourism destinations. *Journal of Travel Research*, *35*(4), 11–15. doi:10.1177/004728759703500402

Banerjee, S. C. (1992). *Tantra in Bengal: A study of its origin, development, and influence*. South Asia Books.

Bansiddhi, P., Brown, J. L., Thitaram, C., Punyapornwithaya, V., & Nganvongpanit, K. (2020). Elephant Tourism in Thailand: A Review of Animal Welfare Practices and Needs. *Journal of Applied Animal Welfare Science*, *23*(2), 164–177. doi:10.1080/10888705.2019.1569522 PMID:30689409

Barlas Bozkuş, Ş. (2014). Kültür ve Sanat İletişimi Çerçevesinde Türkiye'de Sanal Müzelerin Gelişimi. *The Journal of Academic Social Science Studies*, *26*(26), 329–344. doi:10.9761/JASSS2408

Barnes, S. J., Mattsson, J., & Sorensen, F. (2014). Destination brand experience and visitor behavior: Testing a scale in the tourism context. *Annals of Tourism Research*, *48*, Vol-88. doi:10.1016/j.annals.2014.06.002

Bascom, W. (1965). The forms of folklore: Prose narratives. *Journal of American Folklore*, *78*(307), 3–20. doi:10.2307/538099

Batabyal, D., Ray, N., Bag, S., & Nag, K. (2023a). Tourists' Harassment During Pilgrimage: A Case Study of TripAdvisor's Review for the Hindu Pilgrim Centers in India. In Resilient and Sustainable Destinations After Disaster: Challenges and Strategies (pp. 139-156). Emerald Publishing Limited.

Batabyal, D., Das, D. K., Sen, S. K., Porya, A., Halam, H., & Chatterjee, P. (2022). Evolution of Shakti Peetham for religious, mythological and dark tourism in Indian sub-continent. *Revista Turismo & Desenvolvimento (Aveiro)*, *39*, 291–300.

Batabyal, D., Das, D. K., Sen, S. K., Porya, A., Halam, H., & Chatterjee, P. (2022). Evolution of Shakti Peetham for religious, mythological and dark tourism in Indian sub-continent. *Revista Turismo & Desenvolvimento*, *39*, 291–300.

Batabyal, D., Mitra, A., & Goswami, S. (2020). Ecotourism Product Formulation for New Normal Tourism Practices: A Substitutability Approach for Selected Areas of Tamil Nadu, India. *Dogo Rangsang Res. J*, *10*(7), 200–218.

Batabyal, D., Rawal, Y. S., Chatterjee, P., & Goswami, S. (2023b). Temple service quality model for future pilgrimages in Indian sub-continent through user-generated contents (UGC) in TripAdvisor. *International Journal of Tourism Anthropology*, *9*(2), 113–135. doi:10.1504/IJTA.2023.130849

Batabyal, D., Ray, N., Bag, S., & Nag, K. (2023). Tourists' Harassment During Pilgrimage: A Case Study of TripAdvisor's Review for the Hindu Pilgrim Centers in India. In J. Kumar, G. E. Bayram, & A. Sharma (Eds.), *Resilient and Sustainable Destinations After Disaster (Tourism Security-Safety and Post Conflict Destinations)* (pp. 139–156). Emerald Publishing Limited. doi:10.1108/978-1-80382-021-720231011

Batuhan, T. (2023). Kültürel Miras ve Müze İlişkisi: Mardin Müzesi. *Journal of Literature and Humanities*, *70*, 85–91. doi:10.5152/AUJFL.2023.22014

Bec, A., Moyle, B., Timms, K., Schaffer, V., Skavronskaya, L., & Little, Ch. (2019). Management of immersive heritage tourism experiences: A conceptual model. *Tourism Management*, *72*, 117–120. doi:10.1016/j.tourman.2018.10.033

Becker, L., & Jaakkola, E. (2020). Customer experience: Fundamental premises and implications for research. *Journal of the Academy of Marketing Science*, *48*(4), 630–648. doi:10.100711747-019-00718-x

Bedjaoui, M. (2004). The Convention for the Safeguarding of the Intangible Cultural Heritage: The legal framework and universally recognized principles. *Museum International*, *56*(1-2), 150–155. doi:10.1111/j.1350-0775.2004.00469.x

Behrens, K. (2009). Tony Yengeni's Ritual Slaughter: Animal Anti-Cruelty vs. Culture. *South African Journal of Philosophy*, *28*(3), 271–289. doi:10.4314ajpem.v28i3.47859

Bello, R.-W., Akpojaro, J., & Olubummo, D. A. (2019). Impact of Information Communication Technology in Enhancing the Management of the Natural and Cultural Heritage Resources in Idanre Hills for E tourism Attraction Design of a Liquid Level Detector with a Digital Readout View project. *Machine Learning and Cloud Computing View project.* Research Gate. https://www.researchgate.net/publication/336020387

Benkari, N. (2018). Archaeological site of Bat-Oman, management and public perception: Community involvement in archaeological heritage management and planning. *Journal of Cultural Heritage Management and Sustainable Development, 8*(3), 293–308. doi:10.1108/JCHMSD-10-2016-0059

Bertella, G. (2014). The Co-creation of Animal-based Tourism Experience. *Tourism Recreation Research, 39*(1), 115–125. https://www.trrworld.org/. doi:10.1080/02508281.2014.11081330

Beyers, J. (2017). Religion and culture: Revisiting a close relative'. *Hervormde Teologiese Studies, 73*(1), a3864. doi:10.4102/hts.v73i1.3864

Bharadwaj, R. M. (2021). *Trees, Flowers and Waters: Nature and Environment in the Hindu Vrata Religious Tradition-AD. 400-1200.* Indraprastha Shodha Sandarsh.

Bhatia, A. K. (2002). *Tourism development: Principles and practices.* Sterling Publishers Pvt. Ltd.

Bhatt, G. P. (2004). The Skanda Purana. Part X. AITM, 58. Motilal Banarasidass Publishing House.

Bhattacharyya, N. N. (1974). *History of the tantric religion.* MunshiramManoharlal Publisher.

Bhavin Shukla, K. B. (2014). *Development in the Context of Urban Conservation: Examining the Interrelationship between Architectural Heritage and Intangible Culture in Historic Settlements.* CEPT Archives.

Bindhu, D., & Panakaje, N. (2023). Adopt a Heritage Scheme: A CSR Initiative for Preserving our Past Glories to Future. [IJCSBE]. *International Journal of Case Studies in Business, IT, and Education, 7*(2), 250–262. doi:10.47992/IJCSBE.2581.6942.0272

Binkhorst, E., & Den Dekker, T. (2009). Agenda for co-creation tourism experience research. *Journal of Hospitality Marketing & Management, 18*(2), 311–327. doi:10.1080/19368620802594193

Birbhum D.O.W. (n.d.). *Brief History of Birbhum.* Birbhum DOW. http://birbhum.gov.in/geog1.htm

Biswas, S. N. (2020). *Odisha: The Land of Chromatic Tourism Destinations.* Mousumi Prokashani.

Biswas, S., & Mohapatra, P. (2022). Analysis of Impact of Covid-19 on Religious Tourism Destinations of Odisha, India. IMPACT: International Journal of Research in Applied. *Natural and Social Sciences, 10*(9), 1–12.

Bitner, M. J. (1992). Servicescapes: Impact of physical surroundings on customers and employees. *Journal of Marketing, 56*(2), 57–71. doi:10.1177/002224299205600205

Bitušíková, A. (2021). Cultural Heritage as a means of heritage tourism development. *Muzeológia a Kultúrne Dedičstvo, 9*(1), 81–95. doi:10.46284/mkd.2021.9.1.5

Björk, P., Prebensen, N., Räikkönen, J., & Sundbo, J. (2021). 20 years of Nordic tourism experience research: A review and future research agenda. *Scandinavian Journal of Hospitality and Tourism, 21*(1), 26–36. doi:10.1080/15022250.2020.1857302

Black, P., & Street, E. (2014). The power of perceptions: Exploring the role of urban design in cycling behaviours and healthy ageing. *Transportation Research Procedia, 4*, 68–79. doi:10.1016/j.trpro.2014.11.006

Boaz, D. N. (2019). The "Abhorrent" Practice of Animal Sacrifice and Religious Discrimination in the Global South. *Religions, 10*(3), 160. doi:10.3390/rel10030160

Boboc, R. G., Băutu, E., Gîrbacia, F., Popovici, N., & Popovici, D. M. (2022). Augmented Reality in Cultural Heritage: An Overview of the Last Decade of Applications. *Applied Sciences, 12*,(19), 9859. doi:10.3390/app12199859

Bolger, R. K. (2021). Finding wholes in the metaverse: Posthuman mystics as agents of evolutionary contextualization. *Religions, 12*(9), 768. doi:10.3390/rel12090768

Bond, N., Packer, J., & Ballantyne, R. (2015). Exploring visitor experiences, activities and benefits at three religious tourism sites. *International Journal of Tourism Research, 17*(5), 471–481. doi:10.1002/jtr.2014

Bowlby, J. (1958). The nature of the child's tie to his mother. *The International Journal of Psycho-Analysis, 39*, 350–373. PMID:13610508

Braojos, J., Benitez, J., Llorens, J., & Ruiz, L. (2020). Impact of IT integration on the firm's knowledge absorption and desorption. *Information & Management, 57*(7), 103290. doi:10.1016/j.im.2020.103290

Brar, R. (2020, August 3). *Heritage Tourism In Gujarat* [Video]. Digital India Learning. YouTube. https://www.youtube.com/watch?v=A4i8q_6pRUQ

Braslavsky, C. Fernǹdez Ana La̓, & Cox Cristiǹ. (2002). A Educaô Secundr̀ia: mudanȧ ouimutabilidade? Brasil: UNESCO Brasil.

Brida, J. G., Osti, L., & Barquet, A. (2010). Segmenting resident perceptions towards tourism—A cluster analysis with a multinomial logit model of a mountain community. *International Journal of Tourism Research, 12*(5), 591–602. doi:10.1002/jtr.778

Brown, L. (2016). Tourism and pilgrimage: Paying homage to literary heroes. *International Journal of Tourism Research, 18*(2), 167–175. doi:10.1002/jtr.2043

Brundtland, G. H. (1987). Our common future—Call for action. *Environmental Conservation, 14*(4), 291–294. doi:10.1017/S0376892900016805

Brundtland, G. H. (1987). World commission on environment and development. *Environmental Policy and Law, 14*(1), 26–30.

Buhalis, D. (2000). Marketing the competitive destination of the future. *Tourism Management, 21*(1), 97–116. doi:10.1016/S0261-5177(99)00095-3

Buhalis, D. (2020). Technology in tourism-from information communication technologies to eTourism and smart tourism towards ambient intelligence tourism: A perspective article. *Tourism Review, 75*(1), 267–272. doi:10.1108/TR-06-2019-0258

Bulut, P., & Uzun, E. B. (2021). Sanal Müzeler ve Yaratıcı Drama için Bir Atölye Önerisi: Müze Sınıfa Geliyor! *Yaratıcı Drama Dergisi, 16*(1), 131–156.

Bunnell, F. L., & McCann, R. (1993). The Brown or Grizzly Bear. In Bears: Majestic Creatures of the Wild. Rodale Press.

Burgess, J. A. (1979). Place-Making: The Contribution of Environmental Perception Studies in Planning. *Geography: Journal of the Geographical Association, 64*(4), 317-326. https://www.jstor.org/stable/40569988

Butler, R., & Hinch, T. (2007). *Tourism and indigenous peoples: Issues and implications*. Routledge. doi:10.4324/9780080553962

Byrne, D. (2022). *Heritage Corridors: transnational flows and the built environment of migration. Journal of ethnic and migration studies, 42 (14)* (1st ed.). Routledge.

Cahn, M. (2006). *Sustainable rural livelihoods, micro-enterprise and culture in the Pacific Islands: case studies from Samoa*. [Thesis, Massey University].

Cai, W., & McKenna, B. (2021). Knowledge Creation in Information Technology and Tourism Research. In Journal of Travel Research, 60(4). doi:10.1177/0047287520903142

Cai, W., McKenna, B., Wassler, P., & Williams, N. (2021). Rethinking Knowledge Creation in Information Technology and Tourism. In Journal of Travel Research, 60(6). doi:10.1177/0047287520946100

Cai, Z., Fang, C., Zhang, Q., & Chen, F. (2021). Joint Development of Cultural Heritage Protection and Tourism: The case of mount lushan cultural landscape heritage site. *Heritage Science*, *9*(1), 86. doi:10.118640494-021-00558-5

Calver, S. J., & Page, S. J. P. (2013). Enlightened hedonism: Exploring the relationship of service value, visitor knowledge and interest, to visitor enjoyment at heritage attractions. *Tourism Management*, *39*, 23–36. doi:10.1016/j.tourman.2013.03.008

Campanella, M. (2013). *Establishing tourist places: between social constructivism and phenomenology*. [Doctoral dissertation, Tartu Ülikoo].

Campos, A. C., Mendes, J., Valle, P. O., & Scott, S. (2018). Co-creation of tourist experiences: A literature review. *Current Issues in Tourism*, *21*(4), 369–400. doi:10.1080/13683500.2015.1081158

Cape Town Declaration on Responsible Tourism. (2002). *International Conference on Responsible Tourism in Destinations,* Cape Town, South Africa.

Carù, A., & Cova, B. (2003). Revisiting consumption experience: A more humble but complete view of the concept. *Marketing Theory*, *3*(2), 267–286. doi:10.1177/14705931030032004

Catterall, M., & Maclaran, P. (2001). Body talk: Questioning the assumptions in cognitive age. *Psychology and Marketing*, *18*(10), 1117–1133. doi:10.1002/mar.1046

Cetin, G., & Bilgihan, A. (2016). Components of cultural tourists' experiences in destinations. *Current Issues in Tourism*, *19*(2), 137–154. doi:10.1080/13683500.2014.994595

Chalana, M., & Sprague, T. S. (2013). Beyond Le Corbusier and the modernist city: Reframing Chandigarh's 'World Heritage' legacy. *Planning Perspectives*, *28*(2), 199–222. doi:10.1080/02665433.2013.737709

Chalmers, A. (2017). Experiencing the multisensory past. In M. Loannides, N. Magnenat-Thalmann, & G. Papagiannakis (Eds.), *Mixed Reality and Gamification for Cultural Heritage*. Springer. doi:10.1007/978-3-319-49607-8_14

Chambers, R., & Conway, G. (1992). *Sustainable rural livelihoods: practical concepts for the 21st century*. Institute of Development Studies.

Chand, S. (2005), Orissa Review Vol. LXII No. 4, Pp- 45-47

Chandralal, K. P. L. (2010). Impacts of tourism and community attitude towards tourism: A case study in Sri Lanka. *South Asian Journal of Tourism and Heritage*, *3*(2), 41–49.

Chang, C. L., McAleer, M., & Ramos, V. (2020). A charter for sustainable tourism after COVID-19. *Sustainability (Basel)*, *12*(9), 3671. doi:10.3390u12093671

Chauhan, E. (2022). Residents' motivations to participate in decision-making for cultural heritage tourism: Case study of New Delhi. *Sustainability (Basel)*, *14*(14), 8406. doi:10.3390u14148406

Cheer, J. M., Belhassen, Y., & Kujawa, J. (2017). The search for spirituality in tourism: Toward a conceptual framework for spiritual tourism. *Tourism Management Perspectives*, *24*, 252–256. doi:10.1016/j.tmp.2017.07.018

Chen, S., Law, R., Xu, S., & Zhang, M. (2020). Bibliometric and visualized analysis of mobile technology in tourism. In Sustainability (Switzerland), 12(19). https://doi.org/ doi:10.3390u12197975

Chen, C. F., & Chen, P. C. (2010). Resident attitudes toward heritage tourism development. *Tourism Geographies*, *12*(4), 525–545. doi:10.1080/14616688.2010.516398

Chen, C.-F., & Chen, F.-S. (2010). Experience quality, perceived value, satisfaction and behavioral intentions for heritage tourists. *Tourism Management*, *31*(1), 29–35. doi:10.1016/j.tourman.2009.02.008

Chen, H., & Rahman, I. (2018). Cultural Tourism: An Analysis of Engagement, Cultural Contact, Memorable Tourism Experience and Destination Loyalty. *Tour. Management Perspective*, *26*, 153–163. doi:10.1016/j.tmp.2017.10.006

Chen, Z. (2022). Visualizing experiencescape–from the Art of intangible cultural heritage. *Current Issues in Tourism*, *25*(4), 559–578. doi:10.1080/13683500.2021.1892040

Chhabra, D., Healy, R., & Sills, E. (2003). Staged authenticity and heritage tourism. *Annals of Tourism Research*, *30*(3), 702–719. doi:10.1016/S0160-7383(03)00044-6

Choi, S., Imon, S. S., & Couto, U. (2020). Negotiating with authenticity to ensure sustainability of traditional cultural festivals: Residents' perspectives about Macao's Drunken Dragon festival. *Sustainability (Basel)*, *12*(3), 885. doi:10.3390u12030885

Christie, M. F., & Mason, P. A. (2003). Transformative Tour Guiding: Training Tour Guides to be Critically Reflective Practitioners. *Journal of Ecotourism*, *2*(1), 1–16. doi:10.1080/14724040308668130

Chung, T. (1970). ANCIENT INDIAN LIFE THROUGH CHINESE EYES. *Proceedings of the Indian History Congress*, *32*, 137–149. https://www.jstor.org/stable/44141059

Cibinskiene, A., & Snieskiene, G. (2015). Evaluation of city tourism competitiveness. *Procedia - Social and Behavioral Sciences, 20th International Scientific Conference Economics and Management - 2015 (ICEM-2015)*, *213*, 105-110. DOI:10.1016/j.sbspro.2015.11.411

Çınar, C., Utkugün, C., & Gazel, A. A. (2021). Sosyal Bilgiler Dersinde Sanal Müze Kullanımı Hakkında Öğrenci Görüşleri. *Uluslararası Sosyal ve Eğitim Bilimleri Dergisi*, *16*(16), 150–170. doi:10.20860/ijoses.1017419

Claiborne, P. (2010). Community participation in tourism development and the value of social capital. The case of Bastimentos, Bocas Del Toro, Panamá. [Master Degree Project, University of Gothenburg].

Cohen, E. (1992). Pilgrimage and tourism: convergence and divergence. In *Sacred journeys: The anthropology of pilgrimage* (pp. 47–61). Greenwood Press.

Cole, S. (2006). Cultural tourism, community participation and empowerment. *Cultural tourism in a changing world: Politics, participation and (re) presentation*, 89-103.

Cole, S. K., Mullor, E. C., Ma, Y., & Sandang, Y. (2020). *Tourism, water, and gender—An international review of an unexplored nexus.* Wiley.

Cole, S. (2012). A political ecology of water equity and tourism: A case study from Bali. *Annals of Tourism Research*, *39*(2), 1221–1241. doi:10.1016/j.annals.2012.01.003

Collins-Kreiner, N. (2016a). Dark tourism as/is pilgrimage. *Current Issues in Tourism*, *19*(12), 1185–1189. doi:10.1080/13683500.2015.1078299

Collins-Kreiner, N. (2016b). The lifecycle of concepts: The case of 'Pilgrimage Tourism'. *Tourism Geographies*, *18*(3), 322–334. doi:10.1080/14616688.2016.1155077

Colton, J. W., & Whitney-Squire, K. (2010). Exploring the relationship between aboriginal tourism and community development. *Leisure/Loisir, 34*(3), 261-278. doi:10.1080/14927713.2010.521321

Cominelli, F., & Greffe, X. (2012). Intangible cultural heritage: Safeguarding for creativity. City. *Cultura e Scuola, 3*(4), 245–250. doi:10.1016/j.ccs.2012.10.003

Conzen, M. P., & Wulfestieg, B. M. (2001). Metropolitan Chicago's Regional Cultural Park: Assessing the Development of the Illinois & Michigan Canal National Heritage Corridor. *The Journal of Geography, 100*(3), 111–117. doi:10.1080/00221340108978426

Correia, A., Kozak, M., & Ferradeira, J. (2013). From tourist motivations to tourist satisfaction. *International Journal of Culture, Tourism and Hospitality Research, 7*(4), 411–424. doi:10.1108/IJCTHR-05-2012-0022

Corridor, K. (2018). India and Pakistan sign a deal on the Sikh temple project. *BBC news*. https://www.bbc.com/news/world-asia-india-50138275.

Craik, J. (2002). The culture of tourism. In *Touring cultures* (pp. 123–146). Routledge.

Cranmer, E. E., tom Dieck, M. C., & Fountoulaki, P. (2020). Exploring the value of augmented reality for tourism. *Tourism Management Perspectives, 35*, 100672. doi:10.1016/j.tmp.2020.100672

Cravidão, F., Nossa, P., Santos, N., Vidal, M., Crous, N., Aulet, S., & Dallari, F. (2018). Tourism at world heritage sites opportunities and challenges. *Tourism Management at UNESCO World Heritage Sites*, 23-30.

Cravidão, F., Nossa, P., Santos, N., Vidal, M., Crous, N., Aulet, S., & Dallari, F. (2018). Tourism at world heritage sites opportunities and challenges. *Tourism Management at UNESCO World Heritage Sites,* 23-30. UNESCO.

Crick, H. Q. (2004). The impact of climate change on birds. *The Ibis, 146*(s1), 48–56. doi:10.1111/j.1474-919X.2004.00327.x

Crompton, J. L. (1979). Motivations for pleasure vacation. *Annals of Tourism Research, 6*(4), 408–424. doi:10.1016/0160-7383(79)90004-5

Cusick, J. G. (2015). *Studies in Culture Contact: Interaction, Culture Change, and Archaeology*. Southern Illinois University Press.

Cutler, S. Q., & Carmichael, B. A. (2010). The dimensions of the tourist experience. In M. Morgan, P. Lugosi, & J. R. B. Ritchie (Eds.), *The tourism and leisure experience: Consumer and managerial perspectives* (pp. 3–26). Channel View. doi:10.21832/9781845411503-004

D'Alba, A. (2012). *Analyzing Visitors' Discourse, Attitudes, Perceptions, and Knowledge Acquisition In An Art Museum Tour After Using A 3d Virtual Environment.* [Unpublished PhD Thesis, University of North Texas].

Dahles, H. (2002). Tourism, small enterprises and community development. In *Tourism and sustainable community development* (pp. 172–187). Routledge.

Dai, Y. (2021). A Behavioral Cultural-Based Development Analysis of Entrepreneurship in China. *Sustainability*.

Dalal, R. (2014). *Hinduism: An Alphabetical Guide*. Penguin.

Damar, M. (2021). *Metaverse Shape of Your Life for Future: A bibliometric snapshot.*

Dann, G. M. (1977). Anomie, ego-enhancement and tourism. *Annals of Tourism Research, 4*(4), 184–194. doi:10.1016/0160-7383(77)90037-8

Das, R. P. (1994). *Terminology of Orissan Temple Architecture*. Lalita Kala Academy, New Delhi, December.

Dasgupta, D. (2011). *Tourism Marketing*. Pearson Education.

Dash, P. K. (2000). *Management of Tourism Sector in Orissa, Problems and Future*. [Thesis, Utakal University].

Daskon, C., & Binns, T. (2009). Culture, tradition and sustainable rural livelihoods: Exploring the culture-development interface in Kandy, Sri Lanka. *Community Development Journal: An International Forum, 45*(4), 494–517. doi:10.1093/cdj/bsp019

Das, M. (2017). Sriniketan: A Model Centre for Rural Reconstruction. *International Journal of Humanities And Social Science Invention, 6*(11), 57–61.

David Throsby, K. P. (2022). *Assessing the Impacts of Heritage-Led Urban Rehabilitation The Case of George Town, Penang, Malaysia.* Research Report - Getty Conservation Institute.

Dawson, E., & Jensen, E. (2011). Towards a contextual turn in visitor studies: Evaluating visitor segmentation and identity-related motivations. *Visitor Studies, 14*(2), 127–140. doi:10.1080/10645578.2011.608001

De Ascaniis, S., Gravari-Barbas, M., & Cantoni, L. (2018). *Tourism management at UNESCO world heritage sites.* Università della Svizzera italiana.

de Bruyn, C., Said, F. B., Meyer, N., & Soliman, M. (2023). Research in tourism sustainability: A comprehensive bibliometric analysis from 1990 to 2022. *Heliyon, 9*(8), e18874. doi:10.1016/j.heliyon.2023.e18874 PMID:37636413

De Luca, V., Marcantonio, G., Barba, M. C., & De Paolis, L. T. (2022). A Virtual Tour for the Promotion of Tourism of the City of Bari. *Information, 13*(7), 339. doi:10.3390/info13070339

Della Corte, V., Piras, A., & Zamparelli, G. (2010). Brand and Image: The Strategic Factors in Destination Marketing. *International Journal of Leisure and Tourism Marketing, 1*(4), 358–377. https://www.inderscienceonline.com/doi/abs/10.1504/IJLTM.2010.032064. doi:10.1504/IJLTM.2010.032064

Desai, M. (2017). *Banaras Reconstructed. Architecture and Sacred Space in a Hindu Holy City.* University of Washington Press.

Dessein, J., Soini, K., Fairclough, G., & Horlings, L. (Eds.). (2015). *Culture in, for and as Sustainable Development. Conclusions from the COST Action IS1007 Investigating Cultural Sustainability.* University of Jyväskylä.

Dhanalekshmi, U., Alam, T., & Khan, S. A. (2022). *The folkloric uses and economic importance of some selected edible medicinal plants native to Oman: a brief overview. Edible Plants in Health and Diseases* (Vol. 1). Cultural, Practical and Economic Value. doi:10.1007/978-981-16-4880-9_1

Dhar, P., Sarkar, A. K. (1998). *Indian Tourism.* Kanishka Publication.

Dhonju, H. K., Xiao, W., Mills, J. P., & Sarhosis, V. (2018). Share Our Cultural Heritage (SOCH): Worldwide 3D heritage reconstruction and visualization via web and mobile GIS. *ISPRS International Journal of Geo-Information, 7*(9), 360. doi:10.3390/ijgi7090360

Digance, J. (2006). Religious and secular pilgrimages: Journeys redolent with meaning. In D. Timothy & D. H. Olsen (Eds.), *Tourism, religion and spiritual journeys* (pp. 36–48). Routledge.

Dikshit, R. (2018). 7.35 crore visit Kashi Vishwanath Dham in opening year. *The Times of India [TOI].* https://timesofindia.indiatimes.com/city/varanasi/7-35-crore-visit-kashi-vishwanath-dham-in-opening-year/articleshow/96211539.cms

Divya, A. (2021, December 14). Explained: What is changing at the ancient Kashi Vishwanath temple complex? *Indian Express.* https://indianexpress.com/article/explained/kashi-vishwanath-temple-narendramodi-varanasi-7669709

Donaldson, T. (1975). Propitious Apotropaic Eroticism in the Art of Orissa. *Artibus Asiae, 37*(1), 75–100. doi:10.2307/3250213

Donlon, J. G., Donlon, J. H., & Agrusa, J. (2010). Cultural tourism, camel wrestling, and the tourism 'bubble' in Turkey. *Anatolia. International Journal of Tourism and Hospitality Research, 21*(1), 29–39.

Donovan, R., & Rypkema, C. C. (2013). *Measuring Economic Impacts of Historic Preservation.* Advisory Council on Historic Preservation by Place Economics.

Dragicevic Curkovic, M. (2021). The Role of the Traditional Crafts as Intangible Heritage on the Global Tourist Market. *SHS Web of Conferences*. Ecosciences. 10.1051hsconf/20219206005

Dredge, D., & Jenkins, J. (2007). *Tourism policy and planning*. John Wiley & Sons.

Drot-Delange, B., Fleury, S., Lafourcade, P., & More, M. (2018). *Un tour de magie pour introduire la représentation binaire des nombres*. CS Unplugged. https://csunplugged.org

du Cros, H. (2001). A new model to assist in planning for sustainable cultural heritage tourism. *International Journal of Tourism Research, 3*(2), 165–170. doi:10.1002/jtr.297

Duffy, R., & Moore, L. (2011). Global regulations and local practices: The politics and governance of animal welfare in elephant tourism. *Journal of Sustainable Tourism, 19*(4), 589–604. doi:10.1080/09669582.2011.566927

Durgamohan, M. (2015). Cultural and economic significance of Hampi festival. *Worldwide Hospitality and Tourism Themes, 7*(4), 377–392. doi:10.1108/WHATT-03-2015-0013

Dwivedi, O. P., & Tiwari, B. H. (1987). *Environmental Crisis and Hindu Religion.* Gitanjali Publishing House.

Dycka, M. G., & Baydack, R. K. (2004). Vigilance behavior of polar bears (Ursus maritimus) in the context of wildlife-viewing activities at Churchill, Manitoba, Canada. *Biological Conservation, 116*(3), 343–350. doi:10.1016/S0006-3207(03)00204-0

Earp, B. D. (2015). On the Supposed Distinction between Culture and Religion: A Brief Comment on Sir James Munby's Decision in the Matter of B and G (Children). *University of Oxford Practical Ethics blog*. University of Oxford.

Eck, D. L. (1982). *Banares, City of Light.* Columbia University Press.

Edgell, D. L. Sr. (2019). *Managing sustainable tourism: A legacy for the future*. Routledge.

Edmunds, L. O. (2013). Maximizing the Local Economic Impact of Heritage Conservation for Cultural Tourism. *Asia Tourism Forum 2016 – The 12th Biennial Conference of Hospitality and Tourism Industry in Asia.*

Eichler, J. (2020). Intangible cultural heritage, inequalities, and participation: Who decides on heritage? *International Journal of Human Rights, 25*(5), 793–814. doi:10.1080/13642987.2020.1822821

El Menchawy, A., Aly, S. S., & Hakim, M. A. (2011). The impact of urban sprawl on the heritage areas through the urban fabric of cities. *Conference: Sustainable Development and Planning V. WIT Transactions on Ecology and the Environment, 150*, 299–314. doi:10.2495/SDP110261

Elbaz, A. M., Soliman, M., Al-Alawi, A., Al-Romeedy, B. S., & Mekawy, M. (2022). Customer responses to airline companies' service failure and recovery strategies: The moderating role of service failure habit. *Tourism Review, 78*(1), 1–17. doi:10.1108/TR-03-2022-0108

Eliot, T. S. (2010). *Notes towards the Definition of Culture*. Faber & Faber.

Elkington, J. (1998). Cannibals with forks: The triple bottom line of 21st century business. Gabriola Island, BC; Stony Creek, CT: New Society Publishers.

Elkington, J. (1994). Towards the sustainable corporation: Win win win business strategies for sustainable development. *California Management Review*, *36*(2), 90–100. doi:10.2307/41165746

El-Said, O., & Aziz, H. (2022). Virtual Tours a Means to an End: An Analysis of Virtual Tours' Role in Tourism Recovery Post COVID-19. *Journal of Travel Research*, *61*(3), 528–548. doi:10.1177/0047287521997567

Enrico, B., Nuccio, M., & Durio, A. (2021). Proximity tourism and cultural amenities: Evidence from a regional museum card. *Tourism Economics*, *27*(1), 187–204. doi:10.1177/1354816619890230

Erbaş, H. E., & Tuncel, D. (2022). Kültür Varlıklarının Sanal Ortamda Yeniden İşlevlendirilmesi: Yerebatan Sarnıcı Örneği. *European Journal of Science and Technology*, *41*, 362–372. doi:10.31590/ejosat.1179021

Ermertz, A., Kázmér, M., Adolphs, S., Falkenroth, M., & Hoffmann, G. (2019). Geoarchaeological evidence for the decline of the medieval city of Qalhat, Oman. *Open Quaternary*, *5*(8), 1–14. doi:10.5334/oq.56

Erschbamer, M. (2019). Taming of supernatural entities and animal sacrifice, The synthesis of Tibetan Buddhism and local shamanistic traditions in Northern Sikkim (India). *Études mongoles et sibériennes, centrasiatiques et tibétaines*, *50*. . doi:10.4000/emscat.3915

Espiner, S., Stewart, E. J., & Lama, L. T. (2017). Assessing the effectiveness of 'Appreciative Inquiry'(AI) in Nepali pro-poor tourism (PPT) development processes. *Tourism Planning & Development*, *14*(3), 369–388. doi:10.1080/21568316.2016.1243148

Etikan, I., Musa, S. A., & Alkassim, R. S. (2016). Comparison of convenience sampling and purposive sampling. *American Journal of Theoretical and Applied Statistics*, *5*(1), 1–4. doi:10.11648/j.ajtas.20160501.11

Falk, J. H. (2006). An identity-centered approach to understanding museum learning. *Curator (New York, N.Y.)*, *49*(2), 151–166. doi:10.1111/j.2151-6952.2006.tb00209.x

Falk, J. H., & Storksdieck, M. (2010). Science learning in a leisure setting. *Journal of Research in Science Teaching: The Official Journal of the National Association for Research in Science Teaching*, *47*(2), 194–212.

Farsari, I. (2023). Exploring the nexus between sustainable tourism governance, resilience and complexity research. *Tourism Recreation Research, 48*(3), 352-367.

Feather, J. (2006) Managing the documentary heritage: issues for the present and future. In: Gorman, G.E. and Sydney J.Shep. (eds.), Preservation management for libraries, archives and museums. London: Facet, pp. 1-18. doi:10.29085/9781856049214.002

Fedele, A. (2014). Energy and transformation in alternative pilgrimages to Catholic shrines: Deconstructing the tourist/pilgrim divide. *Journal of Tourism and Cultural Change*, *12*(2), 150–165. doi:10.1080/14766825.2014.915091

Fennel, D. A. (2014). Exploring the boundaries of a new moral order for tourism's global code of ethics: An opinion piece on the position of animals in the tourism industry. *Journal of Sustainable Tourism*, *22*(7), 983–996. doi:10.1080/09669582.2014.918137

Fennell, D. A. (2012). Tourism, animals, and utilitarianism. *Tourism Recreation Research*, *37*(3), 239–249. doi:10.1080/02508281.2012.11081712

Fennell, D. A., & Sheppard, V. (2020). Tourism, animals and the scales of justice. *Journal of Sustainable Tourism*, *29*(3), 314–335. doi:10.1080/09669582.2020.1768263

Ferri, C. (2017). Competitiveness of Heritage Sites: a comparison between Scuola Grande della Misericordia in Venice and Oude Kerk in Amsterdam. *Working Papers 05*. Department of Management, Università Ca' Foscari Venezia. https://ideas.repec.org/p/vnm/wpdman/138.html

FICCI. (2015). *Tourism Infrastructure: The role states play*. FICCI.

Florentina-Cristina Merciu, A.-I. P.-L. (2021). *Economic Valuation of Cultural Heritage Using the Travel Cost Method: The Historical Centre of the Municipality of Bucharest as a Case Study*. Heritage.

Flores de León, R.M., Babere, N.J., & Swai, O. (2020). Implications of Cultural Heritage in Urban Regeneration: The CBD of Dar es Salaam. *Revue Interventions économiques, 63*. doi:10.4000/interventionseconomiques.9171

Folklife at the International Level: Intangible Cultural Heritage Defined (2020, January 31). *Folklife Today*. https:////blogs.loc.gov/folklife/2020/01/folklife-at-the-international-level-intangible-cultural-heritage-defined

Foroudi, P., Cuomo, M. T., Foroudi, M. M., Katsikeas, C. S., & Gupta, S. (2020). Linking identity and heritage with image and a reputation for competition. *Journal of Business Research, 113*, 317–325. doi:10.1016/j.jbusres.2019.09.042

Fountain, J., Espiner, S., & Xie, X. (2010). A cultural framing of nature: Chinese tourists' motivations for, expectations of, and satisfaction with, their New Zealand tourist experience. *Tourism Review International, 14*(2-3), 71–83. doi:10.3727/154427211X13044361606298

Francis-Lindsay, J. (2010, June 28). *From fashion to 'tangible-intangible' action: local communities 'culturizing' new tourism development. WIT Transactions on Ecology and the Environment*. WIT Press. doi:. doi:10.2495/st100421

Fraser, N. (1981). Foucault on modern power: Empirical insights and normative confusions. *Praxis International, 3*, 272–287.

Freitag, S. B. (1800). Introduction: The History and Political Economy of Bañaras. Sandria B. Freitag (ed.) Cultural and Power in Banaras. University of California Press. doi:10.1525/9780520313392-005

Frooman, J. (1999). Stakeholder influence strategies. *Academy of Management Review, 24*(2), 191–205. doi:10.2307/259074

Fulkerson, M. (2013). Explaining Multisensory Experience. Consciousness Inside and Out: *Phenomenology, Neuroscience, and the Nature of Experience*, 365–373. doi:10.1007/978-94-007-6001-1_25

Fuller, C. J. (2004). The camphor flame: popular Hinduism and society in India (Revised and Expanded ed.). Princeton University Press.

Fu, X., Sittithai, T., & Chankoson, T. (2023). Factors affecting the development of Lipu Yi costumes: Intangible cultural heritage tourism. *Journal of Cultural Heritage Management and Sustainable Development*. doi:10.1108/JCHMSD-05-2022-0068

Gabor, D., Colombo, U., & King, A. S. (2016). *Beyond the age of waste: a report to the Club of Rome*. Elsevier.

Galav, P., Jain, A., Katewa, S. S., & Nag, A. (2010). Animal healthcare practices by livestock owners at Puskar animal fairs, Rajasthan. *Indian Journal of Traditional Knowledge, 9*(3), 581–584.

Gam, A., Bahri-Ammari, N., & Soliman, M. (2023). Intelligent customer experience and behavioral brand loyalty: Empirical evidence from the commerce environment. *International Journal of Human-Computer Interaction, 39*(10), 2090–2101. doi:10.1080/10447318.2022.2074120

Gannon, C., Lynch, P., & Harrington, D. (2010). *Framework for Enhancing Competitiveness in Culture and Heritage Organisations in the Irish Tourism Industry*. Semantic Scholar. https://www.semanticscholar.org/paper/FRAMEWORK-FOR-ENHANCING-COMPETITIVENESS-IN-CULTURE-Gannon-Lynch/f077c904847ac8641454ea06a0a8a032396f63dc

Gantait, A., Mohanty, P., & Swamy, A. (2018). Conservation and Management of Indian Built-Heritages: Exploring the Issues and Challenges. *South Asian Journal of Tourism and Heritage*, *11*(1), 5–21.

Gaur, S. S., & Chapnerkar, S. M. (2015). Indian festival's contribution to the cultural and economic well-being: A case study of Ganapati festival. *Worldwide Hospitality and Tourism Themes*, *7*(4). doi:10.1108/WHATT-03-2015-0017

Gawlik, A., Apollo, M., Andreychouk, V., & Wengel, Y. (2022). 11 Pilgrimage Tourism and Sacred Places in the Himalaya. *Tourism and Development in the Himalaya: Social, Environmental, and Economic Forces*, 53.

Geertz, C. (1973). *The interpretation of cultures: Selected essays*. Basic Books.

George, B. P., & Swain, S. K. (2005). *Advancements in Tourism Theory and Practices From India* (1st ed.). Abhijeet Publication.

Ghimire, D. K., Gautam, P., Karki, S. K., Ghimire, J., & Takagi, I. (2022). Small business and livelihood: A study of Pashupatinath UNESCO heritage site of Nepal. *Sustainability (Basel)*, *15*(1), 612. doi:10.3390u15010612

Giampiccoli, A., & Kalis, J. H. (2012). Tourism, food, and culture: Community-based tourism, local food, and community development in m pondoland. *Culture, Agriculture, Food and Environment*, *34*(2), 101–123. doi:10.1111/j.2153-9561.2012.01071.x

Giglio, S., Gabriele, L., Tavernise, A., Pantano, P., Bilotta, E., & Bertacchini, F. (2015). *Virtual museums and Calabrian cultural heritage: Projects and challenges.* Digital Heritage. doi:10.1109/DigitalHeritage.2015.7419603

Ginzarly, M., & Jordan Srour, F. (2022). Cultural heritage through the lens of COVID-19. *Poetics*, *92*, 101622. doi:10.1016/j.poetic.2021.101622 PMID:34725533

Ginzarly, M., Roders, A. P., & Teller, J. (2019). Mapping historic urban landscape values through social media. *Journal of Cultural Heritage*, *36*, 1–11. doi:10.1016/j.culher.2018.10.002

Gjerde, M. (2010). Visual Aesthetic Perception and Judgement Of Urban Streetscapes. *Paper for Building a Better World: CIB World Congress*, 12-22. https://scholar.google.co.nz/citations?view_op=view_citation&hl=en&user=_Iw_Wf4AAAAJ&citation_for_view=_Iw_Wf4AAAAJ:u-x6o8ySG0sC

Gnoth, J. (1997). Tourism Motivation and Expectation Formation. *Annals of Tourism Research*, *24*(2), 283–304. doi:10.1016/S0160-7383(97)80002-3

Gnoth, J., & Zins, A. H. (2013). Developing Tourism Cultural Contact Scale. *Journal of Business Research*, *66*(1), 738–744. doi:10.1016/j.jbusres.2011.09.012

Goffi, G., Cucculelli, M., & Masiero, L. (2019). Fostering tourism destination competitiveness in developing countries: The role of sustainability. *Journal of Cleaner Production*, *209*, 101–115. doi:10.1016/j.jclepro.2018.10.208

Gomez-Oliva, A., Alvarado-Uribe, J., Parra-Meroño, M. C., & Jara, A. J. (2019). Transforming communication channels to the co-creation and diffusion of intangible heritage in smart tourism destination: Creation and testing in ceutí (spain). *Sustainability (Basel)*, *11*(14), 3848. doi:10.3390u11143848

Gómez-Vega, M., & Picazo-Tadeo, J. (2019). Ranking world tourist destinations with a composite indicator of competitiveness: To weigh or not to weigh? *Tourism Management*, *72*, 281–291. doi:10.1016/j.tourman.2018.11.006

González-Rodríguez, M. R., Díaz-Fernández, M. C., & Pino-Mejías, M. Á. (2020). The impact of virtual reality technology on tourists' experience: A textual data analysis. *Soft Computing*, *24*(18), 13879–13892. doi:10.100700500-020-04883-y

Goodey, B., & Gold, J. R. (1987). Environmental Perception: The Relationship with Urban Design. *Progress in Human Geography*, *11*(1), 126–133. doi:10.1177/030913258701100106

Goodman, R. (2002). Pastoral Livelihoods in Tanzania: Can the Maasai Benefit from Conservation? *Current Issues in Tourism, 5*(3-4), 280–286. doi:10.1080/13683500208667924

Goodwin, C. J. (2005). *Research in Psychology: Methods and Design.* John Wiley & Sons, Inc.

Goodwin, H. (2007). *Indigenous tourism and poverty reduction* (Vol. 104). Elsevier.

Goodwin, H. (2016). *Responsible tourism: using tourism for sustainable development.* Goodfellow. doi:10.23912/978-1-910158-84-5-3101

Goodwin, H. (2021). *Responsible Tourism in India. Government of India Institute of Secretariat Training & Management.* ISTM.

Goossens, C. (2000). Tourism information and pleasure motivation. *Annals of Tourism Research, 27*(2), 301–321. doi:10.1016/S0160-7383(99)00067-5

Gortan-Carlin, I. P., KRAJNOVIČ, A., & Raguž, I. V. (2022). Study programs in culture and tourism in higher education in the Republic of Croatia – A critical analysis. In *Proceedings of FEB Zagreb International Odyssey Conference on Economics and Business* (Vol. 4, No. 1, pp. 462-484). University of Zagreb, Faculty of Economics and Business.

Graham, B. (2000). The past in place: historical geographies of identity. In B. Graham & C. Nash (Eds.), *Modern Historical Geographies* (pp. 70–79). Pearson.

Graham, B. (2002). Heritage as Knowledge: Capital of Culture. *Urban Studies (Edinburgh, Scotland), 39*(56), 1003–1017. doi:10.1080/00420980220128426

Gravari-Barbas, M., Cominelli, F., Condevaux, A., Jacquot, S., & Conti, A. (2018). Chapter: Concepts, Doctrines and Lists. *Tourism Management at UNESCO World Heritage Sites*, 7.

Gray, B. (1989). Collaborating: Finding Common Ground for Multiparty Problems. *Overcoming obstacles for successful collaboration.* https://www.ojp.gov/ncjrs/virtual-library/abstracts/collaborating-finding-common-ground-multiparty-problems#:~:text=This%20book%20presents%20an%20innovative,%2C%20national%2C%20and%20international%-20problems

Gray, P. B., & Young, S. M. (2011). Human–pet dynamics in cross-cultural perspective. *Anthropos, 24*(1), 17–30.

Green, T. A. (Ed.). (1997). *Folklore: an encyclopedia of beliefs, customs, tales, music, and art* (Vol. 1). Abc-clio.

Griffiths, M. (2011). Those who come to pray and those who come to look: Interactions between visitors and congregations. *Journal of Heritage Tourism, 6*(1), 63–72. doi:10.1080/1743873X.2010.536234

Gupta, S., & Singh, A. (2019). Measurement Scale of Tourism Destination competitiveness: Supply Side Perspectives. *SCMS Journal of Indian Management, 16*(1), 105-116. https://www.proquest.com/scholarly-journals/measurement-scale-tourism-destination/docview/2231860053/se-2?accountid=132709

Gupta, V. (1999). Sustainable tourism: learning from Indian religious traditions. *International Journal of Contemporary Hospitality Management, 11*, 91–95. //. doi:10.1108/09596119910250751

Gursoy, D., Jurowski, C., & Uysal, M. (2002). Resident attitudes: A structural modeling approach. *Annals of Tourism Research, 29*(1), 79–105. doi:10.1016/S0160-7383(01)00028-7

Gustafsson, C. (2015). Sonic branding: A consumer-oriented literature review. *Journal of Brand Management, 22*(1), 20–37. doi:10.1057/bm.2015.5

Gutierrez-Montes, I., Emery, M., & Fernandez-Baca, E. (2009). The sustainable livelihoods approach and the community capitals framework: The importance of system-level approaches to community change efforts. *Community Development (Columbus, Ohio)*, *40*(2), 106–113. doi:10.1080/15575330903011785

Guttentag, D. A. (2010). Virtual reality: Applications and implications for tourism. *Tourism Management*, *31*(5), 637–651. doi:10.1016/j.tourman.2009.07.003

Guzmán, A. P. (2016). Measuring links between cultural heritage management and sustainable urban development: An overview of global monitoring tools. *Cities (London, England)*.

Halder, S., & Sarda, R. (2021). Promoting intangible cultural heritage (ICH) tourism: Strategy for socioeconomic development of snake charmers (India) through geoeducation, geotourism and geoconservation. *International Journal of Geoheritage and Parks*, *9*(2), 212–232. doi:10.1016/j.ijgeop.2021.02.008

Hall, C. M. (2005). Rural wine and food tourism cluster and network development. *Rural tourism and sustainable business*, *26*, 149-164.

Hall , C.M., & McArthur, S. (1998). Integrated heritage management: Principles and practice. The Stationery Office.

Hall, C. M. (2008). *Tourism Planning: Policies, Processes, and Relationships*. Pearson Education.

Hall, D., & Richards, G. (2002). *Tourism and sustainable community development*. Routledge. doi:10.4324/9780203464915

Hall, M. (2016). Cultural heritage as civilising mission: From decay to recovery. *International Journal of Heritage Studies*, *23*(2), 176–177. doi:10.1080/13527258.2016.1218917

Hall, M. C., & Zeppel, H. (1990). History, Architecture, Environment: Cultural Heritage and Tourism. *Journal of Travel Research*, *29*(2), 54–55. doi:10.1177/004728759002900212

Hamdan, H. A., Andersen, P. H., & De Boer, L. (2021). Stakeholder collaboration in sustainable neighborhood projects—A review and research agenda. *Sustainable Cities and Society*, *68*, 102776. doi:10.1016/j.scs.2021.102776

Hammersley, M., & Atkinson, P. (2007). *Ethnography: Principles in practice*. Tavistock.

Hanzaee, K. H., & Khonsari, Y. (2011). A review of the role of hedonic and utilitarian values on customer's satisfaction and behavioural intentions. *Interdisciplinary Journal of Research in Business.*, *1*(5), 34–35.

Harkness, T. & Sinha, A. (2004). Taj Heritage Corridor: Intersection between History and culture on the Yamuna riverfront. *Places*, *16*(2).

Harrison, R. (2013). *Heritage: Critical approaches*. Routledge., doi:10.4324/9780203108857

Hassan, S. B., & Soliman, M. (2021). COVID-19 and repeat visitation: Assessing the role of destination social responsibility, destination reputation, holidaymakers' trust and fear arousal. *Journal of Destination Marketing & Management*, *19*, 100495. doi:10.1016/j.jdmm.2020.100495

Hede, A. M., & Kellett, P. (2017). Religious tourism and cultural sustainability. *Journal of Sustainable Tourism*, *25*(1), 1–16.

Henderson, J. (2000). Attracting tourists to Singapore's Chinatown: A case study in conservation and promotion. *Tourism Management*, *21*(5), 525–534. doi:10.1016/S0261-5177(99)00108-9

Hernández-Rojas, R. D., del Río, J. A. J., Fernández, A. I., & Vergara-Romero, A. (2021). The cultural and heritage tourist, SEM analysis: The case of The Citadel of the Catholic King. *Heritage Science*, *9*(1), 1–19. doi:10.118640494-021-00525-0

He, Z., Wu, L., & Li, X. (2018). When art meets tech: The role of augmented reality in enhancing museum experiences and purchase intentions. *Tourism Management*, *68*, 127–139. doi:10.1016/j.tourman.2018.03.003

Higgins-Desbiolles, F., Bigby, B. C., & Doering, A. (2022). Socialising tourism after COVID-19: Reclaiming tourism as a social force? *Journal of Tourism Futures*, *8*(2), 208–219. doi:10.1108/JTF-03-2021-0058

Hiltebeitel, A. (1980). The Sacrifices of the Water Buffalo and the Bull of Heaven. *History of Religions*, *19*(3), 187–223. doi:10.1086/462845

Hirschman, E. C., & Holbrook, M. B. (1982). Hedonic consumption: Emerging concepts, methods and propositions. *Journal of Marketing*, *46*(3), 92–101. doi:10.1177/002224298204600314

Holloway, J., & Valins, O. (2002). Pacing Religion and Spirituality in Geography. *Social & Cultural Geography*, *3*(1), 5–9. doi:10.1080/14649360120114107

Holst, A., &Martens, P. (2016). Determinants of animal protection policy: a cross country empirical study. *Politics and Animals*, 1-14. Olju.

Hosany, S., Sthapit, E., & Björk, P. (2022). Memorable tourism experience: A review and research agenda. *Psychology and Marketing*, *39*(8), 1467–1486. doi:10.1002/mar.21665

Hsu, F. C., Zhang, S., Zhang, Y., & Lee, T. J. (2022). Decision-making behavior in the sustainable development of intangible cultural heritage tourism. *International Journal of Tourism Research*, *24*(6), 800–812. doi:10.1002/jtr.2546

Huang, S. (2021). Tourist motivation: a critical overview. In R. Sharpley (Ed.), *Routledge handbook of the tourist experience* (pp. 200–212). Taylor & Francis. doi:10.4324/9781003219866-18

Huang, W., Xiang, H., & Li, S. (2019). The application of augmented reality and unity 3D in interaction with intangible cultural heritage. *Evolutionary Intelligence*, *1*, 1–1. doi:10.100712065-019-00314-6

Huffer, E. (2006). Regionalism and cultural identity: Putting the Pacific back into the plan. *Globalisation, governance and the Pacific islands: State, society and governance in Melanesia*, 43-55.

Hughes, M., & Carlsen, J. (2010). The business of cultural heritage tourism: Critical success factors. *Journal of Heritage Tourism*, *5*(1), 17–32. doi:10.1080/17438730903469805

Hung, K., Yang, X., Wassler, P., Wang, D., Lin, P., & Liu, Z. (2017). Contesting the Commercialization and Sanctity of Religious Tourism in the Shaolin Monastery, China. *International Journal of Tourism Research*, *19*(2), 145–159. doi:10.1002/jtr.2093

IG Group. (2021). *Economic and social impacts of cultural and creative sectors*. OECD.

Iliev, D. (2020). The evolution of religious tourism: Concept, segmentation and development of new identities. *Journal of Hospitality and Tourism Management*, *45*, 131–140. doi:10.1016/j.jhtm.2020.07.012

Ismagilova, G., Safiullin, L., & Gafurov, I. (2015). Using historical heritage as a factor in tourism development. *Procedia: Social and Behavioral Sciences*, *188*, 157–162. doi:10.1016/j.sbspro.2015.03.355

Ismail, N., Masron, T., & Ahmad, A. (2014). Cultural heritage tourism in Malaysia: Issues and challenges. SHS Web of Conferences, Iso-Ahola, S. E. (1982). Toward a social psychological theory of tourism motivation: A rejoinder. *Annals of Tourism Research*, *9*(2), 256–262.

Jacobsen, J. K. S., & Munar, A. M. (2012). Tourist informationsearch and destination choice in a digital age. *Tourism Management Perspectives*, *1*(1), 39–47. doi:10.1016/j.tmp.2011.12.005

Jafari, J. (1996). Tourism and culture: an inquiry into paradoxes. In *Culture, Tourism Development: Crucial Issues for the XXIst Century*. UNESCO.

Jain, R., & Bagdare, S. (2011). Music and consumption experience: A review. *International Journal of Retail & Distribution Management, 39*(4), 289–302. doi:10.1108/09590551111117554

Jalilvand, M. R., & Heidari, A. (2017). Comparing face-to-face and electronic word-of-mouth in destination image formation: The case of Iran. *Information Technology & People, 30*(4), 710–735. doi:10.1108/ITP-09-2016-0204

Jantzen, C., Firchett, J., Ostergaard, P., & Vetner, M. (2012). Just for fun? The emotional regime of experiential consumption. *Marketing Theory, 12*(2), 137–15. doi:10.1177/1470593112441565

Jasrotia, S. S., & Koul, S. (2020). Do young consumers buy country made products? Evidence from India. *International Journal of Management and Enterprise Development, 19*(4), 378–397. doi:10.1504/IJMED.2020.110830

Jenkins, T. (2000). Putting postmodernity into practice: Endogenous development and the role of traditional cultures in the rural development of marginal regions. *Ecological Economics, 34*(3), 301–313. doi:10.1016/S0921-8009(00)00191-9

Jeong, M., & Shin, H. H. (2019). Tourists' Experiences with Smart Tourism Technology at Smart Destinations and Their Behavior Intentions. *Sage, 59*(8), 1464–1477. https://doi.org/ doi:10.1177/0047287519883034

Jiang, S., Moyle, B., Yung, R., Tao, L., & Scott, N. (2023). Augmented reality and the enhancement of memorable tourism experiences at heritage sites. *Current Issues in Tourism, 26*(2), 242–257. doi:10.1080/13683500.2022.2026303

Johnson, C. L. (2017). *Millennials' Perception of Destination Attractiveness*. HDL. https://hdl.handle.net/1805/15114

Johnston, J., Weiler, A., & Baumann, S. (2022). The cultural imaginary of ethical meat: A study of producer perceptions. *Journal of Rural Studies, 89*, 186–198. doi:10.1016/j.jrurstud.2021.11.021

Jokilehto, J. (1986). *A History of Architectural Conservation: The Contribution Of English, French, German And Italian Thought Towards An International Approach To The Conservation Of Cultural Property* [D.Phil. Thesis, The University Of York]. https://vk.com/doc277688559_437435201?hash=204ac4f1d4c699116f&dl=39b65a8b182586a0cd

Jørgensen, N. N. K. D., Eade, J., Ekeland, T. J., & Lorentzen, C. A. N. (2020). *The processes, effects and therapeutics of pilgrimage walking the St. Olav Way*.

Jovanović, A., & Milosavljević, A. (2022). VoRtex Metaverse Platform for Gamified Collaborative Learning. *Electronics (Basel), 11*(3), 317. doi:10.3390/electronics11030317

Kahraman, Z. (2021). Sanal Müzecilikte Yeni Yaklaşımlar. *Güzel Sanatlar Fakültesi Dergisi, 3*(2), 145–160.

Kamal, A. M. (2019, February 28). Intangible Cultural Heritage as a Tool for Community Empowerment: A Case Study of the Date Palm Festival in Siwa Oasis, Egypt. *Ottoman Journal of Tourism and Management Research, 4*(1), 361–377. doi:10.26465/ojtmr.2018339516

Kamariotou, V., Kamariotou, M., & Kitsios, F. (2021). Strategic planning for virtual exhibitions and visitors' experience: A multidisciplinary approach for museums in the digital age. *Digital Applications in Archaeology and Cultural Heritage, 21*, 1–11. doi:10.1016/j.daach.2021.e00183

Kamboj, S., & Joshi, R. (2021). Examining the factors influencing smartphone apps use at tourism destinations: A UTAUT model perspective. *International Journal of Tourism Cities, 7*(1), 135–157. doi:10.1108/IJTC-05-2020-0094

Kandari O.P. & Ashish, C. (2004). *Tourism Development Principles and Practices*. Shree Publishers: New Delhi.

Kasapoğlu Akyol, P. (2020). COVID-19 Küresel Salgın Dönemi Ve Sonrası Müze Etkinlikleri. *Milli Folklor, 16*(127), 72–86.

Kauppinen-Räisänen, H., Gummerus, J., & Lehtola, K. (2013). Remembered eating experiences described by the self, place, food, context and time. *British Food Journal, 115*(5), 666–685. doi:10.1108/00070701311331571

Kaur, P., Jawaid, A., & Othman, N. B. A. (2016). The impact of community-based tourism on community development in Sarawak. *Journal of Borneo Kalimantan, 2*(1), 15–26.

Kavrečič, P., & Virloget, K. H. (2020). Traditional Festivities in North Istria in Terms of Authenticity and (dis) continuity. *Academica Turistica-Tourism and Innovation Journal, 12*(2), 109–120.

Kerala. (2008). *Kerala declaration on responsible tourism.* 2nd International Conference on responsible tourism in destinations, Kerala, India.

Kermani, S. (2018). Kartarpur corridor: A road to peace between India and Pakistan. *BBC News.* https://www.bbc.com/news/world-asia-46382657

Kersten, T. P., Tschirschwitz, F., & Deggim, S. (2017). DEVELOPMENT OF A VIRTUAL MUSEUM INCLUDING A 4D PRESENTATION OF BUILDING HISTORY IN VIRTUAL REALITY. The *International Archives of the Photogrammetry, Remote Sensing and Spatial Information Sciences, XLII-2-W3*(2W3), 361–367. doi:10.5194/isprs-archives-XLII-2-W3-361-2017

Ketmuang, S. (2005). *Local Residents and Tourists Attitudes towards Tourism Development at Patong Beach, Phuket, Thailand.* [Unpublished Ph.D. Thesis, University Putra Malaysia, Serdang].

Kim, K., Seo, B. K., Han, J. H., & Park, J. Il. (2009). Augmented reality tour system for immersive experience of cultural heritage. *Proceedings - VRCAI 2009: 8th International Conference on Virtual Reality Continuum and Its Applications in Industry*, (pp. 323–324). IEEE. 10.1145/1670252.1670325

Kim, B., & Kim, S. (2018). Hierarchical value map of religious tourists visiting the vatican city/rome. *Tourism Geographies.*

Kim, B., Kim, S. S., & King, B. (2016). The sacred and the profane: Identifying pilgrim traveler value orientations using means-end theory. *Tourism Management, 56*, 142–155. doi:10.1016/j.tourman.2016.04.003

Kim, B., Kim, S., & King, B. (2020). Religious tourism studies: Evolution, progress, and future prospects. *Tourism Recreation Research, 45*(2), 185–203. doi:10.1080/02508281.2019.1664084

Kim, S., Whitford, M., & Arcodia, C. (2019, February 19). Development of intangible cultural heritage as a sustainable tourism resource: The intangible cultural heritage practitioners' perspectives. *Journal of Heritage Tourism, 14*(5–6), 422–435. doi:10.1080/1743873X.2018.1561703

King, T. F. (2014). *Cultural heritage and tourism: An introduction.* Channel View Publications. https://books.google.co.in/books?hl=en&lr=&id=oADPBQAAQBAJ&oi=fnd&pg=PR5&dq=King,+T.+F.+(2014).+Cultural+heritage+and+tourism:+An+introduction.+Channel+View+Publications.&ots=uT7c721YmT&sig=WHafF3i2jRwtJ0yit56fMmYn7bg&redir_esc=y#v=onepage&q&f=false

Kolay, S. (2016). Cultural heritage preservation of traditional Indian Art through virtual new-media. *Procedia: Social and Behavioral Sciences, 225*, 309–320. doi:10.1016/j.sbspro.2016.06.030

Kõmmus, H. (2019). How the Concepts of Folk Music Emerge: The Terminology of Folk Music Festivals. *Etnomusikologian vuosikirja [Etnomusicology Yearbook], 31,* 112-142.

Kondo, Y., Miki, T., Kuronuma, T., Hayakawa, Y. S., Kataoka, K., & Oguchi, T. (2016). Concurrent and sustainable development of a local-scale digital heritage inventory through action research at Bat, Oman. *Journal of Cultural Heritage Management and Sustainable Development*, *6*(2), 195–212. doi:10.1108/JCHMSD-01-2016-0005

Kongjian, Y., Li, W., Li, D., Li, C.B., Huang, G. & Liu, H.L. (2005). Suitability analysis of heritage corridor in the rapidly urbanising region: A case study of Taizhou City. *Geographical Research.* Pecking University, Beijing.

Koster, R. L., & Lemelin, R. H. (2009). Appreciative inquiry and rural tourism: A case study from Canada. *Tourism Geographies*, *11*(2), 256–269. doi:10.1080/14616680902827209

Kovačević, N. D., Kovačević, L., Stankov, U., Dragićević, V., & Miletić, A. (2018). Applying destination competitiveness model to strategic tourism development of small destinations: The case of South Banat district. *Journal of Destination Marketing & Management*, *8*, 114–124. doi:10.1016/j.jdmm.2017.01.002

Kozinets, R. (2010). *Netnography: Doing ethnographic research online.* Sage.

Kozinets, R. (2015). *Netnography: Redefined.* Sage.

Kozinets, R. V. (2002). The field behind the screen: Using netnography for marketing research in online communities. *JMR, Journal of Marketing Research*, *39*(1), 61–72. doi:10.1509/jmkr.39.1.61.18935

Krantz, L. (2001). The sustainable livelihood approach to poverty reduction. SIDA. *Division for Policy and Socio-Economic Analysis, 44.*

Kumar, R. A., & Singh, L. (2014). Pilgrimage Tourism in Kurukshetra (Haryana): A Sustainable Development Approach. *African Journal of Hospitality, Tourism and Leisure*, *3*(2), 1–10.

Kumar, S. (2015). Rural Society and Rural Economy in the Ganga Valley during the Gahadavalas. *Social Scientist*, *43*(5/6), 29–45.

Kumar, S., & Dhir, A. (2020). Associations between travel and tourism competitiveness and culture. *Journal of Destination Marketing & Management*, *18*, 100501. doi:10.1016/j.jdmm.2020.100501

Kurtz, L. R. (2015). *Gods in the global village: the world's religions in sociological perspective.* Sage Publications.

Kusuma, P. A., Mutiarin, D., & Damanik, J. (2021). STRATEGI PEMULIHAN DAMPAK WABAH COVID PADA SEKTOR PARIWISATA DI DAERAH ISTIMEWA YOGYAKARTA. *Journal of Tourism and Economic*, *4*(1), 47–59. doi:10.36594/jtec.v4i1.110

Kuznets, S. (1955). Economic growth and income inequality. *The American Economic Review*, *45*, 1–28.

Kye, B., Han, N., Kim, E., Park, Y., & Jo, S. (2021). Educational applications of metaverse: Possibilities and limitations. In Journal of Educational Evaluation for Health Professions, 18. doi:10.3352/jeehp.2021.18.32

Land, N. (2002). *The thirst for annihilation: Georges Bataille and virulent nihilism.* Routledge. https://books.google.co.in/books?hl=en&lr=&id=d1SIAgAAQBAJ&oi=fnd&pg=PP1&dq=68.%09S%C3%B6rlin,+S.+(2009).+The+thirst+for+annihilation:+Georges+Bataille+and+virulent+nihilism.+Routledge.&ots=T5Lg6GHulz&sig=3DmvbuP2hZKvJpv--iMHqoF5rg8&redir_esc=y#v=onepage&q&f=false

Landorf, C. (2009). Managing for sustainable tourism: A review of six cultural World Heritage Sites. *Journal of Sustainable Tourism*, *17*(1), 53–70. doi:10.1080/09669580802159719

Lapointe, D. (2020). Reconnecting tourism after COVID-19: The paradox of alterity in tourism areas. *Tourism Geographies*, *22*(3), 633–638. doi:10.1080/14616688.2020.1762115

Lee, U. K. (2022). Tourism Using Virtual Reality: Media Richness and Information System Successes. *Sustainability, 14*(7), 3975. doi:10.3390/su14073975

Leipämaa-Leskinen, H., Syrjälä, H., Kauppinen-Räisänen, H., & Jaskari, M. M. (2020). Saunascape in the Nexus of Socio-Material Sauna Bathing Practices. In *Research in consumer culture theory*. Consumer culture theory consortium.

Lemon, K. N., & Verhoef, P. C. (2016). Understanding customer experience throughout the customer journey. *Journal of Marketing, 80*(6), 69–96. doi:10.1509/jm.15.0420

Lenzerini, F. (2011). Intangible cultural heritage: The living culture of peoples. *European Journal of International Law, 22*(1), 101–120. doi:10.1093/ejil/chr006

Lerner, P., & Rabello, A. M. (2007). The Prohibition of Ritual Slaughtering (Kosher Shechita and Halal) and Freedom of Religion of Minorities. *The Journal of Law and Religion, 22*(1), 1–6. doi:10.1017/S0748081400003210

Liang, X., Lu, Y., & Martin, J. (2021). A review of the role of social media for the cultural heritage sustainability. *Sustainability (Basel), 13*(3), 1055. doi:10.3390u13031055

Li, J., Krishnamurthy, S., Roders, A. P., & Van Wesemael, P. (2020). State-of-the-practice: Assessing community participation within Chinese cultural World Heritage properties. *Habitat International, 96*, 102107. doi:10.1016/j.habitatint.2019.102107

Lima, J., Serra, J., Borges, M. D. R., & Marujo, N. (2023). Is gastronomy important for UNESCO world heritage sites' tourists? A case study from Portugal. *Anatolia*, 1-13.

Little, C., Bec, A., Moyle, B. D., & Patterson, D. (2020). Innovative methods for heritage tourism experiences: Creating windows into the past. *Journal of Heritage Tourism, 15*(1), 1–13. doi:10.1080/1743873X.2018.1536709

Liutikas, D. (2017). The manifestation of values and identity in travelling: The social engagement of pilgrimage. *Tourism Management Perspectives, 24*, 217–224. doi:10.1016/j.tmp.2017.07.014

Li, Y., Lau, C., & Su, P. (2020). Heritage tourism stakeholder conflict: A case of a World Heritage Site in China. *Journal of Tourism and Cultural Change, 18*(3), 267–287. doi:10.1080/14766825.2020.1722141

Li, Y., Xiao, L., Ye, Y., Xu, W., & Law, A. (2016). Understanding tourist space at a historic site through space syntax analysis: The case of Gulangyu, China. *Tourism Management, 52*, 30–43. doi:10.1016/j.tourman.2015.06.008

Li, Z., Wang, D., Abbas, J., Hassan, S., & Mubeen, R. (2022). Tourists' health risk threats amid COVID-19 era: Role of technology innovation, Transformation, and recovery implications for sustainable tourism. *Frontiers in Psychology, 12*, 769175. doi:10.3389/fpsyg.2021.769175 PMID:35465147

Lonardi, S., & Unterpertinger, Y. (2022). The Relevance of Intangible Cultural Heritage and Traditional Languages for the Tourism Experience: The Case of Ladin in South Tyrol. *Sustainability, 14*(5), 2729. doi:10.3390u14052729

López-Guzmán, T., Torres Naranjo, M., Pérez Gálvez, J. C., & Carvache Franco, W. (2019). Segmentation and motivation of foreign tourists in world heritage sites. A case study, Quito (Ecuador). *Current Issues in Tourism, 22*(10), 1170–1189. doi:10.1080/13683500.2017.1344625

Loulanski, T., & Loulanski, V. (2011). The sustainable integration of cultural heritage and tourism: A meta-study. *Journal of Sustainable Tourism, 19*(7), 837–862. doi:10.1080/09669582.2011.553286

Lubbe, B. A., Preez, E. A., Douglas, A., & Wessels, F. F. (2017). The impact of rhino poaching on tourist experiences and future visitation to National Parks in South Africa. *Current Issues in Tourism, 22*(1), 8–15. doi:10.1080/13683500.2017.1343807

Luke, W. (2008). *Correa's Artist's Village Today - The incremental house*. Blogpost. http://incrementalhouse.blogspot.co.nz/2008/04/correas-artists-village-today.html

Luo, Y. (2020, June 15). Safeguarding intangible heritage through edutainment in China's creative urban environments. *International Journal of Heritage Studies*, *27*(2), 170–185. doi:10.1080/13527258.2020.1780463

Lussetyowati, T. (2015). Preservation and conservation through cultural heritage tourism. Case study: Musi Riverside Palembang. *Procedia: Social and Behavioral Sciences*, *184*, 401–406. doi:10.1016/j.sbspro.2015.05.109

Lv, X., Li, S. C., & McCabe, S. (2020). Expanding theory of tourists' destination loyalty: The role of sensory impressions. *Tourism Management*, *77*, 77. doi:10.1016/j.tourman.2019.104026

Lynch, J. J. (1977). *The broken heart: The medical consequences of loneliness*. Basic Books.

Lynch, J. J. (2000). *A cry unheard: New insights into the medical consequences of loneliness*. Bancroft Press.

Lyon, S. (2013). Coffee tourism and community development in Guatemala. *Human Organization*, *72*(3), 188–198. doi:10.17730/humo.72.3.00l1g4xj1m83x68j

M., A. (2019). The Effect of Cultural Heritage Sites in Local Economic Development - The Case Study of Syria. *IUPR (Russian Journal)*.

MacCannell, D. (1973). Staged authenticity: Arrangements of social space in tourist settings. *American Journal of Sociology*, *793*(3), 589–603. doi:10.1086/225585

Madandola, M., & Boussaa, D. (2023). Cultural Heritage Tourism as a catalyst for sustainable development; the case of Old Oyo Town in Nigeria. *International Journal of Heritage Studies*, *29*(1–2), 21–38. doi:10.1080/13527258.2023.2169332

Mahmoud Abdulkader Alghafri, V. I. (2020). Cultural Heritage Management as a Generator of Socioeconomic Development (The Case Study of Syria). *SHS Web of Conferences*.

Mair, H., & Reid, D. G. (2007). Tourism and community development vs. tourism for community development: Conceptualizing planning as power, knowledge, and control. *Leisure/Loisir, 31*(2), 403-425.

Malchrowicz-M, Ewa., Munsters, W., Korzeniewska-Nowakowska, P., & Gravelle, F. (2020). Controversial Animal Tourism Considered from a Cultural Perspective. *Tourism / Turyzm*, *30*(1,5), 21-30. . doi:10.18778/0867-5856.30.1.14

Mandić, A. (2020). Structuring challenges of sustainable tourism development in protected natural areas with driving force–pressure–state–impact–response (DPSIR) framework. *Environment Systems & Decisions*, *40*(4), 560–576. doi:10.100710669-020-09759-y

MariottiA. (2012). Local System, Networks and International Competitiveness: from Cultural Heritage to Cultural Routes. *Journal of Tourism, Culture and Territorial Development, 3*(5), 81-95. doi:10.6092/issn.2036-5195/3208

Martina Bosone, P. D. (2021). Indicators for Ex-Post Evaluation of Cultural Heritage Adaptive Reuse Impacts in the Perspective of the Circular Economy. *Sustainability*.

Martin, J., Dominica, W., Klara, Ł., & John, A. (2021). Development of the My Cult-Rural Toolkit. *Sustainability*, *13*(1), 7128. doi:10.3390u13137128

Mason, R. (2002). Assessing Values in Conservation Planning: Methodological Issues and Choices. In Assessing the Values of Cultural Heritage Los Angeles: Research Report. The Getty Conservation Institute.

Mason, P., Augustyn, M., & Seakhoa-King, A. (2010). Exploratory study in tourism: Designing an initial, qualitative phase of sequenced, mixed methods research. *International Journal of Tourism Research*, *12*(5), 432–448. doi:10.1002/jtr.763

Masud, M., Aldakhil, A., Nassani, A., & Azam, M. (2017). Community-based ecotourism management for sustainable development of marine protected areas in Malaysia. *Ocean and Coastal Management, 136*, 104–112. doi:10.1016/j.ocecoaman.2016.11.023

Mathew, V., & Soliman, M. (2021). Does digital content marketing affect tourism consumer behavior? An extension of t echnology acceptance model. *Journal of Consumer Behaviour, 20*(1), 61–75. doi:10.1002/cb.1854

Matlegiewicz, M. (2020). Motivation in Cultural Tourism: The Case of the Traditional and Regional Food Products of West Pomerania. *Zarządzanie w kulturze, 21*(3), 219-237.

Matteucci, X., Koens, K., Calvi, L., & Moretti, S. (2022). Envisioning the futures of cultural tourism. *Futures, 142*, 103013. doi:10.1016/j.futures.2022.103013 PMID:36061420

Ma, X., Wang, R., Dai, M., & Ou, Y. (2020). The influence of culture on the sustainable livelihoods of households in rural tourism destinations. *Journal of Sustainable Tourism, 29*(8), 1–18.

McCamley, C. (2016). Achieving Competitiveness In An Emerging Heritage Region; Examining The Role Of Trust As A Critical Antecedent To Heritage Tourism Development. *Global Marketing Conference 2016: Bridging Asia and the World: Global Platform for Interface Between Marketing and Management.* HUD. https://pure.hud.ac.uk/en/activities/achieving-competitiveness-in-an-emerging-heritage-region-examinin

McCool, S. F., & Martin, S. R. (1994). Community attachment and attitudes toward tourism development. *Journal of Travel Research, 32*(3), 29–34. doi:10.1177/004728759403200305

McIntosh, A. J. (1999). Into the tourist's mind: Understanding the value of the heritage experience. *Journal of Travel & Tourism Marketing, 8*(1), 41–64. doi:10.1300/J073v08n01_03

McKercher, B. (2002). Towards a classification of cultural tourists. *International Journal of Tourism Research, 4*(1), 29–38. doi:10.1002/jtr.346

McKercher, B., & du Cros, H. D. (2002). *Cultural tourism: the partnership between tourism and cultural heritage management.* Haworth Hospitality Press.

Meha, A., Tahiri, A., & Zhubi, M. (2020). *Importance of Cultural Heritage in tourism development: The case of Kosovo.* ResearchGate. https://www.researchgate.net/publication/354223359_Importance_of_Cultural_Heritage_in_Tourism_Development_The_Case_of_Kosovo

Mehmood, F., & Aazem, G. (2002). Shilpgram in Udaipur: Organising craftspeople. *Economic and Political Weekly, 37.* https://architexturez.net/doc/10-2307/4412740

Mehta, K., Rawal, Y. S., & Pal, S. (2022). The Role of Smart City Mission in Tourism Development and Sustainability in Udaipur. *Atithya: A Journal of Hospitality, 8*(2).

Mele, C. (2000). *Selling the lower east side: Culture, real estate, and resistance in New York City* (Vol. 5). U of Minnesota Press.

Melubo, K., & Lovelock, B. (2020). Living inside a UNESCO world heritage site: The perspective of the Maasai community in Tanzania. In *Sustainable Tourism Policy and Planning in Africa* (pp. 89–108). Routledge., doi:10.4324/9781003038832-6

Meskell, L. (Ed.). (2002). *Archaeology under fire: nationalism, politics and heritage in the Eastern Mediterranean and Middle East.* Routledge. https://books.google.co.in/books?hl=en&lr=&id=QaWEAgAAQBAJ&oi=fnd&pg=PP1&dq=Archaeology+under+fire:+Nationalism,+politics,+and+heritage+in+the+Eastern+Mediterranean+and+Middle+East&ots=TIOCryURJK&sig=LheXmbPirhLtWhglV-e9vxw5w-U&redir_esc=y#v=onepage&q=Archaeology%20under%20fire%3A%20Nationalism%2C%20politics%2C%20and%20heritage%20in%20the%20Eastern%20Mediterranean%20and%20Middle%20East&f=false doi:10.4324/9780203029817

Meyer, D. (2004). *Tourism routes and getaways: key issues for the development of tourism routes and gateways and their potential for pro-poor tourism.*

Ministry of Finance. Govt. of India. (2019). *India Economic Survey, 2018.* Ministry of France. http://mofapp.nic.in:8080/economicsurvey/

Ministry of Tourism, Government of India. (2022). *Ministry of Tourism, Government of India, Annual Report 2022.* Ministry of Tourism, Government of India, New Delhi. www.indiatourism.com.

Ministry of Tourism. Govt. of India. (2019). *India Tourism Statistics, 2019.* (pp. 7-10). Ministry of Tourism. https://tourism.gov.in/sites/default/files/Other/ITS_Glance_2018_Eng_Version_for_Mail.pdf

Minkiewicz, J., Evans, J., & Bridson, K. (2013). How do consumers co-create their experiences? An exploration in the heritage sector. *Journal of Marketing Management, 30*, 1–2, 30–59. doi:10.1080/0267257X.2013.800899

Mishra, R. (2000). *Pilgrimage Tourism – Case Study of Brajmandal.* [Unpublished Ph.D. Thesis, Kurukshetra University, Kurukshetra].

Mitra, P. (2023, July, 2). In pictures: Sonajhuri Haat — a slice of Santiniketan in the heart of New Town, Saris, jewellery, trinkets, paintings – this Friday-to-Sunday (af)fair with a 'mela' vibe is Kolkata's newest weekend shopping destination. *My Kolkata/ Telegraph India/ The Telegraph Online.* https://www.telegraphindia.com/my-kolkata/places/kolkatas-sonajhuri-haat-a-slice-of-santiniketan-in-the-heart-of-new-town-photogallery/cid/1949245?slide=1

Mkono, M., & Markwell, K. (2014). The application of netnography in tourism studies. *Annals of Tourism Research, 48*, 266–291. doi:10.1016/j.annals.2014.07.005

Moaven, Z. (2017). Globalization of Culture and Religious Tourism. *European Online Journal of Natural and Social Sciences, 6*(1). www.european-science.com

Mohammadi, M., Khalifah, Z., & Hosseini, H. (2010). Local People Perceptions toward Social, Economic and Environmental Impacts of Tourism in Kermanshah (Iran). *Asian Social Science, 6*(11), 220–225. doi:10.5539/ass.v6n11p220

Mohanty, G. N., & Mohanty, S. N. (2004). Living with Snakes: The Lifestyle of the Snake Charmers (SapuaKela). *Adivasi, 44*(12), 24–40.

Moodie, D. (2019). On blood, power, and public interest: The concealment of Hindu sacrificial rites under Indian law. *The Journal of Law and Religion, 34*(2), 1–18. doi:10.1017/jlr.2019.24

Moore, D.J., & Lee, S. P. 9 2012). How advertising influences consumption impulses. *Journal of Advertising, 41*(3), 107-120

Moorhouse, T. P., D'Cruze, N. C., & Macdonald, D. W. (2016). Unethical use of wildlife in tourism: What is the problem, who is responsible, and what can be done? *Journal of Sustainable Tourism*, 1–12. doi:10.1080/09669582.2016.1223087

Morris, M. C. (2013). Improved animal welfare is more related to income equality than it is to income. *Journal of Applied Animal Science, 16*, 272–293. doi:10.1080/10888705.2013.768921 PMID:23795689

Moshin, A., Brochado, A., & Rodrigues, H. (2020). Halal tourism is traveling fast: Community perceptions and implications. *Journal of Destination Marketing & Management, 18*, 100503. doi:10.1016/j.jdmm.2020.100503

Motichandra. (1962). *Kashi Ka Itihas*. Mumbai: Hindi Granth Ratnakar Pvt. Ltd.

Mrđa, A., & Carić, H. (2019). *Cultural Urban Heritage*. The Urban Book Series. doi:10.1007/978-3-030-10612-6

Mukhambetov, M. O. (2021). Cluster Approach. In *Cultural Heritage Tourism (Case Of The Central Asian Section Of Silk Road)*. Central Asian Journal of Social Sciences and Humanities.

Mulder, D. C. (1985). Het vak godsdienstwetenskap. In D. J. Hoens, J. H. Kamstra, & D. C. Mulder (Eds.), *Inleiding tot de studie van godsdiensten* (pp. 35–40). J.H. Kok, Kampen.

Munjal, P. (2015). Potential of participatory approach in sustaining fairs and festivals of small towns: The case of Sohna, Haryana. *Worldwide Hospitality and Tourism Themes, 7*(4), 347–366. doi:10.1108/WHATT-03-2015-0008

Munjal, P. G. (2021). Charting the role of digital platforms for cultural heritage tourism in India. *Worldwide Hospitality and Tourism Themes, 13*(2), 186–198. doi:10.1108/WHATT-10-2020-0126

Munjal, S., & Jauhari, V. (2015). Leveraging the cultural and economic value of fairs and festivals in India. *Worldwide Hospitality and Tourism Themes, 7*(4), 417–423. doi:10.1108/WHATT-03-2015-0006

Murphy, P., Pritchard, M. P., & Smith, B. (2000). The destination product and its impact on traveller perceptions. *Tourism Management, 21*(1), 43–52. doi:10.1016/S0261-5177(99)00080-1

Muštra, V., Perić, B. Š., & Pivčević, S. (2023). Cultural heritage sites, tourism and regional economic resilience. *Papers in Regional Science, 102*(3), 465–482. doi:10.1111/pirs.12731

Nadalipour, Z., Imani Khoshkhoo, M. H., & Eftekhari, A. R. (2019). An integrated model of destination sustainable competitiveness. *Competitiveness Review, 29*(4), 314–335. doi:10.1108/CR-12-2017-0086

Nag, A. (2020). Health and Housing for Urban Poor in India Post Covid-19. *International Journal of Recent Technology and Engineering, 9*(3), 208–216. doi:10.35940/ijrte.C4350.099320

Nag, A. (2022). Industrial infrastructure development of cottage industries for inclusive economic growth in a sustainable manner: Case study of the urban growth centre in Bishnupur, West Bengal, India. *International Journal of Indian Culture and Business Management, 26*(2), 204–233. doi:10.1504/IJICBM.2022.123591

Nag, A., & Mishra, S. (2023a). Destination Competitiveness and Sustainability: Heritage Planning From the Perspective of the Tourism Industry Stakeholders. In I. G. I. Giuseppe Catenazzo (Ed.), *Cases on Traveler Preferences, Attitudes, and Behaviors: Impact in the Hospitality Industry* (pp. 1–32). Global. doi:10.4018/978-1-6684-6919-4.ch001

Nag, A., & Mishra, S. (2023b). Stakeholders' perception and competitiveness of heritage towns: A systematic literature review. *Tourism Management Perspectives, 48*(101156), 101156. doi:10.1016/j.tmp.2023.101156

Nag, A., & Sarkar, S. (2022). Developing an integrated model for assessment of tourism, transportation and economy in small heritage sites. *International Journal of Technological Learning. Innovation and Development, 14*(3), 272–296. doi:10.1504/IJTLID.2022.125694

Narain, A. K., & Roy, T. N. (1977). *Excavations at Rājghāt, 1957-1958, 1960-1965: The pottery (Vol. 2)*. Department of Ancient Indian History, Culture and Archaeology, Banaras Hindu University.

Nasar, J. L. (1989). Perception, Cognition, and Evaluation of Urban Places. *Public Places and Spaces* [Advances in Theory and Research]. *Human Behavior and Environment, 10*, 31–56. doi:10.1007/978-1-4684-5601-1_3

Nastase, C. E., Lucaci, A., & Chaşovschi, C. (2020). Circular Economy: A Perspective for the Transformation of Rural Built Heritage in Europe. *Competitiveness and sustainable development, Economic International Conference*. UTM. http://repository.utm.md/handle/5014/17575

NDTV. (2019). PM Modi Lays the Foundation stone of kasha Vishwanath Temple Corridor. *NDTV Corridor*. www.ndtv.com

Negi, P. C., Sondhi, S., Asotra, S., Mahajan, K., & Mehta, A. (2019). Current status of rheumatic heart disease in India. *Indian Heart Journal*, *71*(1), 85–90. doi:10.1016/j.ihj.2018.12.007 PMID:31000189

Nestor Shpak, O. M.-K. (2021). Simulation of the Influence of External Factors on the Level of Use of the Regional Tourism Potential: A Practical Aspect. *Administrative Sciences*.

Neto, F. (2003, August). A new approach to sustainable tourism development: Moving beyond environmental protection. []). Oxford, UK: Blackwell Publishing Ltd.]. *Natural Resources Forum*, *27*(3), 212–222. doi:10.1111/1477-8947.00056

Nguyen, T. H. H., & Cheung, C. (2016). Chinese heritage tourists to heritage sites: What are the effects of heritage motivation and perceived authenticity on satisfaction? *Asia Pacific Journal of Tourism Research*, *21*(11), 1155–1168. doi:10.1080/10941665.2015.1125377

Ng, W.-K., Hsu, F.-T., Chao, C.-F., & Chen, C.-L. (2023). Sustainable competitive advantage of cultural heritage sites: Three destinations in East Asia. *Sustainability (Basel)*, *15*(11), 8593. doi:10.3390u15118593

Nilsson, M., & Tesfahuney, M. (2016). Performing the "post-secular" in Santiago de Compostela. *Annals of Tourism Research*, *57*, 18–30. doi:10.1016/j.annals.2015.11.001

Nilsson, M., & Tesfahuney, M. (2018). The post-secular tourist: Re-thinking pilgrimage tourism. *Tourist Studies*, *18*(2), 159–176. doi:10.1177/1468797617723467

Norgaard, R. B. (1988). Sustainable development: A co-evolutionary view. *Futures*, *20*(6), 606–620. doi:10.1016/0016-3287(88)90003-1

Nunkoo, R., & Ramkissoon, H. (2010). Small island urban tourism: A residents' perspective. *Current Issues in Tourism*, *13*(1), 37–60. doi:10.1080/13683500802499414

Nunkoo, R., & Ramkissoon, H. (2012). Power, trust, social exchange and community support. *Annals of Tourism Research*, *39*(2), 997–1023. doi:10.1016/j.annals.2011.11.017

Nyaupane, G. P., Timothy, D. J., & Poudel, S. (2015). Understanding tourists in religious destinations: A social distance perspective. *Tourism Management*, *48*, 343–353. doi:10.1016/j.tourman.2014.12.009

Nyaupane, G., & Poudel, S. (2012). Application of appreciative inquiry in tourism research in rural communities. *Tourism Management*, *33*(4), 978–987. doi:10.1016/j.tourman.2011.10.009

O'Dell, T. (2005). Experiencescapes: Blurring borders and testing connections. In T. O'Dell & P. Billing (Eds.), *Experiencescapes – Tourism, culture, and economy* (pp. 11–33). Copenhagen Business School Press.

O'Shaughnessy, J., & O'Shaughnessy, J. N. (2007). Reply to criticisms of marketing, the consumer society and hedonism. *European Journal of Marketing*, *41*(1/2), 7–16. doi:10.1108/03090560710718076

Obasola, K. E. (2014). Religious tourism and sustainable development: A study of eyo festival in Lagos, Nigeria. *International J. Soc. Sci. & Education*, *4*(2), 524–534.

Office of the District Magistrate and Collector, Birbhum (2016). *Mega Tourism Project at Tarapeeth, Birbhum*. Office of the District Magistrate and Collector. file:///C:/Users/it/Downloads/12111412101217EWVJGJustificationfortheProjectinforestland.pdf

Oklevik, O., Gössling, S., Hall, C. M., Steen Jacobsen, J. K., Grøtte, I. P., & McCabe, S. (2019, December 02). Overtourism, optimisation, and destination performance indicators: A case study of activities in Fjord Norway. *Journal of Sustainable Tourism*, *27*(12), 1804–1824. doi:10.1080/09669582.2018.1533020

Oktadiana, H., Pearce, P. L., & Chon, K. (2016). Muslim travellers' needs: What don't we know? *Tourism Management Perspectives*, *20*, 124–130. doi:10.1016/j.tmp.2016.08.004

Okulicz-Kozaryn, A. (2009). *Religiosity and life satisfaction (A multilevel investigation across nations)*. Institute for Quantitative Social Science, Harvard University.

Oliver, M. B., & Raney, A. A. (2011). Entertainment as pleasurable and meaningful: Identifying hedonic and eudaimonic motivations for entertainment consumption. *Journal of Communication*, *61*(5), 984–1004. doi:10.1111/j.1460-2466.2011.01585.x

Oliver, R. L., & Westbrook, R. A. (1993). Profiles of consumer emotions and satisfaction in ownership and usage. *Journal of Consumer Satisfaction, Dissatisfaction & Complaining Behavior*, *6*, 12–27.

Olsen, D. H. (2014). A scalar comparison of motivations and expectations of experience within the religious tourism market. *International Journal of Religious Tourism and Pilgrimage*.

Olsen, D. H. (2019). 23. Religion, spirituality, and pilgrimage in a globalising world. Handbook of globalisation and tourism.

Olsen, D. H., & Timothy, D. J. (2006). *Tourism and religious journeys in Tourism, Religion & Spiritual Journeys*. Routledge.

Olya, H., Jung, T. H., Tom Dieck, M. C., & Ryu, K. (2020). Engaging visitors of science festivals using augmented reality: Asymmetrical modelling. *International Journal of Contemporary Hospitality Management*, *32*(2), 769–796. doi:10.1108/IJCHM-10-2018-0820

Ooi, C. S. (2005). A Theory of Tourism Experiences: The Management of Attention. In T. O'Dell & P. Billing (Eds.), *Experiencescapes: Tourism, Culture, and Economy* (pp. 51–68). Copenhagen Business School Press.

OTDC. (2023). *Temples*. OTDC. https://odishatourism.gov.in/content/tourism/en/discover/attractions/temples-monuments.html https://www.orissatourism.org/odisha-destinations/ https://magazines.odisha.gov.in/orissaannualreference/ORA-2004/pdf/listof_temples_of_orissa.pdf

Packer, J., & Ballantyne, R. (2002). Motivational factors and the visitor experience: A comparison of three sites. *Curator (New York, N.Y.)*, *45*(3), 183–198. doi:10.1111/j.2151-6952.2002.tb00055.x

Packer, J., & Ballantyne, R. (2016). Conceptualizing the visitor experience: A review of literature and development of a multifaceted model. *Visitor Studies*, *19*(2), 128–143. doi:10.1080/10645578.2016.1144023

Padmanabhan Rao, R. A. (1999). Tourism; Medium of Social and Cultural Development. *Yojana*, *43*(8), 41.

Pal, S., & Rawal, Y. S. (2020). Perception and attitude of the host community on socio-economic development through tourism-a case study on Udaipur Rajasthan. *International Journal of Advanced Science and Technology*, *29*(3), 9202–9217.

Panda B. (2016). Temples of Orissa. *Global journal of Indian history and culture*, *1*, 11-12.

Pandey, V., Agarwal, M. K., & Singh, S. V. (2021). A tourism prospect of religious destination of Uttar Pradesh with special reference to Naimisaranya (Neemsaar). *ZENITH International Journal of Multidisciplinary Research*, *11*(7), 42–54.

Panigrahi, K. C. (1993). History of Orissa. *Kitab Mahal, Cuttack, 1993*, 1–9.

Panneerselvam, R. (2014). *Research methodology*. PHI Learning Pvt. Ltd.

Pansiri, J. (2014). Tourist Motives and Destination Competitiveness: A Gap Analysis Perspective. *International Journal of Hospitality & Tourism Administration, 15*(3), 217–247. doi:10.1080/15256480.2014.925718

Papa, A., Mital, M., Pisano, P., & Del Giudice, M. (2020). E-health and wellbeing monitoring using smart healthcare devices: An empirical investigation. *Technological Forecasting and Social Change, 153*, 119226. doi:10.1016/j.techfore.2018.02.018

Park, S. M., & Kim, Y. G. (2022). A Metaverse: Taxonomy, Components, Applications, and Open Challenges. *IEEE Access : Practical Innovations, Open Solutions, 10*, 4209–4251. doi:10.1109/ACCESS.2021.3140175

Park, S., & Kim, S. (2022). Identifying World Types to Deliver Gameful Experiences for Sustainable Learning in the Metaverse. *Sustainability (Basel), 14*(3), 1361. doi:10.3390u14031361

Partha Sarathi Mishra, S. M. (2021). Value assessment of existing architectural heritage for future generation using criteria importance through inter-criteria correlation and grey relational analysis method: A case of Odisha temple architecture in India. *Current Science*.

Pearce, D. G. (1997). *The roles of the Public sector in conservation and tourism planning. Tourism and Heritage Management*. Gadjahmada University Press.

Pearce, J. (1997). Commentary by J. Pearce: Reality check—but how real should we get? *Bulletin de l'Institut Pasteur, 3*(95), 182. doi:10.1016/S0020-2452(97)81374-7

Pearce, P. L., & Lee, U.-I. (2005). Developing the travel career approach to tourist motivation. *Journal of Travel Research, 43*(3), 226–237. doi:10.1177/0047287504272020

Pearson, M., & Sullivan, S. (2013). *Looking after heritage places: The basics of heritage planning for managers, landowners and administrators*. Melbourne Univ. Publishing.

Peker, N. (2014). *Sosyal Bilgiler Dersinde Sanal Müze Kullanımı Ve Sosyal Bilgiler Öğretmeni Adaylarının Sanal Müze Kullanımına Yönelik Tutumları*. [Unpublished Master Thesis, Aksaray Üniversitesi].

Pendit, N. S. (2002). Tourism Science: A First Introduction (Latest Edition) with Improvements. Pradnya Paramita Publisher: Jakarta.

Perles-Ribes, J. F., Ramón-Rodríguez, A. B., Moreno-Izquierdo, L., & Such-Devesa, M. J. (2019). Online Reputation and Destination Competitiveness: The Case Of Spain. *Tourism Analysis, 24*(2), 161–176. doi:10.3727/108354219X15525055915518

Pine, B. J., & Gilmore, D. M. (1999). *The experience economy: Work is theater and every business a stage*. Harvard Business School Press.

Pine, B. J., & Gilmore, J. H. (1999). *The experience economy: Work is theatre & every business a stage*. Harvard Business Press.

Pine, B. J., & Gilmore, J. H. (2011). *The experience economy*. Harvard Business Press.

Plate, S. B. (2009). Introduction: The varieties of contemporary pilgrimage. *Cross Currents, 59*(3), 260–267. doi:10.1353/cro.2009.a782452

Polácek, M., & Aroch, R. (1984). Analysis of cultural sights attractiveness for tourism. *Tourism Review, 39*(4), 17–18.

Poria, Y., Reichel, A., & Biran, A. (2006). Heritage site perceptions and motivations to visit. *Journal of Travel Research, 44*(3), 10. doi:10.1177/0047287505279004

Povroznik, N. (2018). *Virtual Museums and Cultural Heritage: Challenges and Solutions*. The Digital Humanities in the Nordic Countries 3rd Conference Helsinki, Helsinki, Finland.

Prasad, N. (2019). Kashi Vishwanath Temple Corridor's Final DPR Ready! PM Narendra Modi's Dream Project is ready Pilgrims. *Financial Express*.

Preservation and promotion of languages, folk dance, Art and culture of tribals. (n.d.). PIB. https://pib.gov.in/PressReleaseIframePage.aspx?PRID=1604892

Puren, K., & Jordaan, T. (2014). Towards integrating built heritage resources in urban development through spatial planning. *Conference: The Sustainable City IX. WIT Transactions on Ecology and the Environment, 191*, 209–220. doi:10.2495/SC140181

Puro. (2021). *Yhteinen polku. Opas elävän perinnön vastuulliseen matkailulliseen tuotteistamiseen*. Visit Finland ja Museovirasto.

Qiu, Q., Zuo, Y., & Zhang, M. (2022). Intangible cultural heritage in tourism: Research review and investigation of future agenda. *Land (Basel), 11*(1), 139. doi:10.3390/land11010139

Raj, R., Griffin, K., & Blackwell, R. (2015). Motivations for religious tourism, pilgrimage, festivals and events. In R. Raj & K. Griffin (Eds.), *Religious tourism and pilgrimage D. Iliev Journal of Hospitality and Tourism Management 45 (2020) 131–140 management: An international perspective* (pp. 103–117). CAB International. doi:10.1079/9781780645230.0103

Rakitovac, K. A., Urošević, N., & Vojnović, N. (2021). Community Participation in Sustainable Valorisation of Cultural Heritage: The Case of the Municipality of Vrsar. *Academica Turistica, 14*(1), 7–22. doi:10.26493/2335-4194.14.7-22

Ramos, C. M. Q., Henriques, C. H. N., & Lanquar, R. (2016). *Augmented Reality for Smart Tourism in Religious Heritage Itineraries: Tourism Experiences in the Technological Age*. IGI Global. Https://Services.Igi-Global.Com/Resolvedoi/Resolve.Aspx?Doi=10.4018/978-1-5225-0435-1.Ch010, 245–272. https://doi.org/ doi:10.4018/978-1-5225-0435-1.ch010

Rangaswami, C. M. (Ed.). (2017). Ramanathaswamy temple. In Ramanathaswamy temple, 201–207. New Delhi, Delhi: Bharti publications.

Rani, W. M., Wan, T., Shakir, M., & Wahab, M. H. (2018). Governance of Heritage Conservation: Overview on Malaysian Practice. *The Journal of Social Sciences Research, 847-851*. doi:10.32861/jssr.spi6.847.851

Ranwa, R. (2022). Impact of tourism on intangible cultural heritage: Case of Kalbeliyas from Rajasthan, India. *Journal of Tourism and Cultural Change, 20*(1-2), 20–36. doi:10.1080/14766825.2021.1900208

Rao, S. (2020, May 12). *Odisha – India's Best Kept Secret!* [Video]. Digital India Learning. YouTube. https://www.youtube.com/watch?v=iKsvlx_Vr5k

Rapoport, A., & Hawkes, R. (1970). The Perception of Urban Complexity. *Journal of the American Institute of Planners, 36*(2), 106–111. doi:10.1080/01944367008977291

Rappaport, R. A. (1999). *Ritual and religion in the making of humanity*. Cambridge University Press. doi:10.1017/CBO9780511814686

Rasoolimanesh, M., Roldán, J. L., Jaafar, M., & Ramayah, T. (2017). Factors Influencing Residents' Perceptions toward Tourism Development: Differences across Rural and Urban World Heritage Sites. *Journal of Travel Research, 56*(6), 760–775. doi:10.1177/0047287516662354

Rasoolimanesh, S. M., Ramakrishna, S., Hall, C. M., Esfandiar, K., & Seyfi, S. (2023). A systematic scoping review of sustainable tourism indicators in relation to the sustainable development goals. *Journal of Sustainable Tourism, 31*(7), 1497–1517. doi:10.1080/09669582.2020.1775621

Rasoolimanesh, S. M., Seyfi, S., Hall, C. M., & Hatamifar, P. (2021). Understanding memorable tourism experiences and behavioural intentions of heritage tourists. *Journal of Destination Marketing & Management, 21*, 100621. doi:10.1016/j.jdmm.2021.100621

Rastegar, R., Zarezadeh, Z. Z., & Gretzel, U. (2021). World heritage and social justice: Insights from the inscription of Yazd, Iran. In *Justice and Tourism*. Routledge. doi:10.4324/9781003143055-22

Rawal, Y. S., Soni, H., Dani, R., & Bagchi, P. (2022).A Review on Service Delivery in Tourism and Hospitality Industry Through Artificial Intelligence. Role of Religious Authorities in Sustainable Tourism Practices. *In Proceedings of Third International Conference on Computing, Communications, and Cyber-Security*. Research Gate.

Redclift, M. (2005). Sustainable development (1987-2005): An oxymoron comes of age. *Sustainable Development, 13*(4), 212–227. doi:10.1002d.281

Redden, G. (2005). The New Age: Towards a market model. *Journal of Contemporary Religion, 20*(2), 231–246. doi:10.1080/13537900500067851

Reed, J. (2006). *Appreciative inquiry: Research for change*. Sage publications.

Reher, G. (2021). Heritage as Action Search. *Sustainability, 14*(1), 282. doi:10.3390u14010282

Rennie, J. K., & Singh, N. C. (1996). *Participatory research for sustainable livelihoods: a guidebook for field projects*. Iisd.

Report, D. V. (2020), Government of Uttar Pradesh. *Varanasi tourism statistics*. https://varanasi.nic.in/

Richards (1996). *Journal of Environment Quality, 25*(3). doi:10.2134/jeq1996.253npx

Richards, G., & Munsters, W. (2010). Cultural tourism research methods. CAB International. Research Gate. doi:10.1079/9781845935184.0000

Richards, G. (1999). Heritage visitor attractions in Europe: A visitor profile. *Interpretation (Tulsa), 4*(3).

Richards, G. (2011). Religious tourism and social change: Expressions of faith in contemporary tourism. *Tourism Recreation Research.*

Richards, G. (2018). Cultural tourism: A review of recent research and trends. *Journal of Hospitality and Tourism Management, 36*, 12–21. doi:10.1016/j.jhtm.2018.03.005

Rinschede, G. (1992). Forms of Religious Tourism. *Annals of Tourism Research, 19*(1), 51–67. doi:10.1016/0160-7383(92)90106-Y

Rivera, M. A., Shani, A., & Severt, D. (2009). Perceptions of service attributes in a religious theme site: An importance satisfaction analysis. *Journal of Heritage Tourism, 4*(3), 227–243. doi:10.1080/17438730902822939

Rizvic, S. (2014). Story Guided Virtual Cultural Heritage Applications. *Journal of Interactive Humanities, 2*(1), 1–15. doi:10.14448/jih.02.0002

Roberts, B. H. (2015). *Tool Kit for Rapid Economic Assessment, Planning, and Development of Cities in Asia*. Asian Development Bank.

Roche, E. (2019). India and Pakistan signed a historic agreement to open the Kartarpur corridor. *Live mint.*

Ron, A. (2007). Book Review: Tourism, Religion & Spiritual Journeys. *Annals of Tourism Research, 34*(2), 547–548. doi:10.1016/j.annals.2006.10.007

Rosengren, K. E. (1993, September 1). From Field to Frog Ponds. *Journal of Communication, 43*(3), 6–17. doi:10.1111/j.1460-2466.1993.tb01271.x

Roy, T. (2005). *Rethinking economic change in India: labour and livelihood.* Routledge. doi:10.4324/9780203023105

Ryan, C., & Montgomery, D. (1994). The Attitudes of Bakewell Residents to Tourism and Issues in Community Responsive Tourism. *Tourism Management, 15*(5), 358–369. doi:10.1016/0261-5177(94)90090-6

Rypkema, D. D. (2008). *Heritage Conservation and the Local Economy.* Global Urban Development.

Saatci İzci, B., & Halaç, H. H. (2021). Kent Sanal Turlarında Kültürel Mirasın Değerlendirilmesi. *Turk Turizm Arastirmalari Dergisi, 5*(2), 985–1004. doi:10.26677/TR1010.2021.749

Sahni, D. R. (1914). *Catalogue of the Museum of Archaeology at Sārnāth.* Superintendent Government printing press.

Saidovna, V. F. (2022). Pilgrimage Tour in Tourism. *Miasto Przyszłości, 28*, 342–345.

Saipradist, A., & Staiff, R. (2008). Crossing the cultural divide: Western visitors and interpretation at Ayutthaya world heritage site, Thailand. *Journal of Heritage Tourism, 2*(3), 211–224. doi:10.2167/jht061.0

Sairinen, R., & Kumpulainen, S. (2006). Assessing social impacts in urban waterfront regeneration. *Environmental Impact Assessment Review, 26*(1), 120–135. doi:10.1016/j.eiar.2005.05.003

Salazar, N. B. (2012). Community-based cultural tourism: Issues, threats and opportunities. *Journal of Sustainable Tourism, 20*(1), 9–22. doi:10.1080/09669582.2011.596279

Salemink, O. (2012). Appropriating culture: The politics of intangible cultural heritage in Vietnam. In *State, society and the market in contemporary Vietnam* (pp. 158–180). Routledge.

Samadi, Z., & Yunus, R. M. (2012). Physical and Spiritual Attributes of Urban Heritage Street's Revitalization. *Procedia: Social and Behavioral Sciences, 36*, 342–349. doi:10.1016/j.sbspro.2012.03.038

Samira, D., Mondal, K., & Basu, K. (2006). Dissemination of cultural heritage and impacts of pilgrim tourism at Gangasagar Island. In *Anthropological Survey of India.* Anthropologist.

Santa-Cruz, F. G., & López-Guzmán, T. (2017). Culture, tourism and world heritage sites. *Tourism Management Perspectives, 24*, 111–116. doi:10.1016/j.tmp.2017.08.004

Santos, M. (2000). Fátima: religious tourism in a sanctuarycity. *World Leisure Congress.* Bilbau.

Saravanan, A., & Rao, D. Y. V. (2012). Equitable Tourism Development: Need for Strategic Partnership. *International Journal of Multidisciplinary Research, 2*(3), 344–356.

Sastri, S. A. (1901). *The Taittiriya Upanishad.* GTA Printing Works.

Savishinsky, J. S. (1983). Pet ideas: The domestication of animals, human behavior, and human emotions. In A. H. Katcher & A. M. Beck (Eds.), *New perspectives on our lives with companion animals* (pp. 112–131). University of Pennsylvania Press.

Schmitt, B. (1999). *Experiential marketing: How to get customers to SENSE, FEEL, THINK, ACT and RELATE to our company and brands.* The Free Press.

Schouten, F. F. (1995). Heritage as historical reality. *Heritage, tourism and society*, 21-31.

Schroeder, T. (1996). The Relationship of Residents' Image of their State as a Tourist Destination and their Support for Tourism. *Journal of Travel Research, 34*(4), 71–73. doi:10.1177/004728759603400411

Scoones, I. (1998). Sustainable rural livelihoods, A framework for analysis [). Brighton, IDS.]. *IDS Working Paper*, 72.

Sébastia, B. (2004). *Like an Indian God, Hinduisation of the cult of Saint Anthony of Padua in Tamil Nadu*, Auxlieuxd'Etre.

Secondi, L., Meseguer-Santamaría, M. L., Mondéjar-Jiménez, J., & Vargas-Vargas, M. (2011). Influence of tourist sector structure on motivations of heritage tourists. *Service Industries Journal, 31*(10), 1659–1668. doi:10.1080/02642069.2010.485641

Selmi, N., Bahri-Ammari, N., Soliman, M., & Hanafi, I. (2021). The impact of festivalscape components on festivalgoers' behavioral intentions: The case of the International Festival of Carthage. *Journal of Convention & Event Tourism.*

Şen, H. (2023). *Sosyal Bilgiler Öğretiminde 5 Ve 6. Sınıf Kültür Ve Miras Öğrenme Alanlarında Sanal Müze Kullanımına Yönelik Öğretmen Görüşleri.* [Unpublished Master Thesis, Necmettin Erbakan Üniversitesi].

Serpell, J. (1987). Pet keeping in non-western societies: Some popular misconceptions. *Anthrozoos, 1*(3), 166–174. doi:10.2752/089279388787058443

Serrat, O. (2017). The sustainable livelihoods approach. In *Knowledge solutions* (pp. 21–26). Springer. doi:10.1007/978-981-10-0983-9_5

Seth, R. (1999). Tourism: Problems and Prospects. *Yojana*, 21–24.

Seybold, K. S., & Hill, P. C. (2001). The role of religion and spirituality in mental and physical health. *Journal of Current Directions in Psychological Science, 10*(1), 21–24. doi:10.1111/1467-8721.00106

Seyfi, S., & Hall, C. M. (2021). Cultural heritage tourism in the MENA: introduction and background. In Hall. C.M. & Seyfi. S (Eds.). Cultural and Heritage Tourism in the Middle East and North Africa. Routledge. doi:10.4324/9780429279065-1

Seyfi, S., Hall, C. M., & Rasoolimanesh, S. M. (2020). Exploring memorable cultural tourism experiences. *Journal of Heritage Tourism, 15*(3), 341–357. doi:10.1080/1743873X.2019.1639717

Seyfi, S., Michael Hall, C., & Fagnoni, E. (2019). Managing world heritage site stakeholders: A grounded theory paradigm model approach. *Journal of Heritage Tourism, 14*(4), 308–324. doi:10.1080/1743873X.2018.1527340

Shackley, M. (1996). Community impact of the camel safari industry in Jaisalmar, Rajasthan. *Tourism Management, 17*(3), 213–218. doi:10.1016/0261-5177(96)00008-8

Shaheer, I., Carr, N., & Insch, A. (2021). Spatial distribution of participation in boycott calls: A study of tourism destination boycotts associated with animal abuse. *Anatolia. Anatolia, 33*(3), 323–334. doi:10.1080/13032917.2021.1931896

Sharma, J. K. (2000). *Tourism Planning and Development.* Kansishka Publisher.

Sharma, S., & Sharma, S. (2015). Understanding the socio-economic impact of Pithoragarh's cultural jewels: Its fairs and festivals. *Worldwide Hospitality and Tourism Themes, 7*(4). doi:10.1108/WHATT-03-2015-0007

Sharpley, R. (2016). Tourism and Spirituality: An Evolving Relationship. *International journal of Tourism & Spirituality.*

Sharpley, R. (2014). Host perceptions of tourism: A review of the research. *Tourism Management, 42*, 37–49. doi:10.1016/j.tourman.2013.10.007

Shastri, J. (2017). *Artist Village.* [Master's thesis, Unitec Institute of Technology].

Shastri, B. (1994). *KalikapuraneMurtivinirdesah.* Motilal Banarsidass Publications.

Shastri, J. L. (1970). *The Siva Puraṇa*. Motilal Banarasidass.

Shenoy, S. (2022). Rajasthani Paintings To Jewellery, Shilpgram Is Udaipur's Best Spot For Budget Shopping. *Curly Tales*. https://curlytales.com/rajasthani-paintings-to-jewellery-shilpgram-is-udaipurs-best-spot-for-budget-shopping/

Shinde, K. A. (2021). The environmental impacts of religious and spiritual tourism. The Routledge handbook of religious and spiritual tourism.

Shinde, K. A., & Olsen, D. H. (2020). The environmental impacts of religious tourism. *Religious tourism and the environment*.

Shinde, K. (2008). Religious tourism: exploring a new form of sacred journey in North India. In J. Cochrane (Ed.), *Asian Tourism: Growth and Change* (pp. 245–257). Elsevier Publishing Ltd. doi:10.1016/B978-0-08-045356-9.50025-4

Shinde, K. A. (2003). Residents' perceptions of pilgrimage tourism. *Sociology of Religion*, *63*(4), 475–496.

Shinde, K., & Singh, R. P. (2023). Still on UNESCO's "Tentative List of World Heritage"? Heritage, Tourism, and Stunted Growth in Sarnath (Varanasi), India. *Heritage*, *6*(7), 5051–5065. doi:10.3390/heritage6070267

Shri Kashi Viswanath Temple Trust. (n.d.). *About.* Shri Kashi Viswanath Temple Trust. https://www.shrikashivishwanath.org/

Shukla, S. (2022). *Reimagining socio-spatial Arrangements of a cultural neighbourhood by tourism: a case of silk weaving community of Varanasi* [Doctoral dissertation, SPA Bhopal].

Sigala, M., & Leslie, D. (2016). *International cultural tourism: management, implications, and cases*. Routledge.

Singer, J. L. (1966). *Daydreaming: An Introduction to the Experimental Study of InnerExperimental Study of Inner Experience*. Random House.

Singh, P.K (1998). *Fifty years of Indian Tourism*. Kanishka Publication.

Singh, R. (1998) *Infrastructure of Tourism in India*. Kanishka Publication.

Singhal, V. (2009). *Perspectives and Potential of Kurukshetra as Tourist Destination*. [Unpublished Ph.D. Thesis, Kurukshetra University, Kurukshetra].

Singh, R. P. B. (2011). *Varanasi, India's cultural heritage city: Contestation, Conservation, and Planning*. Surbhi Publications.

Singh, S. (2002). Tourism in India: Policy Pitfalls. *Asia Pacific Journal of Tourism Research*, *7*(1), 45–59. doi:10.1080/10941660208722109

Singh, S. (2006). Secular pilgrimages and sacred tourism in the Indian Himalayas. *GeoJournal*, *64*(3), 215–223. doi:10.100710708-005-5649-8

Singh, S. (2021). 'Quixotic' tourism? Safety, ease, and heritage in post-COVID world tourism. *Journal of Heritage Tourism*, *16*(6), 716–721. doi:10.1080/1743873X.2020.1835924

Singh, S. V., & Ranjan, R. (2019). Online travel portal and their effect on travel agency: A study on outbound visitors of Varanasi. [IJRAR]. *International Journal of Research and Analytical Reviews*, *6*(2), 387–393.

Sinha, A., Harkness, T., & Harkness, T. (2009). Views of the Taj—Figure in the Landscape. *Landscape Journal*, *28*(2), 198-217. https://www.jstor.org/stable/43323846

Sjöholm, J. (2013). *Built Cultural Heritage in an Urban Planning Context: Literature Review*. Semantic Scholar. https://www.semanticscholar.org/paper/Built-cultural-heritage-in-an-urban-planning-%3A-Sj%C3%B6holm/1849c5ada310878db95d966014d23cc1b3520468

Slater, T. R. (2004). Encountering God: Personal reflections on 'geographer as pilgrim'. *Area*, *36*(3), 245–253. doi:10.1111/j.0004-0894.2004.00221.x

Sliem, M. A. E., Hegazi, Y. S., & El-Alfy, M. M. F. (2019). Smart Heritage Buildings As A Tool To Achieve Tourism Competitiveness. *International Journal Of Eco-Cultural Tourism, Hospitality Planning And Development, 3*(1), 1-9. https://ijecth.journals.ekb.eg/article_181298_b17ef9bab59cbe1477ca2f1b08c4be38.pdf

Smith, L. (2006). *Uses of heritage*. Routledge. https://books.google.co.in/books?hl=en&lr=&id=bYx_AgAAQBAJ&oi=fnd&pg=PP1&dq=Smith,+L.+(2006).+Uses+of+heritage.+Routledge.%5C&ots=a0uN22RtEk&sig=KOMynKkGgRoIpTm3Mrl_W9J1KUY&redir_esc=y#v=onepage&q=Smith%2C%20L.%20(2006).%20Uses%20of%20heritage.%20Routledge.%5C&f=false

Smith, M. K., Pinke-Sziva, I., Berezvai, Z., & Buczkowska-Gołąbek, K. (2022). The changing nature of the cultural tourist: Motivations, profiles and experiences of cultural tourists in Budapest. *Journal of Tourism and Cultural Change*, *20*(1-2), 1–19. doi:10.1080/14766825.2021.1898626

Smith, V. L. (1992). Introduction: The quest in guest. *Annals of Tourism Research*, *19*(1), 1–17. doi:10.1016/0160-7383(92)90103-V

Sofield, T., & Sivan, A. (2003). From cultural tourism to creative tourism: A case study of the Jewish community of Rosh Pina, Israel. *Tourism Recreation Research.*

Soliman, M., & Al Balushi, M. K. (2023). Unveiling destination evangelism through generative AI tools. *ROBONOMICS: The Journal of the Automated Economy, 4*, 54. https://journal.robonomics.science/index.php/rj/article/view/54.

Soliman, M. (2021). Extending the theory of planned behavior to predict tourism destination revisit intention. *International Journal of Hospitality & Tourism Administration*, *22*(5), 524–549. doi:10.1080/15256480.2019.1692755

Soliman, M. S. (2015). Pro-poor tourism in protected areas–opportunities and challenges:"The case of Fayoum, Egypt. *Anatolia*, *26*(1), 61–72. doi:10.1080/13032917.2014.906353

Soliman, M., & Al Balushi, M. (2023). Unveiling destination evangelism through generative AI tools. ROBONOMICS. *The Journal of the Automated Economy*, *4*(54), 1.

Soliman, M., Cardoso, L., de Almeida, G. F., Araújo, A., & Vila, N. (2021). Mapping smart experiences in tourism: A bibliometric approach. *European Journal of Tourism Research*, *28*, 2809–2809. doi:10.54055/ejtr.v28i.2254

Soliman, M., Gulvady, S., Lyulyov, O., & Pimonenko, T. (2023). Research Trends and Themes in the Top-Tier Tourism, Leisure and Hospitality Journals: A Bibliometric and Network Analysis Before and During the COVID-19 Era. *International Journal of Hospitality and Tourism Systems*, *16*(1).

Statistical Abstract of Haryana (2022). *Data Compiled* from *Statistical Abstract of Haryana 2021-22*. Department of Economic and Statistical Analysis, Haryana.

Stausberg, M. (2014). Religion and spirituality in tourism. In A. A. Lew, C. M. Hall, & A. M. Williams (Eds.), *The Wiley Blackwell companion to tourism* (pp. 349–360). John Wiley & Sons, Ltd. doi:10.1002/9781118474648.ch28

Stoddard, J. E., Pollard, C. E., & Evans, M. R. (2012). The triple bottom line: A framework for sustainable tourism development. *International Journal of Hospitality & Tourism Administration*, *13*(3), 233–258. doi:10.1080/15256480.2012.698173

Stoddard, R. H. (1997). Defining and classifying pilgrimages. In R. H. Stoddard & A. Morinis (Eds.), *Sacred places, sacred spaces: The geography of pilgrimages* (pp. 41–60). Louisiana State University: Geoscience Publications.

Styliani, S., Fotis, L., Kostas, K., & Petros, P. (2009). Virtual museums, a survey and some issues for consideration. *Journal of Cultural Heritage*, *10*(4), 520–528. doi:10.1016/j.culher.2009.03.003

Subasinghe, M., Magalage, D., Amadoru, N., Amarathunga, L., Bhanupriya, N., & Wijekoon, J. L. (2020). Effectiveness of artificial intelligence, decentralized and distributed systems for prediction and secure channelling for Medical Tourism. *11th Annual IEEE Information Technology, Electronics and Mobile Communication Conference, IEMCON 2020*, (pp. 314–319). IEEE. https://doi.org/10.1109/IEMCON51383.2020.9284898

Subawa, N. S., Widhiasthini, N. W., Astawa, I. P., Dwiatmadja, C., & Permatasari, N. P. I. (2021). The practices of virtual reality marketing in the tourism sector, a case study of Bali, Indonesia. *24*(23), 3284–3295. doi:10.1080/13683500.2020.1870940

Suen, L., Huang, H., & Lee, H. (2014). A comparison of convenience sampling and purposive sampling. *Hu li za zhi The journal of nursing, 61*(3), 105.

Suh, W., & Ahn, S. (2022). Utilizing the Metaverse for Learner-Centered Constructivist Education in the Post-Pandemic Era: An Analysis of Elementary School Students. *Journal of Intelligence*, *10*(1), 17. doi:10.3390/jintelligence10010017 PMID:35324573

Su, J. (2019). Understanding the changing intangible cultural heritage in tourism commodification: The music players' perspective from Lijiang, China. *Journal of Tourism and Cultural Change*, *17*(3), 247–268. doi:10.1080/14766825.2018.1427102

Sunarminto, T., & Mijiarto, J. (2022). Cultural Heritage Tourism Development Strategy in bogor regency. *Jurnal Pariwisata Pesona*, *7*(1), 63–74. doi:10.26905/jpp.v7i1.6643

Sungur, T., & Bülbül, H. (2019). Sınıf Öğretmeni Adaylarının Sanal Müze Uygulamalarına Yönelik Görüşleri. *Abant İzzet Baysal Üniversitesi Eğitim Fakültesi Dergisi*, *19*(2), 652–666. doi:10.17240/aibuefd.2019.19.46660-492112

Suntikul, W., & Dorji, U. (2016). Local Perspectives on the Impact of Tourism on Religious Festivals in Bhutan. *Asia Pacific Journal of Tourism Research*, *21*(7), 741–762. doi:10.1080/10941665.2015.1069365

Sürücü, O., & Başar, M. E. (2016). Kültürel Mirası Korumada Bir Farkındalık Aracı Olarak Sanal Gerçeklik. *Artium*, *4*(1), 13–26.

Surugiu, M.-R., & Surugiu, C. (2015). Heritage tourism entrepreneurship and social media: Opportunities and challenges. *Procedia: Social and Behavioral Sciences*, *188*, 74–81. doi:10.1016/j.sbspro.2015.03.340

Susana Mourato, M. M. (2002). *Economic Valuation of Cultural Heritage: Evidence and Prospects.* The Getty Convention.

Swain, M. (1989). Developing ethnic tourism in Yunnan, China. *Tourism Recreation Research*, *14*(1), 33–39. doi:10.1080/02508281.1989.11014541

Swain, S. K., & Batabyal, D. (2016). Sustainable livelihood and its diversification through community-based ecotourism at Pichavaram and surroundings: An investigative study. *South Asian Journal of Tourism and Heritage*, *9*(2), 82–97.

Swarbrooke, J. (2020). *Postscript-The COVID-19 Pandemic.*

Tan, S. K., Tan, S. H., Kok, Y. S., & Choon, S. W. (2018). Sense of place and sustainability of intangible cultural heritage – The case of George Town and Melaka. *Tourism Management*, *67*, 376–387. doi:10.1016/j.tourman.2018.02.012

Tao, T., & Wall, G. (2009). Tourism as a sustainable livelihood strategy. *Tourism Management, 30*(1), 90–98. doi:10.1016/j.tourman.2008.03.009

Teker, N., & Özer, A. (2016). Sanal Müze Sanal Tur Memnuniyet Ölçeğinin Türkçe'ye Uyarlanması: Geçerlik Ve Güvenirlik Çalışması. *Milli Eğitim Dergisi, 209*, 314–335.

Teo, P., & Huang, S. (2000). Singapore: Tourism development in a planned context. In M. Hall & S. Page (Eds.), *Tourism in South and Southeast Asia: Issues and Cases*. Butterworth Heinemann.

Terzidou, M., Scarles, C., & Saunders, M. N. (2017). Religiousness as tourist performances: A case study of Greek orthodox pilgrimage. *Annals of Tourism Research, 66*, 116–129. doi:10.1016/j.annals.2017.06.009

Thoresen, C. E., & Harris, A. H. (2004). *Spirituality, religion, and health: A scientific perspective*. Research Gate.

Tilden, F. (1957). *Interpreting our heritage: Principles and practices for visitor services in parks, museums, and historic places*. University of North Carolina Press.

Timothy, D. J. (2000). Sacred Journeys: Religious heritage and tourism. *Tourism Recreation Research.*

Timothy, D. J. (2014). Contemporary cultural heritage and tourism: Development issues and emerging trends. *Public Archaeology, 13*(1-3), 30–47. doi:10.1179/1465518714Z.00000000052

Timothy, D. J., & Boyd, S. W. (2003). *Heritage tourism*. Pearson Education.

Timothy, D. J., & Boyd, S. W. (2003). *Heritage Tourism*. Prentice Hall.

Timothy, D. J., & Boyd, S. W. (2006). Heritage tourism in the 21st century: Valued traditions and new perspectives. *Journal of Heritage Tourism, 1*(1), 1–16. doi:10.1080/17438730608668462

Timothy, D. J., & Conover, P. J. (2006). Nature religion, self-spirituality and New Age tourism. In D. J. Timothy & D. H. Olsen (Eds.), *Tourism, religion and spiritual journeys* (pp. 139–155). Routledge. doi:10.4324/9780203001073-21

Timothy, D., & Olsen, D. (Eds.). (2006). *Tourism, religion and spiritual journeys*. Routledge. doi:10.4324/9780203001073

Tiwari, S., Rosak-Szyrocka, J., & Żywiołek, J. (2022). Internet of Things as a Sustainable Energy Management Solution at Tourism Destinations in India. *Energies, 15*(7), 2433. doi:10.3390/en15072433

Tiwari, A., Dhiman, V., Iesa, M. A. M., Alsarhan, H., Mehbodniya, A., & Shabaz, M. (2021). Patient Behavioral Analysis with Smart Healthcare and IoT. *Behavioural Neurology, 2021*, 1–9. doi:10.1155/2021/4028761 PMID:34900023

Tiwari, S. (2015). Culture based development: Empirical evidence for Shilpgram Fair, Rajasthan. *Worldwide Hospitality and Tourism Themes, 7*(4), 393–402. doi:10.1108/WHATT-03-2015-0005

Tnn. (2016, December 27). Shilpgram Utsav: Real festival of craftsmen, folk artists from far off villages. *The Times of India*. https://timesofindia.indiatimes.com/city/jaipur/shilpgram-utsav-real-festival-of-craftsmen-folk-artists-from-far-off-villages/articleshow/56190759.cms

Töppel, M., & Reichel, C. (2021). Qualitative Methods and Hybrid Maps for Spatial Perception with an Example of Security Perception. *Urban Planning, 6*(1), 105–119. Advance online publication. doi:10.17645/up.v6i1.3614

Toubes, D. R., Vila, N. A., & Fraiz Brea, J. A. (2021). Changes in consumption patterns and tourist promotion after the covid-19 pandemic. *Journal of Theoretical and Applied Electronic Commerce Research, 16*(5), 1332–1352. doi:10.3390/jtaer16050075

Tripathy, M. (1998). *Patta painting society and economy of the chitrakaras of Orissa a study of change and development.* [Thesis, Utkal University]. https://shodhganga.inflibnet.ac.in/handle/10603/189442

Tsaur, S. H., & Tu, J. H. (2019). Cultural competence for tour leaders: Scale development and validation. *Tourism Management, 71*, 9–17. doi:10.1016/j.tourman.2018.09.017

Tscheu, F., & Buhalis, D. (2016). Augmented Reality at Cultural Heritage sites. *Information and Communication Technologies in Tourism, 2016*, 607–619. doi:10.1007/978-3-319-28231-2_44

Tsironis, C. N. (2022). *Pilgrimage and religious tourism in society, in the wake of the COVID-19 pandemic: A paradigmatic focus on 'St. Paul's Route'in the Central Macedonia Region.* Religions.

Tuan, Y. F. (1978). Sacred Space: explorations of an idea. In K. W. Butzer (Ed.), *Dimensions of Human Geography: Essays on some familiar and neglected themes* (pp. 84–99). University of Chicago Press.

Tul, H. (2015). Kali's Tongue: Shame, Disgust, and the Rejection of Blood and Violence in Vedic and Hindu Thought. *International Journal of Hindu Studies, 19*(3), 301–332. https://doi.org/. doi:10.1007/s11407-015-9181-2

Tunbridge, J. E., & Ashworth, G. J. (1996). *Dissonant heritage: the management of the past as a resource in conflict.* Wiley.

Tunbridge, J. E., & Ashworth, G. J. (n.d.). *(1996). Dissonant Heritage: The Management of the Past as a UNESCO - Intangible Heritage Home.* Press., https://ich.unesco.org/en/home.University

Tung, V. W. S., & Ritchie, J. B. (2011). Exploring the essence of memorable tourism experiences. *Annals of Tourism Research, 38*(4), 1367–1386. doi:10.1016/j.annals.2011.03.009

Turner, V., & Turner, E. (1969). *The ritual process.* Routledge.

Twain, M. (1897). *Following the Equator: A journey around the world.* Archive. https://archive.org/details/following-equator00twaiuoft

U.S. National Science Foundation. (2000). *Towards a comprehensive geographical perspective on urban sustainability.* New Brunswick, NJ: Rutgers—The State University of New Jersey: http://web.worldbank.org/WBSITE/EXTERNAL/TOPICS/EXTPOVERTY/EXTPA/0,contentMDK:20208959~menuPK:443290~pagePK:148956~piPK:216618~theSitePK:430367~isCURL:Y,00.html

UN. (2023). *The 17 goals.* United Nations. https://sdgs.un.org/goals

UNESCO (n.d.) *Guidelines for the Establishment of National "Living Human Treasures" Systems.* Paris: UNESCO.

UNESCO. (1989). *Recommendation on the Safeguarding of Traditional Culture and Folklore.* In Proceedings of the General Conference of the United Nations Educational, Scientific and Cultural Organization, Paris, France. http://portal.unesco.org/en/ev.php-URL_ID=13141&URL_DO=DO_TOPIC&URL_SECTION=201.html

UNESCO. (2003). *Convention for the safeguarding of the intangible Cultural Heritage.* UNESCO.

UNESCO. (2003). *International convention for the safeguarding of the intangible cultural heritage.* UNESCO.

UNESCO. (2013). *World Heritage Resource Manual.* UNESCO. https://whc.unesco.org/en/managing-cultural-world-heritage/

UNESCO. (2014). Heritage Sustainability. *CDIS Methodology Manual V7.* UNESCO. https://en.unesco.org/creativity/sites/creativity/files/cdis/heritage_dimension.pdf

UNESCO. (2020). *Heritage in Urban Contexts: Impacts of Development Projects on World Heritage properties in Cities.* UNESCO. https://whc.unesco.org/en/events/1516/#:~:text=It%20proposes%20new%20policies%2C%20standards,national%20and%20urban%20development%20policies

UNESCO. (2021). *Cutting edge: Bringing cultural tourism back in the game*. UNESCO. https://www.unesco.org/en/articles/cutting-edge-bringing-cultural-tourism-back-game

UNESCO. (2023). *What is Intangible Cultural Heritage?* UNESCO. https://ich.unesco.org/en/what-is-intangible-heritage-00003

UNESCO. (2023). *World Heritage List: States Parties*. UNESCO. *Oman*

UNESCO. (2023a). *Kaustinen fiddle playing and related practices and expressions*. UNESCO. https://ich.unesco.org/en/RL/kaustinen-fiddle-playing-and-related-practices-and-expressions-01683

UNESCO. (2023a). *What is Intangible Cultural Heritage?* UNESCO. https://ich.unesco.org/en/what-is-intangible-heritage-00003

UNESCO. (2023b). *Sustainable development and living heritage*. UNESCO. https://ich.unesco.org/en/sustainable-development-and-living-heritage

UNESCO. (2023c). *Browse the Lists of Intangible Cultural Heritage and the Register of good safeguarding practices, Finland.* UNESCO. https://ich.unesco.org/en/lists?text=&country[]=00076&multinational=3#tabs

UNESCO. (2023d). *Browse the Lists of Intangible Cultural Heritage and the Register of good safeguarding practices.* UNESCO. https://ich.unesco.org/en/lists

UNWTO. (2017). *International Congress on Religious Tourism and Pilgrimage.* UNESCO. http://europe.unwto.org/event/international-congress-religious-tourism-and-pilgrimage

UNWTO. (2017). *Tourism and culture*. UNESCO. https://www.unwto.org/tourism-and-culture.

UNWTO. (2018). *Tourism and Culture Synergies*. United Nations World Tourism Organization. doi:10.18111/9789284418978

UNWTO. (2020). Tourism and Culture. UNWTO. https://www.unwto.org/tourism-and-culture

Urban, H. B. (2016). Desire, Blood and Power– Georges Bataille and the Study of Hindu Tantra in Northeast India. *RevistaCientíficaArbitrada de la Fundación MenteClara, 1*(3), 26–48. doi:10.32351/rca.v1.3.22

Urban, H. B. (2019). 'The Cradle of Tantra': Modern transformations of a Tantric Centre in Northeast India from Nationalist Symbol to Tourist Destination. *South Asia, 42*(2), 256–277. doi:10.1080/00856401.2019.1570609

Ursache, M. (2015). Tourism – significant driver shaping a destinations heritage. *Procedia: Social and Behavioral Sciences, 188*, 130–137. doi:10.1016/j.sbspro.2015.03.348

Valentina Vasile, M. R. (2016). *Heritage Entrepreneurship and Eco-tourism - A new vision on ecosystem protection and in-situ specific activities for cultural heritage consumption.* Economics and Finance.

Varanasi tourism report. (2020) UP Tourism. http://www.varanasi.gov.in/en/destination

Vargün, Ö. (2020). *Kültürel Miras Eğitiminde BİT'in Rolü ve AR Mobil Uygulama Projesi* [Unpublished PhD Thesis, Yıldız Teknik Üniversitesi].

Vargün, Ö., & Nuhoğlu, M. (2019). Kültürel Miras Eğitiminde Bilgi İletişim Teknolojilerinin Rolü ve Mobil Uygulamalar. *UNIMUSEUM, 2*(2), 45–53.

Vayanou, M., Katifori, A., Chrysanthi, A., & Antoniou, A. (2020). *Cultural Heritage and Social Experiences in the Times of COVID 19.* CHI '05 Extended Abstracts on Human Factors in Computing Systems, Island of Ischia, Italy. https://dl.acm.org/doi/10.1145/1056808.1056881

Veenhoven, R. (2003). Hedonism and Happiness. *Journal of Happiness Studies*, *4*(4), 437–457. doi:10.1023/B:JOHS.0000005719.56211.fd

Vegheş, C. (2018). Cultural Heritage, Sustainable Development and Inclusive Growth: Global Lessons for the Local Communities Under a Marketing Approach. *European Journal of Sustainable Development*, *7*(4). doi:10.14207/ejsd.2018.v7n4p349

Velasco, C., & Obrist, M. (2020). *Multisensory experiences: Where the senses meet technology*. Oxford University Press. doi:10.1093/oso/9780198849629.001.0001

Velasco, C., & Obrist, M. (2021). Multisensory Experiences: A Primer. *Frontiers of Computer Science*, *3*, 614524. doi:10.3389/fcomp.2021.614524

Veldpaus, L. (2015). Reconnecting the city: The historic urban landscape approach and the future of urban heritage. *Cultural Trends*, *24*(4), 340–342. doi:10.1080/09548963.2015.1106036

Vengesayi, S. (2003). A conceptual model of tourism destination competitiveness and attractiveness. *ANZMAC 2003 Conference Proceedings Adelaide*. ANZMAC. https://anzmac.org/conference_archive/2003/

Verma, L. (2021). *PM Modi to inaugurate Kashi Vishwanath corridor in Varanasi today. The New Indian Express*. Express Publications. https://indianexpress.com/article/india/pm-modi-to-inaugurate-kashi-vishwanath-corridor-in-varanasi-7669259/#:~:text=%E2%80%9CTomorrow%2C%2013th%20December%20is%20a,temple%20town%2C%20tweeted%20on%20Sunday

Vijayanand, S. (2013). Pilgrimage Tourism and its Economic Dimensions in Tourism Management. *Radix International Journal of Economics & Business Management*, *2*(1), 1–31.

Vistad, O. I., Øian, H., Williams, D. R., & Stokowski, P. (2020). Long-distance hikers and their inner journeys: On motives and pilgrimage to Nidaros, Norway. *Journal of Outdoor Recreation and Tourism*, *31*, 100326. doi:10.1016/j.jort.2020.100326

Vosinakis, S., Nikolakopoulou, V., Stavrakis, M., Fragkedis, L., Chatzigrigoriou, P., & Koutsabasis, P. (2020). Co-Design of a Playful Mixed Reality Installation: An Interactive Crane in the Museum of Marble Crafts. *Heritage, 3*(4), 1496–1519. https://doi.org/ doi:10.3390/heritage3040083

Vukonic, B. (2002). Religion, tourism and economics: A convenient symbiosis. *Tourism Recreation Research*, *27*(2), 59–64. doi:10.1080/02508281.2002.11081221

Wall, G., & Mathiesom, A. (2006). *Tourism: Change, Impacts and Opportunities*. Pearson.

Wanda George, E. (2010). Intangible cultural heritage, ownership, copyrights, and tourism. *International Journal of Culture, Tourism and Hospitality Research*, *4*(4), 376–388. doi:10.1108/17506181011081541

Wang, C. H. (2020). The necessity of taking a community approach in a historical cultural landscape conservation: A case of the Jianan Irrigation System in Taiwan. *GeoJournal*, *85*(1), 107–126. doi:10.100710708-018-9950-8

Wang, M.-Y., Li, Y.-Q., Ruan, W.-Q., Zhang, S.-N., & Li, R. (2023). Cultural inheritance-based innovation at heritage tourism destinations: Conceptualization, structural dimensions and scale development. *Journal of Hospitality and Tourism Management*, *55*, 118–130. doi:10.1016/j.jhtm.2023.03.009

Wang, N., Fang, M., Beauchamp, M., Jia, Z., & Zhou, Z. (2021). An indigenous knowledge-based sustainable landscape for mountain villages: The Jiabang rice terraces of Guizhou, China. *Habitat International*, *111*, 102360. doi:10.1016/j.habitatint.2021.102360

Wang, Y. C., Chen, P. J., Shi, H., & Shi, W. (2021). Travel for mindfulness through Zen retreat experience: A case study at Donghua Zen Temple. *Tourism Management, 83*, 104211. doi:10.1016/j.tourman.2020.104211 PMID:32895588

WEF TTCR. (2013). *The Travel & Tourism Competitiveness Report 2013*. We Forum. https://reports.weforum.org/travel-and-tourism-competitiveness-report-2013/

WEF TTCR. (2018). *The Travel & Tourism Competitiveness Report 2018*. We Forum. https://reports.weforum.org/global-competitiveness-report-2018/downloads/

WEF TTCR. (2019). *The Travel & Tourism Competitiveness Report 2019*. We Forum. https://www.weforum.org/reports/the-travel-tourism-competitiveness-report-2019

Weiler, B., & Ham, S. H. (2002). Tour Guide Training: A Model for Sustainable Capacity Building in Developing Countries. *Journal of Sustainable Tourism, 10*(1), 52–69. doi:10.1080/09669580208667152

Weismann, N., Dziamski, P., & Haar, L. (2019). The kambārī in the museum of the frankincense land, Salalah, Oman. *International Journal of Nautical Archaeology, 48*(2), 342–359. doi:10.1111/1095-9270.12372

Wei, Y., Liu, H., & Park, K.-S. (2021). Examining the structural relationships among heritage proximity, perceived impacts, attitude and residents' support in Intangible Cultural Heritage Tourism. *Sustainability (Basel), 13*(15), 8358. doi:10.3390u13158358

West, A. (1985). *Heritage. Heritage*. Coronet. www.collinsdictionary.com

Willson, G. B., McIntosh, A. J., & Zahra, A. L. (2013). Tourism and spirituality: A phenomenological analysis. *Annals of Tourism Research, 42*, 150–168. doi:10.1016/j.annals.2013.01.016

Wilson, H., & Venes, J. (2001). *Factors for Success in Rural Promotion and Development*. Deep Publications.

Winter, C. (2020). A review of research into animal ethics in tourism: Launching the annals of tourism research curated collection on animal ethics in tourism. *Annals of Tourism Research, 84*(10), 29–89. doi:10.1016/j.annals.2020.102989

Wong, C. U. I., McIntosh, A., & Ryan, C. (2016). Visitor management at a Buddhist sacred site. *Journal of Travel Research, 55*(5), 675–687. doi:10.1177/0047287514563164

Wongmasa, N. (2020, May 9). *AEN E-learning 05 on Covid-19: Adaptation in Thai Ecotourism sharing cases from TEATA* [Video]. Asian Ecotourism Network AEN. YouTube. https://www.youtube.com/watch?v=oGrWWDCq0gQ

World Tourism Organization. (2011). Religious tourism in Asia and the Pacific. World Tourism Organization.

World Tourism Organization. (2015).United Nation World Tourism Highlights, 2015 Edition. Published by United Nation World Tourism Organization Publication, Madrid Spain.

WTO. (2016). *Tourism sector highlighted as important contributor to trade and development* (pp. 1-2). UNWTO. Retrieved from https://www.wto.org/english/news_e/news16_e/bus_13jul16_e.htm

WTTC. (2018). *Travel and Tourism Economic Impact 2018: India* (pp. 1-3). United Kingdom: WTTC. https://www.wttc.org/-/media/files/reports/economic-impact-research/countries-2018/india2018.pdf

WTTC. (2020). *Economic Impact Reports*. WTTC. https://wttc.org/Research/Economic-Impact

Wu, D., Shen, C., Wang, E., Hou, Y., & Yang, J. (2019). Impact of the Perceived Authenticity of Heritage Sites on Subjective Well-Being: A Study of the Mediating Role of Place Attachment and Satisfaction. *Sustainability, 11*(21). https://doi.org/ doi:10.3390/su11216148

Wu, S. T., Chiu, C. H., & Chen, Y. S. (2020). *The influences of innovative technological introduction on interpretive experiences of exhibition:a discussion on the intention to use augmented reality*. Taylor & Francis. doi:10.1080/10941665.2020.1752754

Wu, G.-M., Chen, S.-R., & Xu, Y.-H. (2023). Generativity and inheritance: Understanding Generation Z's intention to participate in cultural heritage tourism. *Journal of Heritage Tourism, 18*(4), 1–18. doi:10.1080/1743873X.2023.2178927

Wu, H. C., Li, T., & Li, M. Y. (2016). A Study of Behavioral Intentions, Patient Satisfaction, Perceived Value, Patient Trust and Experiential Quality for Medical Tourists. *Journal of Quality Assurance in Hospitality & Tourism, 17*(2), 114–150. doi:10.1080/1528008X.2015.1042621

Wu, M. Y., & Pearce, P. L. (2014). Appraising netnography: Towards insights about new markets in the digital tourist era. *Current Issues in Tourism, 17*(5), 463–474. doi:10.1080/13683500.2013.833179

Wu, X., & Lai, I. K. W. (2021). The acceptance of augmented reality tour app for promoting film-induced tourism: The effect of celebrity involvement and personal innovativeness. *Journal of Hospitality and Tourism Technology, 12*(3), 454–470. doi:10.1108/JHTT-03-2020-0054

Xu, C., Jones, C., & Munday, M. (2019). Tourism inward investment and regional economic development effects: perspectives from tourism satellite accounts. *Regional Studies, 54*(9), 1226–1237. doi:10.1080/00343404.2019.1696954

Yang, W., & Kang, J. (2005). Soundscape and sound preferences in urban squares: A case study in Sheffield. *Journal of Urban Design, 10*(1), 61–80. doi:10.1080/13574800500062395

Yang, Y., Shafi, M., Song, X., & Yang, R. (2018). Preservation of Cultural Heritage Embodied in Traditional Crafts in the Developing Countries. A Case Study of Pakistani Handicraft Industry. *Sustainability, 10*(5), 1336. doi:10.3390u10051336

Yang, Y., Zhang, H., Chen, M., Jiang, Y., & Chai, H. (2021). An Inheritance Mode of Rural Cultural Heritage Based on Virtual Museum in China. *International Journal of Computer Games Technology, 2021*, 1–14. doi:10.1155/2021/4787991

Yekinni, O. B., & Mercy, B. B. (2017). Exploring the Relationship between Religion Tourism and Economic Development of a Host Community. *Int. J. Bus Manag Invent, 6*(9), 41–51.

Yıldırım, M. (2021). Sanal Mekânda Kültür Coğrafyası Çalışmaları ve Artırılmış Gerçeklik Teknolojisinin Kullanılması. *Afyon Kocatepe Üniversitesi Sosyal Bilimler Dergisi, 23*(3), 861–874. doi:10.32709/akusosbil.887804

Yıldırım, T., & Tahiroğlu, M. (2012). Sanal Ortamda Gerçekleştirilen Müze Gezilerinin İlköğretim Öğrencilerinin Sosyal Bilgiler Dersine Yönelik Tutumlarına Etkisi. *Elektronik Sosyal Bilimler Dergisi, 11*(39), 104–114.

Yolal, M., & Negrusa, A. L. (2012). Analysis of cultural tourism motivation: the case of Turkish students. *Chinese Business Review, 11*(3). ISSN 1537-1506

Yoon, Y. (2002). Development of a Structural Model for Tourism Destination Competitiveness from Stakeholders' Perspectives. *ETDs: Virginia Tech Electronic Theses and Dissertations, Theses and Dissertations, Hospitality and Tourism Management (HTM)*. http://hdl.handle.net/10919/29174

Yoshimura, Y., He, S., Hack, G., Nagakura, T., & Ratti, C. (2020). Quantifying memories: Mapping urban perception. *Mobile Networks and Applications, 25*(4), 1275–1286. doi:10.100711036-020-01536-0

Young, L. (2021). A Study on Metaverse Hype for Sustainable Growth. *International Journal of Advanced Smart Convergence, 10*(3).

Yuan, C., Gan, L., & Zhuo, H. (2022). Coupling mechanisms and development patterns of revitalizing Intangible Cultural Heritage by integrating cultural tourism: The case of Hunan Province, China. *Sustainability (Basel), 14*(12), 6994. doi:10.3390u14126994

Zahra, A., & McIntosh, A. J. (2016). Volunteer Tourism: Evidence of Cathartic Tourist Experiences. *Tourism Recreation Research.*

Zainuddin, Z., Mohd Radzi, S., Mohd Zahari, M. S., & Azman Ong, M. H. (2020). Perceived destination competitiveness and response behavior: A test of mediation. *Journal of Tourism, Hospitality & Culinary Arts, 12*(1), 279-288. https://fhtm.uitm.edu.my/images/jthca/Vol12Issue1/Chap_19.pdf

Zainuddin, Z., Radzi, S. M., & Zahari, M. S. M. (2016). Perceived Destination Competitiveness of Langkawi Island, Malaysia. *Procedia: Social and Behavioral Sciences, 222*, 390–397. doi:10.1016/j.sbspro.2016.05.190

Zargham Boroujeny, H. (2017). Tourism: A tool for spirituality development. *International Journal of Tourism & Spirituality.*

Zehrer, A., & Hallmann, K. (2015). A stakeholder perspective on policy indicators of destination competitiveness. *Journal of Destination Marketing & Management, 4*(2), 120–126. doi:10.1016/j.jdmm.2015.03.003

Zhang, Y., Kang, S., & Koo, J. H. (2021). Perception Difference and Conflicts of Stakeholders in the Urban Regeneration Project: A Case Study of Nanluoguxiang. *Sustainability, 13*(5). doi:10.3390/su13052904

Zhang, Q., Liu, X., Li, Z., & Tan, Z. (2021). Multi-experiences in the art performance tourism: Integrating experience economy model with flow theory. *Journal of Travel & Tourism Marketing, 38*(5), 491–510. doi:10.1080/10548408.2021.1952148

Zhang, S., Liang, J., Su, X., Chen, Y., & Wei, Q. (2023). Research on global cultural heritage tourism based on Bibliometric analysis. *Heritage Science, 11*(1), 139. doi:10.118640494-023-00981-w

Zhao, Y., & Agyeiwaah, E. (2023). Exploring value-based motivations for culture and heritage tourism using the means-end chain and laddering approach. *Journal of Heritage Tourism, 18*(5), 1–23. doi:10.1080/1743873X.2023.2215933

Zhou, W. W. (2022). *Management effectiveness evaluation of world cultural landscape heritage: a case from China.* Heritage Science.

Ziff, B., & Rao, P. V. (Eds.). (1997). Borrowed Power: Essays on Cultural Appropriation. Rutgers.

Živanović, M.L. (2016). Culture and heritage tourism – destination competitiveness and regional clusters review. *Horizon A*, 69-77. doi:10.20544/HORIZONS.A.20.1.17.P05

Zouni, G., & Digkas, D. (2019). Marketing suggestions for multi-religious tourism development: The case of Thessaloniki. *J Tour Herit Serv Mark., 5*(2), 36–42.

About the Contributors

Yashwant Rawal has 15 years of experience in both industry and academics. He has worked with the Oberoi Group of Hotels and Carnival Cruise Line USA. His Doctoral Research work was in the area of Staff Turnover Intention in the Hotel industry. He has published more than 57 research papers in Hospitality, Tourism, and Management and presented more than 20 papers at International and National conferences. He was awarded the best role model teacher at the National level in 2019. He also has been awarded with high work engagement award four times at the university level.

Debasish Batabyal, PhD, is currently teaching at Amity University, Kolkata, India. Dr. Debasish regularly speaks at academic conferences in India and abroad on tourism and hospitality management, specifically on eTourism, ecotourism, destination development and management. He has authored two books and edited a book 'Managing Sustainable Tourism Resources'. He has 32 publications in various journals of repute and book chapters at national and international level. As a key member, he also worked with various research institutes and team members, including committee for the production of development reports by UNDP, District Gazetteer Development committee, Board of study, syllabus committee etc. and designed job oriented courses with industry-academia liaison.

***

Musallam Al-Aamri is a PhD student, Taylor's University, Malaysia & University of Technology and Applied Sciences, Salalah, Oman

Hamed Almuhrzi is Dean, Oman Tourism College, Muscat, Oman & Assistant Professor, Sultan Qaboos University, Muscat, Oman.

Sreeparna Bhattacharyya, Assistant Professor (HRM-OB), Bharatiya Vidya Bhavan, Institute of Management Science, Kolkata & Ph.D. Scholar, SOIM, Silver Oak University, Ahmedabad. Her Ph.D. research area covers sustainable human resource management. Her interest area also covers talent management, change management, and well-being. She has published in various reputed national, and international journals, conference proceedings, and book chapters. She has presented her papers at international and national conferences. She has done her M.A. in Psychology (I/O) and M.B.A. (HRM and Marketing) with 1st Class. She has worked as a resource person in various workshops and courses. She is a Sub-Editor of 'The Kolkata Kindle' Committee, a journal published by the National Human

Resource Development Network, Kolkata Chapter. She is a Life-Member at NHRDN, Kolkata Chapter ORCID: 0000-0002-2093-4519

Sarani Bhaumik is a well-recognized tourism academician, having fifteen years of professional experience including working in the tourism industry, teaching, training and research. She is pursuing her Ph.D. in Management from Aliah University, Kolkata. She also obtained few professional courses like a Post-Graduate Diploma in Teaching and Research In Management, a Post-Graduate Diploma in Instructional Design, a Post-Graduate Diploma in Instructional Design, a graduate Diploma in Educational Administration and Supervision, Post-Graduate Diploma in Event Management and Promotions, Post-Graduate Diploma in Public Relations, Post Graduate Diploma in Entrepreneurship, Post Graduate Diploma in International Business and Operations, Post Graduate Diploma in Human Resource Management, Post Graduate Diploma in Marketing Management, Diploma in Training & Development, Diploma in IATA/UFTAA Foundation and EBT Course, Enter the world of Amadeus (ETWA-AMA), Enter the world of Automated Ticketing (ETWAT-AMA) from various recognised institutions. She has published many papers in international and national journals and also contributed many book chapters. She is a lifetime member of the Indian Society for Training & Development (ISTD), All India Management Association (AIMA), Local Management Association: Calcutta Management Association, National Human Resource Development Network (NHRDN), Responsible Tourism Society of India (RTSOI), Indian Society for Technical Education (ISTE) & Associate member of Public Relation Society of India (PRSI). Her current research interests are Intangible Cultural Heritage (ICH) based Tourism, Responsible Tourism and Community Development.

Sarbari Bhowmick has completed her Bachelor's in Business Administration (BBA) in Travel & Tourism from School of Management Sciences, Maulana Abul Kalam Azad University of Technology, West Bengal, India with 1st class. She is a novice researcher. She has a keen interest in the field of travel and tourism with a special interest to ICH. She has a few published articles.

Namrata Chakrabortty is an Assistant Professor of English, associated with Guru Nanak Institute of Technology. She has published research articles in internationally reputed journals (Scopus Index), as well as in national reputed journals (UGC Care Journal). She also attended and presented paper in three national conferences .She has seven years of teaching experience coupled with great desire for research and for contributing to the society.

Samskrati Gulvady is serving as an Assistant Professor in the department of Mass Communication at the University of Technology and Applied Sciences-Salalah, Oman. She has published papers in journals, book chapters and presented papers in international conferences. She is also serving as a reviewer of some journals, and has been a guest editor for the Omani Journal of Applied Sciences. Her research interests include social media, interpersonal communication, mass communication, ICT, youth, culture, tourism, marketing and bibliometric studies.

Viana Hassan is actively involved in academic and research work in Tourism & Cultural Management in Lebanon and Malta. She got her Ph.D. in Tourism Management and Cultural from Saint Joseph University, Beirut. Her Thesis was entitled "Medical Tourism in Lebanon". She has more than 19 years of tourism experience in many related Travel and Tourism Fields, varying from airline sales experience

(MEA), to Co-management several Lebanese Travel agencies. She is an Experienced Lecturer with more than 16 years' experience at several universities in Lebanon (Lebanese University, A.U. L, Lebanese International University, Islamic University), the Institute of Tourism Studies, STC higher education and University of Malta (Malta), Scholars School system (UK) with a demonstrated history of working in the Higher education industry, Skilled in Tourism Management, Customer service, Branding, Event Management, Airline Business Management, Niche Tourism, Business operations management, Customer acquisition and Medical Tourism. Dr. Viana has participated in international conferences (Expo Dubai 2020, 7th Medical Tourism World Summit, etc…) including Athens, Paris, Vienna, Dubai, Beirut, Valencia, Amman, Lisbon, Istanbul and has published papers in peer review journals.

Minna-Maarit Jaskari is a University Lecturer (PhD) at the University of Vaasa, Finland. Introduction to marketing and sales, marketing analytics, design thinking, and research methods are among her award-winning courses. Her current research interests include memorable consumer experiences, music in branding, online brand communities, temporality and spatiality of market spaces, and experiential and innovative pedagogies in higher marketing education. She has published a number of articles in academic journals such as Qualitative Market Research, the Journal of Marketing Education, and the Nordic Journal of Business. She serves on the editorial board of the Journal of Marketing Education.

Rashmi Jha is a UGC-JRF qualified Research Scholar at the Department of History of Art, Banaras Hindu University. She previously worked as a Research Assistant in an IoE Incentive Grant, B.H.U. (MoE, Govt. of India) sponsored project, "Developing a Digital Archive of the Art Heritage of Varanasi (Newly Discovered Group of Ancient Temples and Musical Traditions of Varanasi)". Her research interests are Indian vernacular (religious and non-religious) architecture, and ancient and contemporary craft traditions of Varanasi

Saumya Kapil is currently a PhD Scholar in Tourism Management at CHRIST (Deemed to be) University, Bangalore, where her research focuses on the multi-stakeholder approach in lake conservation and its impact on community well-being. She has successfully completed her Masters in Travel & Tourism Management from Amity University, Rajasthan. She has one year of industry experience working in a leading tour operation company. Her scholarly contributions are evident through her publications in areas such as homestay tourism, animal welfare practices, and environmental conservation, highlighting her dedication to responsible and ethical tourism. Additionally, her research interests extend to encompass broader subjects, including circular economy, sustainability, and conservation studies.

Jyothi Kumar is presently Dean - CHRIST University, Bannerghatta Road Campus. She is a United Board fellow and has attended a summer course at the Graduate School of Education, Harvard University, USA. She has a teaching Experience of 22 years in the areas of General Management, Finance, Taxation and Strategic Management, both at the Under Graduate and Post Graduate level. She is a member of the Academic Council, Member- Board of Management and Member - Planning and Monitoring Board at Christ University. Her PhD is in the area of Strategy and her work is titled Organizational Sustainability: A Study of corporate Organizations in the Indian Context. Her Areas of Research apart from Sustainability is in area of Skill Development for Students in the Higher Educational Context. She has presented various papers at National and International conferences and received Best Paper awards. She has also delivered various lectures in her areas of Interest and was invited to be a part of the inaugural address during the conferences and seminars organized by Educational Institutions. She has widely traveled on Academic purposes and Internationalization initiatives.

Smriti Mishra is an Architect and Urban Planner and is currently working as an Associate Professor in the Department of Architecture and Planning, Birla Institute of Technology, Mesra, Ranchi, India. She has done her research in the field of urban and regional planning and has numerous publications in her records.

Aditi Nag is an Architect and Urban Planner, currently pursuing PhD in the Department of Architecture and Planning, Birla Institute of Technology, Mesra, Ranchi, India. She has done research in the field of urban and rural planning and has a few publications in her record. She is the 1st to have been awarded the DST Inspire Fellowship in the field of Architecture and Planning by the Ministry of Science and Technology, Govt. of India.

Murtaza Reshi is a Research Scholar at Lovely Professional University in the Department of Social Sciences and Humanities

Soumyadeep Roy is an Assistant Professor, Amity School of Hospitality, Amity University Kolkata. He has a profound industry experience and was associated with Taj Group of Hotels before joining Amity University.

Supriyo Roy is a management expert with specialization in Supply Chain Modeling / Supply Chain Intelligence.

Satyaki Sarkar is an Architect and Urban Planner who is in the field of academics and research for the last two decades. He has published numerous research papers, and books and has been involved with numerous research projects throughout his career.

Partho Seal is an avid wordsmith who believes that writing is most influential and resourceful medium to disseminate information has authored, four books Computers in Hotels: Concepts and Application, Food & Beverage Management and How to Succeed in Hotel Management Job Interviews and Food Anthropology of India. His interest is in human resource management, entrepreneurship, and tourism. He has published research papers in indexed international journals and is reviewer of international journals.

Amrik Singh is working as Professor in the School of Hotel Management and Tourism at Lovely Professional University, Punjab, India. He obtained his Ph.D. degree in Hotel Management from Kurukshetra University, Kurukshetra. He started his academic career at Lovely Professional University, Punjab, India in the year 2007. He has published more than 30 research papers in UGC and peer-reviewed journals and 10 publications (Scopus/Web of Science). He has published 12 patents and 01 patent has been granted in the inter-disciplinary domain. Dr. Amrik Singh participated and acted as a resource person in various national and international conferences, seminars, research workshops, and industry talks. His area of research interest is accommodation management, ergonomics, green practices, human resource management in hospitality, waste management, AR VR in hospitality, etc. He is currently guiding 8Ph.D. scholars and 2 Ph.D. scholars have been awarded Ph.D.

Lakhvinder Singh is working as Assistant Professor of Tourism in Government College, Kaithal (Haryana), India. He holds Ph.D. Degree in with UGC-NET qualification in Tourism Administration & Management. He has more than three years of teaching experience and has more than 30 research publications in his credit. He is in the Editorial/Advisory Board of different Journals of International and National repute. His areas of interest include tourist behavior, tourism marketing, tourism impact, culture & religious tourism, e-tourism and ICT in tourism.

Sujay Vikram Singh is a Senior Research Fellow currently pursuing a doctorate from Banaras Hindu University Graduated and post-graduated from IHM Lucknow.His research interest includes Hospitality, CRM, Service Marketing, Service Quality, and Systematic Literature reviews. Published papers in various handbooks and journals. His Recent journal publications include International Journal of Market Research (Sage publishing), ABDC-A,Scopus Journal of Global information management (IGI Publishing) Scopus, ABDC-A . Presented papers at various conferences,seminars. Achieved best papers in various conferences and special mention in various seminar sessions. At present working on projects for Shri Kashi Vishwanath corridor And monograph on the heritage relevance of karkotaka vapi

Mohammad Soliman is currently the Head of Research and Consultancy Department at the University of Technology and Applied Sciences, Salalah, Oman. He is also a Full Professor at the Faculty of Tourism & Hotels, Fayoum University, Egypt. He has published multiple papers in reputed journals indexed in WoS and Scopus (e.g., Current Issues in Tourism, Journal of Destination Management & Marketing, Tourism Review, Tourism Management Perspectives, Journal of Service Theory & Practice, International Journal of Human-Computer Interaction, Mindfulness, Current Psychology; Geoheritage, Journal of Tourism Futures, Tourism & Hospitality Research, Journal of Medical Systems, Journal of Consumer Behaviour, etc.). Additionally, he sits on the editorial board of different academic journals (e.g., Journal of Hospitality and Tourism Education; Tourism & Management Studies, Journal of Responsible Tourism Management, Journal of Global Hospitality and Tourism, Journal of Tourism & Hospitality; and others), and serves as a reviewer for several top-tier journals. He has successfully supervised and examined a number of Master and PhD theses. His research interests include tourism management and marketing, tourism behavior, HR in tourism, PLS-SEM, bibliometrics and systematic reviews.

Supina Supina an experienced Head of Hospitality with a demonstrated history of working in the higher education industry. Skilled in Lecturing, Hospitality Industry, Event Management, Curriculum Development, and Social Media. Strong professional with a Master of Management in International Tourism Management focused in International Tourism Management from Burapha University, Thailand.

Bhavna Taneja has a doctorate in Management and an MBA specializing in marketing and finance. She has over 18 years of teaching experience In premium institutes like Xiss, Xaviers College Ranchi, Birsa Agricultural University, etc. Presently she is associated with Amity University Jharkhand. She has several publications to her name. Dr. Taneja is an active Rotarian and is presently on the board of directors of the Rotary Club of Ranchi. She is a Pranic healer.

Atul Tripathi is a Professor and Head of the Department of History of Art and Tourism Management at Banaras Hindu University. Before this, he taught at the Department of History and Culture, Gujarat Vidyapeeth, Ahmedabad, as Lecturer and Reader, for over a decade. He specializes in the history of ancient and early mediaeval Indian architecture, sculpture and iconography. He has received several prestigious awards, completed a number of valuable research projects and is presently engaged in the IoE Incentive Grant, B.H.U. (MoE, Govt. of India) sponsored project entitled "Developing a Digital Archive of the Art Heritage of Varanasi (Newly Discovered Group of Ancient Temples and Musical Traditions of Varanasi)".

Bindi Varghese is a Doctorate in Commerce,specializing in Tourism. As an academician and tourism professional, she has over 18 years of Academic and one year of Industrial experience. Currently, she is affiliated with Christ University, as an Associate Professor and is the Research Coordinator at School of Business Studies and Social Sciences. She has served many educational institutions in South India and has served as a national and international expert, for a decade among the educational institutions of India. Currently, she is actively associated with Indian Tourism Congress (ITC) and Kerala Development Society (KDS), New Delhi. Along with her academic expertise, she is also an Section editor for 'ATNA- Journal of Tourism Studies', published by Christ University, Bengaluru.

Varsha Vinod is an architect and urban planner. She is currently pursuing her PhD from Department of Architecture and Planning, BIT Mesra and working as Assistant Professor (on contract) at Department of Architecture and Planning, IIEST Shibpur.

Index

A

B

C

D

E

F

G

H

I

K

L

www.ingramcontent.com/pod-product-compliance
Lightning Source LLC
LaVergne TN
LVHW060412300926
849162LV00006B/1411
* 9 7 8 1 6 6 8 4 9 9 5 7 3 *